Lecture Notes in Physics

T0238565

The Lecture Notes in Physics

The series Lecture Notes in Physics (LNP), founded in 1969, reports new developments in physics research and teaching – quickly and informally, but with a high quality and the explicit aim to summarize and communicate current knowledge in an accessible way. Books published in this series are conceived as bridging material between advanced graduate textbooks and the forefront of research to serve the following purposes:

• to be a compact and modern up-to-date source of reference on a well-defined topic;

• to serve as an accessible introduction to the field to postgraduate students and nonspecialist researchers from related areas;

• to be a source of advanced teaching material for specialized seminars, courses and schools.

Both monographs and multi-author volumes will be considered for publication. Edited volumes should, however, consist of a very limited number of contributions only. Proceedings will not be considered for LNP.

Volumes published in LNP are disseminated both in print and in electronic formats, the electronic archive is available at springerlink.com. The series content is indexed, abstracted and referenced by many abstracting and information services, bibliographic networks, subscription agencies, library networks, and consortia.

Proposals should be sent to a member of the Editorial Board, or directly to the managing editor at Springer:

Dr. Christian Caron
Springer Heidelberg
Physics Editorial Department I
Tiergartenstrasse 17
69121 Heidelberg/Germany
christian.caron@springer.com

Stefano Bellucci   (Ed.)

Supersymmetric Mechanics – Vol. 1

Supersymmetry, Noncommutativity
and Matrix Models

 Springer

Editor

Stefano Bellucci
Istituto Nazionale di Fisica Nucleare
Via Enrico Fermi, 40
00044 Frascati (Rome), Italy
E-mail: bellucci@lnf.infn.it

S. Bellucci, *Supersymmetric Mechanics – Vol. 1*, Lect. Notes Phys. 698 (Springer, Berlin Heidelberg 2006), DOI 10.1007/b11730286

ISSN 0075-8450

ISBN 978-3-642-06996-3   e-ISBN 978-3-540-33314-2

Springer is a part of Springer Science+Business Media
springer.com
© Springer-Verlag Berlin Heidelberg 2006
Softcover reprint of the hardcover 1st edition 2006

Cover design: *design & production* GmbH, Heidelberg

To Annalisa, with fatherly love and keen anticipation

Preface

This is the first volume in a series of books on the general theme of Supersymmetric Mechanics, which are based on lectures and discussions held in 2005 and 2006 at the INFN-Laboratori Nazionali di Frascati. These schools originated from a discussion among myself, my long-time foreign collaborators, and my Italian students. We intended to organize these schools as an intense week of learning around some specific topics reflecting our current and "traditional" interests. In this sense, the choice of topics was both rather specific and concrete, allowing us to put together different facets related to the main focus (provided by Mechanics). The selected topics include Supersymmetry and Supergravity, Attractor Mechanism, Black Holes, Fluxes, Noncommutative Mechanics, Super-Hamiltonian Formalism, and Matrix Models.

All lectures were meant for beginners and covered only half of each day. The rest of the time was dedicated to training, solving of problems proposed in the lectures, and collaborations. One afternoon session was devoted to short presentations of recent original results by students and young researchers. The interest vigorously expressed by all attendees, as well as the initiative of the Editors at Springer Verlag, Heidelberg, prompted an effort by all lecturers, helped in some cases and to various degrees by some of the students, including myself, to write down the content of the lectures. The lecturers made a substantial effort to incorporate in their write-ups the results of the animated discussion sessions that followed their lectures. In one case (i.e. for the lectures delivered by Sergio Ferrara) the outgrowth of the original notes during the subsequent reworking, for encompassing recent developments, as well as taking into account the results of the discussion sessions, yielded such a large contribution as to deserve a separate volume on its own. This work is published as the second volume in this series, *Lect. Notes Phys. 701* "Supersymmetric Mechanics – Vol. 2: The Attractor Mechanism" (2006), ISBN: 3-540-34156-0. A third volume on related topics is in preparation.

In spite of the heterogeneous set of lecturers as well as topics, the resulting volumes have reached a not so common unity of style and a homogeneous level of treatment. This is in part because of the abovementioned discussions that have been taken into account in the write-ups, as well as due to the pedagogical character that inspired the school on the whole. In practice, no previous knowledge by attendees was assumed on the treated topics.

As a consequence, these books will be suitable for academic instruction and research training on such topics, both at the postgraduate level, as well as for young postdoctoral researchers wishing to learn about supersymmetry, supergravity, superspace, noncommutativity, especially in the specific context of Mechanics.

I warmly thank both lecturers and students for their collective work and strenuous efforts, which helped shaping up these volumes. Especially, I wish to mention Professors Ferrara, Gates, Krivonos, Nair, Nersessian and Sochichiu for their clear teaching, enduring patience, and deep learning, as well as in particular the students Alessio Marrani and Emanuele Orazi for their relentless questioning, sharp curiosity, and thorough diligence. Last, but not least, I wish to express my gratitude to Mrs. Silvia Colasanti, at INFN in Frascati, for her priceless secretarial work and skilled organizing efforts. Finally, I am grateful to my wife Gloria, and my daughters Costanza and Eleonora, for providing me a peaceful and favorable environment for the long hours of work needed to complete these contributions.

Frascati, Italy   *Stefano Bellucci*
December 2005

Contents

List of Contributors

Stefano Bellucci
INFN-Laboratori Nazionali di
Frascati
Via E. Fermi 40, C.P. 13
00044 Frascati, Italy
bellucci@lnf.infn.it

Sergio Ferrara
CERN, Physics Department
1211 Geneva 23, Switzerland
and
INFN-Laboratori Nazionali di
Frascati
Via E. Fermi 40, C.P. 13
00044 Frascati, Italy
and
Department of Physics and
Astronomy
University of California
Los Angeles, CA, USA
Sergio.Ferrara@cern.ch

Sylvester James Gates, Jr.
Physics Department
University of Maryland
Rm. 4125, College Park
MD 20742-4111, USA
gatess@wam.umd.edu

Sergey Krivonos
Bogoliubov Laboratory of
Theoretical Physics
Joint Institute for Nuclear Research
141980 Dubna, Moscow Reg., Russia
krivonos@thsun1.jinr.ru

Alessio Marrani
Centro Studi e Ricerche "E. Fermi"
Via Panisperna 89A,
00184 Roma, Italy
and
INFN-Laboratori Nazionali di
Frascati
Via E. Fermi 40, C.P. 13
00044 Frascati, Italy
Alessio.Marrani@lnf.infn.it

V.P. Nair
City College of the CUNY
New York, NY 10031, USA
vpn@sci.ccny.cuny.edu

Armen Nersessian
Artsakh State University
Stepanakert and Yerevan State
University
Alex Manoogian St., 1
Yerevan, 375025, Armenia
arnerses@yerphi.am

Emanuele Orazi
Dipartimento di Fisica
Università di Roma "Tor Vergata"
Via della Ricerca Scientifica 1
00133 Roma, Italy
and
INFN-Laboratori Nazionali di
Frascati
Via E. Fermi 40, C.P. 13
00044 Frascati, Italy
orazi@lnf.infn.it

Corneliu Sochichiu
Max-Planck-Institut für Physik
Werner-Heisenberg-Institut
Föhringer Ring 6
80805 München, Germany

and
Institutul de Fizică Aplicată AŞ
str. Academiei, nr. 5, Chişinău
MD2028 Moldova
and
Bogoliubov Laboratory of
Theoretical Physics
Joint Institute for Nuclear Research
141980 Dubna, Moscow Reg., Russia
sochichi@mppmu.mpg.de

1

A Journey Through Garden Algebras

S. Bellucci,[1] S.J. Gates Jr.,[2] and E. Orazi[1,3]

[1] INFN-Laboratori Nazionali di Frascati, Via E. Fermi 40, C.P. 13, 00044 Frascati,
Italy
`bellucci@lnf.infn.it`
[2] Physics Department, University of Maryland, Rm. 4125, College Park, MD
20742-4111
`gatess@wam.umd.edu`
[3] Dipartimento di Fisica, Università di Roma "Tor Vergata", Via della Ricerca
Scientifica 1, 00133 Roma, Italy
`orazi@lnf.infn.it`

Abstract. The main purpose of these lectures is to give a pedagogical overview on
the possibility to classify and relate off-shell linear supermultiplets in the context
of supersymmetric mechanics. A special emphasis is given to a recent graphical
technique that turns out to be particularly effective for describing many aspects of
supersymmetric mechanics in a direct and simplifying way.

1.1 Introduction

Sometimes problems in mathematical physics go unresolved for long periods
of time in mature topics of investigation. During this World Year of Physics,
which commemorates the pioneering efforts of Albert Einstein, it is perhaps
appropriate to note the irreconcilability of the symmetry group of Maxwell
equations with that of Newton's equation (via his second law of motion) was
one such problem. The resolution of this problem, of course, led to one of the
greatest revolutions in physics. This piece of history suggests a lesson on what
can be the importance of problems that large numbers of physicists regard as
unimportant or unsolvable.

In light of this episode, the presentation which follows hereafter is focused
on a problem in supersymmetry that has long gone unresolved and seems generally regarded as one of little importance. While there is no claim or preten-
sion that this problem has the importance of the one resolved by the brilliant
genius of Einstein, it is a problem that perhaps holds the key to a more mathematically complete understanding of the area known as "supersymmetry."

The topic of supersymmetry is over 30 years old now. It has been vigorously researched by both mathematicians and physicists. During this entire
time, this subject has been insinuated into a continuously widening array of

S. Bellucci et al.: *A Journey Through Garden Algebras*, Lect. Notes Phys. **698**, 1–47 (2006)
DOI 10.1007/3-540-33314-2_1   © Springer-Verlag Berlin Heidelberg 2006

increasingly sophisticated mathematical models. At the end of this stream of development lies the mysterious topic known as "M-theory." Accordingly, it may be thought that all fundamental issues regarding this area have already a satisfactory resolution.

However, as surprising as it may seem, in fact very little is known about the representation theory of supersymmetry required for the classification of irreducible superfield theories in a manner that allows for quantization consistent with a manifest realization of supersymmetry.

Superspace is to supersymmetry as Minkowski space is to the Lorentz group. Superspace provides the most natural geometrical setting in which to describe supersymmetrical theories. Almost no physicist would utilize the component of Lorentz four-vectors or higher rank tensor to describe relativistic physics. Yet, the analog of this is a common practice in describing supersymmetrical theory. This is so because "component fields" are the predominant language by which most discussions of supersymmetry are couched.

One fact that hides this situation is that much of the language used to describe supersymmetrical theories appears to utilize the superspace formalism. However, this appearance is deceiving. Most often what appears to be a superspace presentation is actually a component presentation in disguise. A true superspace formulation of a theory is one that uses "unconstrained" superfields as their fundamental variables. This is true of a tiny subset of the discussions of supersymmetrical theories and is true of none of the most interesting such theories involving superstrings.

This has led us to the belief that possibly some important fundamental issues regarding supersymmetry have yet to be properly understood. This belief has been the cause of periodic efforts that have returned to this issue. Within the last decade this investigation has pointed toward two new tools as possibly providing a fresh point of departure for the continued study (and hopefully ultimate resolution) of this problem. One of these tools has relied on a totally new setting in which to understand the meaning of supersymmetry. This has led to the idea that the still unknown complete understanding of the representation theory of supersymmetry lies at the intersection of the study of Clifford algebras and K-theory. In particular, a certain class of Clifford algebras (to which the moniker $\mathcal{GR}(d, N)$ have been attached) provides a key to making such a connection. Within the confines of an interdisciplinary working group that has been discussing these problems, the term "garden algebra" has been applied to the symbolic name $\mathcal{GR}(d, N)$. It has also been shown that these Clifford algebras naturally lead to a graphical representation somewhat akin to the root and weight spaces seen in the classification of compact Lie algebras. These graphs have been given the name "Adinkras." The topic of this paper will be introducing these new tools for the study of supersymmetry representation theory.

1.2 $\mathcal{GR}(d, N)$ Algebras

1.2.1 Geometrical Interpretation of $\mathcal{GR}(d, N)$ Algebras

In a field theory, boson and fermions are to be regarded as diffeomorphisms generating two different vector spaces; the supersymmetry generators are nothing but sets of linear maps between these spaces. Following this picture we can include a supersymmetric theory in a more general geometrical framework defining the collection of diffeomorphisms

$$\phi_i : R \to R^{d_L}, \quad i = 1, \ldots, d_L \tag{1.1}$$

$$\psi_{\hat{\alpha}} : R \to R^{d_R}, \quad i = 1, \ldots, d_R , \tag{1.2}$$

where the one-dimensional dependence reminds us that we restrict our attention to mechanics. The free vector spaces generated by $\{\phi_i\}_{i=1}^{d_L}$ and $\{\psi_{\hat{\alpha}}\}_{\hat{\alpha}=1}^{d_R}$ are respectively $\mathcal{V}_L$ and $\mathcal{V}_R$, isomorphic to R^{d_L} and R^{d_R}. For matrix representations in the following, the two integers are restricted to the case $d_L = d_R = d$. Four different linear mappings can act on $\mathcal{V}_L$ and $\mathcal{V}_R$

$$\begin{aligned} \mathcal{M}_L : \mathcal{V}_L \to \mathcal{V}_R, \quad & \mathcal{M}_R : \mathcal{V}_R \to \mathcal{V}_L \\ \mathcal{U}_L : \mathcal{V}_L \to \mathcal{V}_L, \quad & \mathcal{U}_R : \mathcal{V}_R \to \mathcal{V}_R , \end{aligned} \tag{1.3}$$

with linear maps space dimensions

$$\begin{aligned} \dim\mathcal{M}_L = \dim\mathcal{M}_R = d_R d_L = d^2 , \\ \dim \mathcal{U}_L = {d_L}^2 = d^2, \quad \dim \mathcal{U}_R = {d_R}^2 = d^2 , \end{aligned} \tag{1.4}$$

as a consequence of linearity. To relate this construction to a general real ($\equiv \mathcal{GR}$) algebraic structure of dimension d and rank N denoted by $\mathcal{GR}(d, N)$, two more requirements need to be added.

1. Let us define the generators of $\mathcal{GR}(d, N)$ as the family of $N + N$ linear maps[1]

$$\begin{aligned} L_I \in \{\mathcal{M}_L\}, \quad & I = 1, \ldots, N \\ R_K \in \{\mathcal{M}_R\}, \quad & K = 1, \ldots, N \end{aligned} \tag{1.5}$$

such that for all $I, K = 1, \ldots, N$, we have

$$\begin{aligned} L_I \circ R_K + L_K \circ R_I = -2\delta_{IK} I_{\mathcal{V}_R} , \\ R_I \circ L_K + R_K \circ L_I = -2\delta_{IK} I_{\mathcal{V}_L} , \end{aligned} \tag{1.6}$$

where $I_{\mathcal{V}_L}$ and $I_{\mathcal{V}_R}$ are identity maps on $\mathcal{V}_L$ and $\mathcal{V}_R$. Equations (1.6) will later be embedded into a Clifford algebra but one point has to be emphasized, we are working with real objects.

---

[1] Notice that in previous works on the subject [1, 2], the maps L_I and R_K were exchanged, so that $L_I \in \{\mathcal{M}_R\}$ and $R_K \in \{\mathcal{M}_L\}$.

2. After equipping $\mathcal{V}_L$ and $\mathcal{V}_R$ with euclidean inner products $\langle\cdot,\cdot\rangle_{\mathcal{V}_L}$ and $\langle\cdot,\cdot\rangle_{\mathcal{V}_R}$, respectively, the generators satisfy the property

$$\langle\phi, R_I(\psi)\rangle_{\mathcal{V}_L} = -\langle L_I(\phi), \psi\rangle_{\mathcal{V}_R}, \quad \forall(\phi,\psi)\in\mathcal{V}_L\oplus\mathcal{V}_R. \tag{1.7}$$

This condition relates L_I to the hermitian conjugate of R_I, namely $R_I{}^\dagger$, defined as usual by

$$\langle\phi, R_I(\psi)\rangle_{\mathcal{V}_L} = \langle R_I^\dagger(\phi), \psi\rangle_{\mathcal{V}_R} \tag{1.8}$$

so that

$$R_I^\dagger = R_I^t = -L_I. \tag{1.9}$$

The role of $\{\mathcal{U}_L\}$ and $\{\mathcal{U}_R\}$ maps is to connect different representations once a set of generators defined by conditions (1.6) and (1.7) has been chosen. Notice that $(R_I L_J)_i{}^j \in \mathcal{U}_L$ and $(L_I R_J)_{\hat{\alpha}}{}^{\hat{\beta}} \in \mathcal{U}_R$. Let us consider $\mathcal{A} \in \{\mathcal{U}_L\}$ and $\mathcal{B} \in \{\mathcal{U}_R\}$ such that

$$\mathcal{A} : \phi \rightarrow \phi' = \mathcal{A}\phi$$
$$\mathcal{B} : \psi \rightarrow \psi' = \mathcal{B}\psi \tag{1.10}$$

then, taking the $\mathcal{V}_L$ sector as example, we have

$$\begin{aligned}\langle\phi, R_I(\psi)\rangle_{\mathcal{V}_L} &\rightarrow \langle\mathcal{A}\phi, R_I\mathcal{B}(\psi)\rangle_{\mathcal{V}_L}\\ &= \langle\phi, \mathcal{A}^\dagger R_I\mathcal{B}(\psi)\rangle_{\mathcal{V}_L}\\ &= \langle\phi, R_I'(\psi)\rangle_{\mathcal{V}_L}\end{aligned} \tag{1.11}$$

so a change of representation transforms the generators in the following manner:

$$\begin{aligned}L_I &\rightarrow L_I' = \mathcal{B}^\dagger L_I\mathcal{A}\\ R_I &\rightarrow R_I' = \mathcal{A}^\dagger R_I\mathcal{B}.\end{aligned} \tag{1.12}$$

In general (1.6) and (1.7) do not identify a unique set of generators. Thus, an equivalence relation has to be defined on the space of possible sets of generators, say $\{L_I, R_I\} \sim \{L_I', R_I'\}$ if and only if there exist $\mathcal{A} \in \{\mathcal{U}_L\}$ and $\mathcal{B} \in \{\mathcal{U}_R\}$ such that $L' = \mathcal{B}^\dagger L_I\mathcal{A}$ and $R' = \mathcal{A}^\dagger R_I\mathcal{B}$.

Now, we want to show how a supersymmetric theory arises. Algebraic derivations are defined by

$$\begin{aligned}\delta_\epsilon\phi_i &= i\epsilon^I(R_I)_i{}^{\hat{\alpha}}\psi_{\hat{\alpha}}\\ \delta_\epsilon\psi_{\hat{\alpha}} &= -\epsilon^I(L_I)_{\hat{\alpha}}{}^i\partial_\tau\phi_i,\end{aligned} \tag{1.13}$$

where the real-valued fields $\{\phi_i\}_{i=1}^{d_L}$ and $\{\psi_{\hat{\alpha}}\}_{\hat{\alpha}=1}^{d_R}$ can be interpreted as bosonic and fermionic respectively. The fermionic nature attributed to the $\mathcal{V}_R$ elements implies that $\mathcal{M}_L$ and $\mathcal{M}_R$ generators, together with supersymmetry transformation parameters ϵ^I, anticommute among themselves. Introducing the $d_L + d_R$ dimensional space $\mathcal{V}_L \oplus \mathcal{V}_R$ with vectors

$$\Psi = \begin{pmatrix} \phi \\ \psi \end{pmatrix} , \tag{1.14}$$

Equation (1.13) reads

$$\delta_\epsilon(\Psi) = \begin{pmatrix} i\epsilon R\psi \\ \epsilon L \partial_\tau \phi \end{pmatrix} \tag{1.15}$$

so that

$$[\delta_{\epsilon_1}, \delta_{\epsilon_2}]\Psi = i\epsilon_1^I \epsilon_2^J \begin{pmatrix} R_I L_J \partial_\tau \phi \\ L_I R_J \partial_\tau \psi \end{pmatrix} - i\epsilon_2^J \epsilon_1^I \begin{pmatrix} R_J L_I \partial_\tau \phi \\ L_J R_I \partial_\tau \psi \end{pmatrix} = -2i\epsilon_1^I \epsilon_2^I \partial_\tau \Psi , \tag{1.16}$$

utilizing that we have classical anticommuting parameters and that (1.6) hold. It is important to stress that components of (1.23) can be interpreted as superfield components, so it is as if we were working with a particular superfield multiplet containing only these physical bosons and fermions. From (1.16) it is clear that δ_ϵ acts as a supersymmetry generator, so that we can set

$$\delta_Q \Psi := \delta_\epsilon \Psi = i\epsilon^I Q_I \Psi \tag{1.17}$$

which is equivalent to writing

$$\delta_Q \phi_i = i \left(\epsilon^I Q_I \psi\right)_i , $$
$$\delta_Q \psi_{\hat{a}} = i \left(\epsilon^I Q_I \phi\right)_{\hat{a}} , \tag{1.18}$$

with

$$Q_I = \begin{pmatrix} 0 & R_I \\ L_I H & 0 \end{pmatrix} , \tag{1.19}$$

where $H = i\partial_\tau$. As a consequence of (1.16) a familiar anticommutation relation appears

$$\{Q_I, Q_J\} = -2i\delta_{IJ}H , \tag{1.20}$$

confirming that we are talking about genuine supersymmetry. Once the supersymmetry is recognized, we can associate to the algebraic derivations (1.13), the variations defining the scalar supermultiplets. However, the choice (1.13) is not unique: one can check that

$$\delta_Q \xi_{\hat{a}} = \epsilon^I (L_I)_{\hat{a}}^{\phantom{\hat{a}}i} F_i , $$
$$\delta_Q F_i = -i\epsilon^I (R_I)_i^{\hat{a}} \partial_\tau \xi_{\hat{a}} , \tag{1.21}$$

is another proposal linked to ordinary supersymmetry as the previous one. In this case we will refer to the supermultiplet defined by (1.21) as the spinorial one.

1.2.2 Twisted Representations

The construction outlined above suffers from an ambiguity in the definition of superfield components $(\phi_i, \psi_{\hat{\alpha}})$ and $(\xi_{\hat{\alpha}}, A_i)$ due to the possibility of exchanging the role of R and L generators, giving rise to the new superfields $(\phi_{\hat{\alpha}}, \psi_i)$ and $(\xi_i, A_{\hat{\alpha}})$ with the same supersymmetric properties of the previous ones. The variations associated to these twisted versions are, respectively

$$\delta_Q \phi_{\hat{\alpha}} = i\epsilon^I (L_I)_{\hat{\alpha}}^{\ i} \psi_i$$
$$\delta_Q \psi_i = -\epsilon^I (R_I)_i^{\ \hat{\alpha}} \partial_\tau \phi_{\hat{\alpha}} \,, \tag{1.22}$$

and

$$\delta_Q \xi_i = \epsilon^I (R_I)_i^{\ \hat{\alpha}} F_{\hat{\alpha}} \,,$$
$$\delta_Q F_{\hat{\alpha}} = -i\epsilon^I (L_I)_{\hat{\alpha}}^{\ i} \partial_\tau \xi_i \,. \tag{1.23}$$

The examples mentioned above are just some cases of a wider class of inequivalent representations, referred to as "twisted" ones. The possibility to pass from a supermultiplet to its twisted version is realized by the so called "mirror maps." Moreover, it is possible to define superfields in a completely different manner by parameterizing the supermultiplet using component fields which take value in the algebra vector space. We will refer to these objects as Clifford algebraic superfields. An easy way to construct this kind of representations is tensoring the superspace $\{\mathcal{V}_L\} \oplus \{\mathcal{V}_R\}$ with $\{\mathcal{V}_L\}$ or $\{\mathcal{V}_R\}$. For instance, if we multiply from the right by $\{\mathcal{V}_L\}$ then we have

$$(\{\mathcal{V}_L\} \oplus \{\mathcal{V}_R\}) \otimes \{\mathcal{V}_L\} = \{\mathcal{U}_L\} \oplus \{\mathcal{M}_L\} \tag{1.24}$$

whose fields content is

$$\phi_i^{\ j} \in \{\mathcal{U}_L\} \,,$$
$$\psi_{\hat{\alpha}}^{\ i} \in \{\mathcal{M}_L\} \,, \tag{1.25}$$

with supersymmetry transformations

$$\delta_Q \phi_i^{\ j} = -i\epsilon^I (R_I)_i^{\ \hat{\alpha}} \psi_{\hat{\alpha}}^{\ j} \,,$$
$$\delta_Q \psi_{\hat{\alpha}}^{\ i} = \epsilon^I (L_I)_{\hat{\alpha}}^{\ j} \partial_\tau \phi_j^{\ i} \,, \tag{1.26}$$

still defining a scalar supermultiplet. An analogous structure can be assigned to $\{\mathcal{U}_R\} \oplus \{\mathcal{M}_L\}$, $\{\mathcal{U}_L\} \oplus \{\mathcal{M}_R\}$, and $\{\mathcal{U}_R\} \oplus \{\mathcal{M}_R\}$ type superspaces. Even in these cases, twisted versions can be constructed applying considerations similar to those stated above. The important difference between the Clifford algebraic superfields approach and the $\mathcal{V}_R \oplus \mathcal{V}_L$ superspace one, resides in the fact that in the latter case the number of bosonic fields (which actually describe coordinates) increases with the number of supersymmetric charges, while in the first case there is a way to make this not happen, allowing for a description of arbitrary extended supersymmetric spinning particle systems, as it will be shown in the third section.

1.2.3 $\mathcal{GR}(d, N)$ Algebras Representation Theory

It is time to clarify the link with real Clifford Γ-matrices of Weyl type ($\equiv$ block skew diagonal) space which is easily seen to be

$$\Gamma_I = \begin{pmatrix} 0 & R_I \\ L_I & 0 \end{pmatrix} . \tag{1.27}$$

In fact, due to (1.6), Γ-matrices in (1.27) satisfy

$$\{\Gamma_I, \Gamma_J\} = -2i\delta_{IJ}I, \quad \forall I, J = 1, \ldots, N , \tag{1.28}$$

which is the definition of Clifford algebras. One further Γ-matrix, namely

$$\Gamma_{N+1} = \begin{pmatrix} I & 0 \\ 0 & -I \end{pmatrix} , \tag{1.29}$$

can be added. Therefore, the complete algebra obeys the relationships

$$\{\Gamma_A, \Gamma_B\} = -2i\eta_{AB}I, \quad \forall A, B = 1, \ldots, N+1 , \tag{1.30}$$

where

$$\eta_{AB} = \operatorname{diag}(\underbrace{1, \ldots, 1}_{N}, -1) . \tag{1.31}$$

In the following we assume that A, B indices run from 1 to $N + 1$ while I, J run from 1 to N. The generator (1.29), that has the interpretation of a fermionic number, allow us to construct the following projectors on bosonic and fermionic sectors:

$$P_\pm = \frac{1}{2}(I \pm \Gamma_{N+1}) , \tag{1.32}$$

which are the generators of the usual projectors algebra

$$P_a P_b = \delta_{ab} P_a . \tag{1.33}$$

Commutation properties of $P_\pm$ with Γ-matrices are easily seen to be

$$P_\pm \Gamma_I = \Gamma_I P_\mp ,$$
$$P_\pm \Gamma_{N+1} = \pm \Gamma_{N+1} P_\pm . \tag{1.34}$$

The way to go back to $\mathcal{GR}(d, N)$ from a real Clifford algebra is through

$$R_I = P_+ \Gamma_I P_- ,$$
$$L_I = P_- \Gamma_I P_+ , \tag{1.35}$$

that yield immediately the condition (1.6)

$$R_{(I} L_{J)} = P_+ \Gamma_{(I} P_- \Gamma_{J)} P_+ = -2\delta_{IJ} P_+ \equiv -2\delta_{IJ} \mathbf{I}_+ , \tag{1.36}$$
$$L_{(I} R_{J)} = P_- \Gamma_{(I} P_+ \Gamma_{J)} P_- = -2\delta_{IJ} P_- \equiv -2\delta_{IJ} \mathbf{I}_- . \tag{1.37}$$

In this way, we have just demonstrated that representations of $\mathcal{GR}(d, N)$ are in one-to-one correspondence with the real-valued representations of Clifford algebras, which will be classified in the following using considerations of [3]. To this end, let M be an arbitrary $d \times d$ real matrix and let us consider

$$S = \sum_A \Gamma_A^{-1} M \Gamma_A , \tag{1.38}$$

then

$$\forall \ \Gamma_B \in C(p, q), \ \ \Gamma_B^{-1} S \Gamma_B = \sum_A (\Gamma_B \Gamma_A)^{-1} M \Gamma_A \Gamma_B = \sum_C \Gamma_C^{-1} M \Gamma_C = S , \tag{1.39}$$

where we have used the property of Γ-matrices

$$\Gamma_A \Gamma_B = \epsilon_{AB} \Gamma_C + \delta_{AB} I . \tag{1.40}$$

Equation (1.39) tells us that for all $\Gamma_A \in C(p, q)$ there exists at least one S such that $[\Gamma_A, S] = 0$. Thus, by Shur's lemma, S has to be invertible (if not vanishing). It follows that any set of such M matrices defines a real division algebra. As a consequence of a Frobenius theorem, three possibilities exist that we are going to analyze.

1. **Normal representations (N)**. The division algebra is generated by the identity only
$$S = \lambda I, \quad \lambda \in R . \tag{1.41}$$

2. **Almost complex representations (AC)**. There exists a further division algebra real matrix J such that $J^2 = -I$ and we have
$$S = \mu I + \nu J, \quad \mu, \nu \in R . \tag{1.42}$$

3. **Quaternionic representations (Q)**. Three elements E_1, E_2, and E_3 satisfying quaternionic relations
$$E_i E_j = -\delta_{ij} E + \sum_{k=1}^{3} \epsilon_{ijk} E_k, \quad i, j = 1, 2, 3 , \tag{1.43}$$
are present in this case. Thus it follows
$$S = \mu I + \nu E_1 + \rho E_2 + \sigma E_3, \quad \mu, \nu, \rho, \sigma \in R . \tag{1.44}$$

The results about irreducible representations obtained in [3] for $C(p, q)$ are summarized in Table 1.1.

The dimensions of irreducible representations are referred to faithful ones except the $p - q = 1, 5$ cases where exist two inequivalent representations of the same dimension, related to each other by $\bar{\Gamma}_A = -\Gamma_A$. To obtain faithful representations, the dimensions of those cases should be doubled defining

Table 1.1. Representation dimensions for $C(p,q)$ algebras

$p - q =$	0	1	2	3	4	5	6	7
Type	N	N	N	AC	Q	Q	Q	AC
Rep. dim.	2^n	2^n	2^n	2^{n+1}	2^{n+1}	2^{n+1}	2^{n+1}	2^{n+1}

It appears $n = \left[\frac{p+q}{2}\right]$ with [.] denoting here and in the following, the integer part.

$$\tilde{\Gamma}_A = \begin{pmatrix} \Gamma_A & 0 \\ 0 & -\Gamma_A \end{pmatrix} . \tag{1.45}$$

Once the faithfulness has been recovered, we can say that a periodicity theorem holds, asserting that

$$C\left(p + 8,\ 0\right) = C\left(p,\ 0\right) \otimes M_{16}\left(R\right), \tag{1.46}$$
$$C\left(0,\ q + 8\right) = C\left(0,\ q\right) \otimes M_{16}\left(R\right), \tag{1.47}$$

where $M_r(R)$ stands for the set of all $r \times r$ real matrices. Furthermore we have

$$C\left(p,p\right) = M_r\left(R\right), \quad r = 2^n . \tag{1.48}$$

The structure theorems (1.47) and (1.48) justify the restriction in Table 1.1 to values of $p - q$ from 0 to 7. As mentioned in [4], the dimensions reported in Table 1.1 can be expressed as functions of the signature (p, q) introducing integer numbers k, l, m, and n such that

$$q = 8k + m, \quad 0 \le m \le 7 ,$$
$$p = 8l + m + n, \quad 1 \le n \le 8 , \tag{1.49}$$

where n fix $p - q$ up to $l - k$ multiples of eight as can be seen from

$$p - q = 8(l - k) + n , \tag{1.50}$$

while m encode the p, q choice freedom keeping $p - q$ fixed. Obviously, k and l take into account the periodicity properties. The expression of irreducible representation dimensionalities reads

$$d = 2^{4k+4l+m} F(n) , \tag{1.51}$$

where $F(n)$ is the Radon–Hurwitz function defined by

$$F(n) = 2^r, \quad [\log_2 n] + 1 \ge r \ge [\log_2 n], \quad r \in N . \tag{1.52}$$

Turning back to $\mathcal{GR}(d, N)$ algebras, from (1.31) we deduce that we have to deal only with $C(N, 1)$ case which means that irreducible representation dimensions depend only on N in the following simple manner:

$$d = 2^{4a} \mathcal{F}(b) \tag{1.53}$$

Table 1.2. Representation dimensions for $\mathcal{GR}(d, N)$ algebras

$b =$	1	2	3	4	5	6	7	8
Type	N	AC	Q	Q	Q	AC	N	N
Rep. dim.	2^{4a}	2×2^{4a}	4×2^{4a}	4×2^{4a}	8×2^{4a}	8×2^{4a}	8×2^{4a}	8×2^{4a}

where $N = 8a + b$ with a and b integer running respectively from 1 to 8 and from 0 to infinity. This result can be obtained straightforwardly setting $p = 1$ and $q = N$ in (1.49). Representation dimensions obtained adapting the results of Table 1.1 to the $C(N, 1)$ case are summarized in Table 1.2.

In what follows we focus our attention to the explicit representation's construction. First of all we enlarge the set of linear mappings acting between $\mathcal{V}_L$ and $\mathcal{V}_R$, namely $\mathcal{M}_L \oplus \mathcal{M}_R$ (i.e., $\mathcal{GR}(d, N)$), to $\mathcal{U}_L \oplus \mathcal{U}_R$ defining the enveloping general real algebra

$$\mathcal{EGR}(d, N) = \mathcal{M}_L \oplus \mathcal{M}_R \oplus \mathcal{U}_L \oplus \mathcal{U}_R . \tag{1.54}$$

As noticed before, we have the possibility to construct elements of $\mathcal{U}_L$ and $\mathcal{U}_R$ as products of alternating elements of $\mathcal{M}_L$ and $\mathcal{M}_R$ so that

$$\begin{aligned} L_I R_J, L_I R_J L_K R_L, \ldots & \in \mathcal{U}_R , \\ R_I L_J, R_I L_J R_K L_L, \ldots & \in \mathcal{U}_L , \end{aligned} \tag{1.55}$$

but L_I and R_J come from $C(N, 1)$ through (1.35) so all the ingredients are present to develop explicit representation of $\mathcal{EGR}(d, N)$ starting from Clifford algebra.

We focus now on the building of enveloping algebras' representations starting from Clifford algebras. Indeed, we need to divide the representations into three cases.

1. **Normal representations.** In this case basic definition of Clifford algebra (1.30) suggests a way to construct a basis $\{\Gamma\}$ by wedging Γ matrices

$$\{\Gamma\} = \{I, \Gamma^I, \Gamma^{IJ}, \Gamma^{IJK}, \ldots, \Gamma^{N+1}\}, \quad I < J < K \ldots , \tag{1.56}$$

where $\Gamma^{I,\ldots,J}$ are to be intended as the antisymmetrization of $\Gamma^I \cdots \Gamma^J$ matrices otherwise denoted by $\Gamma^{[I} \cdots \Gamma^{J]}$ or $\Gamma^{[N]}$ if the product involve N elements. Dividing into odd and even products of Γ we obtain the sets

$$\begin{aligned} \{\Gamma_e\} &= \{I, \ \Gamma^{N+1}, \ \Gamma^{IJ}, \ \Gamma^{IJ}\Gamma^{N+1}, \ \Gamma^{IJKL}, \ldots\} , \\ \{\Gamma_o\} &= \{\Gamma^I, \ \Gamma^I\Gamma^{N+1}, \ \Gamma^{IJK}, \ \Gamma^{IJK}\Gamma^{N+1} \ldots\} , \end{aligned} \tag{1.57}$$

respectively related to $\{\mathcal{M}\}$ and $\{\mathcal{U}\}$ spaces. Projectors (1.32) have the key role to separate left sector from right sector. In fact, for instance, we have

$$P_+\Gamma_{IJ}P_+ = P_+\Gamma_{[I}\Gamma_{J]}P_+P_+ = P_+\Gamma_{[I}P_-\Gamma_{J]}P_+$$
$$= P_+\Gamma_{[I}P_-P_-\Gamma_{J]}P_+ = R_{[I}L_{J]} \in \{\mathcal{U}_L\}\,, \qquad (1.58)$$

and in a similar way $P_-\Gamma^{IJ}P_- = L^{[I}R^{J]} \in \{\mathcal{U}_R\}$, $P_+\Gamma_{IJK}P_- = R_{[I}L_JR_{K]} \in \{\mathcal{M}_R\}$, and so on. Those remarks provide the following solution

$$\{\mathcal{U}_R\} = \{\, P_-, P_-\Gamma^{IJ}P_-, \dots, P_-\Gamma^{[N]}P_-\} \equiv \{\, I_{\mathcal{V}_R}, L^{[I}R^{J]}, \dots\}\,,$$
$$\{\mathcal{M}_R\} = \{P_+\Gamma^I P_-, \dots, P_+\Gamma^{[N-1]}P_-\} \equiv \{R^I, R^{[I}L^JR^{K]}, \dots\}\,,$$
$$\{\mathcal{U}_L\} = \{\, P_+, P_+\Gamma_{IJ}P_+, \dots, P_+\Gamma_{[N]}P_+\} \equiv \{\, I_{\mathcal{V}_L}, R_{[I}L_{J]}, \dots\}\,,$$
$$\{\mathcal{M}_L\} = \{P_-\Gamma_I P_+, \dots, P_-\Gamma_{[N-1]}P_+\} \equiv \{L_I, L_{[I}R_JL_{K]}, \dots\}\,,$$
$$(1.59)$$

which we will denote as $\wedge\mathcal{GR}(d, N)$ to remember that it is constructed by wedging L_I and R_J generators. Clearly enough, from each $\Gamma_{[I,\dots,J]}$ matrix we get two elements of $\mathcal{EGR}(d, N)$ algebra as a consequence of the projection. Thus, we can say that in the normal representation case, $C(N, 1)$ is in 1–2 correspondence with the enveloping algebra which can be identified by $\wedge\mathcal{GR}(d, N)$. By the wedging construction in (1.59) naturally arise p-forms that is useful to denote

$$f_I = L_I, \qquad \hat{f}^I = R^I\,,$$
$$f_{IJ} = R_{[I}L_{J]}, \qquad \hat{f}^{IJ} = L^{[I}R^{J]}\,,$$
$$f_{IJK} = L_{[I}R_JL_{K]}, \qquad \hat{f}^{IJK} = R^{[I}L^JR^{K]}\,,$$
$$\vdots \qquad\qquad \vdots \qquad (1.60)$$

The superfield components for the $\{\mathcal{U}_L\} \oplus \{\mathcal{M}_L\}$ type superspace introduced in (1.25), can be expanded in terms of this normal basis as follows

$$\phi_i{}^j = \phi\,\delta_i{}^j + \phi^{IJ}(f_{IJ})_i{}^j + \cdots \in \{\mathcal{U}_L\}\,,$$
$$\psi_{\dot\alpha}{}^i = \psi^I(f_I)_{\dot\alpha}{}^i + \psi^{IJK}(f_{IJK})_{\dot\alpha}{}^i + \cdots \in \{\mathcal{M}_L\} \qquad (1.61)$$

according to the fact that $f_{[\text{even}]} \in \{\mathcal{U}_L\}$ and $f_{[\text{odd}]} \in \{\mathcal{M}_L\}$. We will refer to this kind of supefields as bosonic Clifford algebraic ones because of the bosonic nature of the level zero field. Similar expansion can be done for the $\{\mathcal{U}_R\} \oplus \{\mathcal{M}_R\}$ type superspace where $\hat{f}_{[\text{even}]} \in \{\mathcal{U}_R\}$ and $\hat{f}_{[\text{odd}]} \in \{\mathcal{M}_R\}$

$$\phi_{\dot\alpha}{}^{\dot\beta} = \phi\,\delta_{\dot\alpha}{}^{\dot\beta} + \phi^{IJ}(\hat{f}_{IJ})_{\dot\alpha}{}^{\dot\beta} + \cdots \in \{\mathcal{U}_R\}\,,$$
$$\psi_i{}^{\dot\alpha} = \psi^I(\hat{f}_I)_i{}^{\dot\alpha} + \psi^{IJK}(\hat{f}_{IJK})_i{}^{\dot\alpha} + \cdots \in \{\mathcal{M}_R\}\,. \qquad (1.62)$$

In the (1.62) case we deal with a fermionic Clifford algebraic superfield because the component ϕ is a fermion. For completeness we include the remaining cases, namely $\{\mathcal{U}_R\} \oplus \{\mathcal{M}_L\}$ superspace

$$\phi_{\hat{\alpha}}{}^{\hat{\beta}} = \phi\, \delta_{\hat{\alpha}}^{\hat{\beta}} + \phi^{IJ}(\hat{f}_{IJ})_{\hat{\alpha}}{}^{\hat{\beta}} + \cdots \ \in \{\mathcal{U}_R\}\,,$$
$$\psi_{\hat{\alpha}}{}^{i} = \psi^{I}(\hat{f}_I)_{\hat{\alpha}}^{i} + \psi^{IJK}(\hat{f}_{IJK})_{\hat{\alpha}}^{i} + \cdots \ \in \{\mathcal{M}_L\}\,, \qquad (1.63)$$

and $\{\mathcal{U}_L\} \oplus \{\mathcal{M}_R\}$ superspace

$$\phi_i{}^{j} = \phi\, \delta_i{}^{j} + \phi^{IJ}(\hat{f}_{IJ})_i{}^{j} + \cdots \ \in \{\mathcal{U}_L\}\,,$$
$$\psi_i{}^{\hat{\alpha}} = \psi^{I}(\hat{f}_I)_i{}^{\hat{\alpha}} + \psi^{IJK}(\hat{f}_{IJK})_i{}^{\hat{\alpha}} + \cdots \ \in \{\mathcal{M}_R\}\,. \qquad (1.64)$$

2. **Almost complex representations.** As already pointed out, those kind of representations contain one more generator J with respect to normal representations so that to span all the space, the normal part, which is generated by wedging, is doubled to form the basis for the Clifford algebra

$$\{\Gamma\} = \{I, J, \Gamma^I, \Gamma^I J, \Gamma^{IJ}, \Gamma^{IJ} J, \ldots, \Gamma^{N+1}, \Gamma^{N+1} J\}\,. \qquad (1.65)$$

Starting from (1.65) it is straightforward to apply considerations from (1.57) to (1.59) to end with a $\mathcal{EGR}(d, N)$ almost complex representation in 1–2 correspondence with the previous. Concerning almost complex Clifford algebra superfields, it is important to stress that we obtain irreducible representations only restricting to the normal part.

3. **Quaternionic representations.** Three more generators E^α satisfying

$$\left[E^\alpha, E^\beta\right] = 2\epsilon^{\alpha\beta\gamma} E^\gamma \qquad (1.66)$$

have to be added to the normal part to give the following quaternionic Clifford algebra basis

$$\{\Gamma\} = \{I, E^\alpha, \Gamma^I, \Gamma^I E^\alpha, \Gamma^{IJ}, \Gamma^{IJ} E^\alpha, \ldots, \Gamma^{N+1}, \Gamma^{N+1} E^\alpha\}\,, \qquad (1.67)$$

which is four times larger than the normal part. Again, repeating the projective procedure presented above, generators of Clifford algebra (1.67) are quadrupled to produce the $\mathcal{EGR}(d, N)$ quaternionic representation. Even in this case only the normal part gives irreducible representations for the Clifford algebra superfields.

Notice that from the group manifold point of view, the presence of the generator J for the almost complex case and generators E^α for the quaternionic one, separate the manifold into sectors which are not connected by left or right group elements multiplication giving rise to intransitive spaces. Division algebra has the role to link those different sectors.

Finally we explain how to produce an explicit matrix representation using a recursive procedure mentioned in [1] that can be presented in the following manner for the case $N = 8a + b$ with $a \geq 1$:

$$\begin{aligned}
L_1 &= i\sigma^2 \otimes I_b \otimes I_{8a} = R_1\,, \\
L_I &= \sigma^3 \otimes (L_b)_I \otimes I_{8a} = R_I, \quad 1 \leq I \leq b-1\,, \\
L_J &= \sigma^1 \otimes I_b \otimes (L_{8a})_J = R_J, \quad 1 \leq I \leq 8a-1\,, \\
L_N &= I_2 \otimes I_b \otimes I_{8a} = -R_N\,,
\end{aligned} \qquad (1.68)$$

where I_n stands for the n-dimensional identity matrix while L_b and L_{8a} are referred respectively to the cases $N = b$ and $N = 8a$. Expressions for the cases where $N \leq 7$ which are the starting points to apply the algorithm in (1.68), can be found in appendix A of [5].

1.3 Relationships Between Different Models

It turns out that apparently different supermultiplets can be related to each other using several operations.

1. Leaving N and d unchanged, one can increase or decrease the number of physical bosonic degrees of freedom (while necessarily and simultaneously to decrease or increase the number of auxiliary bosonic degrees of freedom) within a supermultiplet by shifting the level of the superfield θ-variables expansion by mean of an automorphism on the superalgebra representation space, commonly called automorphic duality (AD).
2. It is possible to reduce the number of supersymmetries maintaining fixed representation dimension (reduction).
3. The space-time coordinates can be increased preserving the supersymmetries (oxidation).
4. By a space-time compactification, supersymmetries can be eventually increased.

These powerful tools can be combined together to discover new supermultiplets or to relate the known ones. The first two points will be analyzed in the following paragraphs while for the last two procedures, we remind to [6, 7] and references therein.

1.3.1 Automorphic Duality Transformations

Until now, we encountered the following two types of representations: the first one defined on $\mathcal{V}_L$ and $\mathcal{V}_R$ superspace complemented with the second one, Clifford algebraic superfields. In the latter case we observed that in order to obtain irreducible representations, it is needed a restriction to normal representations or to their normal parts. If we consider irreducible cases of Clifford algebraic superfields then there exists the surprising possibility to transmute physical fields into auxiliary ones changing the supermultiplet degrees of freedom dynamical nature. The best way to proceed for an explanation of the subject is to begin with the $N = 1$ example which came out to be the simplest. In this case only two supermultiplets are present:

- the scalar supermultiplet (X, ψ) respectively composed of one bosonic and one fermionic field arranged in the superfield

$$X(\tau, \theta) = X(\tau) + i\theta\psi(\tau) , \qquad (1.69)$$

with transformation properties

$$\delta_Q X = i\epsilon\psi \,,$$
$$\delta_Q \psi = \epsilon\partial_\tau X \,; \tag{1.70}$$

- the spinor supermultiplet (ξ, A) respectively composed of one bosonic and one fermionic field arranged in the superfield

$$Y(\tau,\theta) = \xi(\tau) + \theta A(\tau) \,, \tag{1.71}$$

with transformation properties

$$\delta_Q A = i\epsilon\partial_\tau\xi \,,$$
$$\delta_Q \xi = \epsilon A \,. \tag{1.72}$$

The invariant Lagrangian for the scalar supermultiplet transformations (1.70)

$$\mathcal{L} = \dot{X}^2 + ig\psi\dot{\psi} \,, \tag{1.73}$$

gives to the fields X and ψ a dynamical meaning and offers the possibility to perform an automorphic duality map that at the superfield level reads

$$Y(\tau,\theta) = -i\mathcal{D}X(\tau,\theta) \,, \tag{1.74}$$

where $\mathcal{D} = \partial_\theta + i\theta\partial_\tau$ is the superspace covariant derivative. At the component level (1.74) corresponds to the map upon bosonic components

$$X(\tau) = \partial_\tau^{-1}A(\tau) \,, \tag{1.75}$$

and identification of fermionic ones. The mapping (1.96) is intrinsically not local but it can be implemented in a local way both in the transformations (1.70) and in the Lagrangian (4.220) producing respectively (1.72) and the Lagrangian

$$\mathcal{L} = A^2 + ig\psi\dot{\psi} \,. \tag{1.76}$$

As a result we get that automorphic duality transformations map $N = 1$ supermultiplets into each other in a local way, changing the physical meaning of the bosonic field X from dynamical to auxiliary A (not propagating) as is showed by the Lagrangian (1.76) invariant for (1.72) transformations. Note that the auxiliary meaning of A is already encoded into (1.72) transformations that enlighten on the nature of the fields and consequently of the supermultiplet.

Let us pass to the analysis of the $N = 2$ case making a link with the considerations about representation theory discussed above. At the $N = 2$ level, we deal with an AC representation so, in order to implement AD transformations, we focus on the normal part, namely $\wedge\mathcal{GR}(d, N)$, defining the Clifford algebraic bosonic superfield

$$\phi_j{}^i = \phi\delta_j{}^i + \phi^{IJ}(f_{IJ})_j{}^i \ ,$$
$$\psi_{\dot\alpha}{}^i = \psi^I(f_I)_{\dot\alpha}{}^i \ , \tag{1.77}$$

constructed with the forms (1.60). Notice that if we work in an N-dimensional space then the highest rank for the forms is N. This is the reason why, writing (1.77), we stopped at f^{IJ} level. Some comments about transformation properties. By comparing each level of the expansion, it is straightforward to prove that superfields (1.61) transform according to (1.26) if the component field transformations are recognized to be

$$\delta_Q\phi^{I_1\cdots I_{p_{\rm even}}} = -i\epsilon^{[I_1}\psi^{I_2\cdots I_{p_{\rm even}}]} + i(p_{\rm even}+1)\epsilon_J\psi^{I_1\cdots I_{p_{\rm even}}J} \ ,$$
$$\delta_Q\psi^{I_1\cdots I_{p_{\rm odd}}} = -\epsilon^{[I_1}\dot\phi^{I_2\cdots I_{p_{\rm odd}}]} + i(p_{\rm odd}+1)\epsilon_J\dot\phi^{I_1\cdots I_{p_{\rm odd}}J} \ . \tag{1.78}$$

Therefore (1.78) for the $N=2$ case read

$$\delta_Q\phi = i\epsilon_I\psi^I \ ,$$
$$\delta_Q\psi^I = -\epsilon^I\dot\phi + 2\epsilon_J\dot\phi^{IJ} \ ,$$
$$\delta_Q\phi^{IJ} = -i\epsilon^{[I}\psi^{J]} \ . \tag{1.79}$$

Once again, transformations (1.79) admit local AD maps between bosonic fields. To discuss this in a way that brings this discussion in line with that of [8], we adhere to a convention that list three numbers (PB, PF, AB) where PB denotes the number of "propagating" bosonic fields, AB denotes the number of "auxiliary" bosonic fields, and PF denotes the number of "fermionic" fields.

We briefly list the resulting supermultiplets arising from the dualization procedure.

- The AD map involving ϕ field

$$\phi(\tau) = \partial_\tau^{-1}A(\tau) \ , \tag{1.80}$$

  yield a $(1,2,1)$ supermultiplet whose transformation properties are

$$\delta_Q A = i\epsilon_I\partial_\tau\psi^I \ ,$$
$$\delta_Q\psi^I = -\epsilon^I A + 2\epsilon_J\dot\phi^{IJ} \ ,$$
$$\delta_Q\phi^{IJ} = -i\epsilon^{[I}\psi^{J]} \ . \tag{1.81}$$

- By redefining the ϕ^{IJ} field

$$\phi^{IJ} = \partial_\tau^{-1}B^{IJ} \ , \tag{1.82}$$

  another $(1,2,1)$ supermultiplet is obtained. Accordingly, we have

$$\delta_Q\phi = i\epsilon_I\psi^I \ ,$$
$$\delta_Q\psi^I = -\epsilon^I\dot\phi + 2\epsilon_J A^{IJ} \ ,$$
$$\delta_Q A^{IJ} = -i\epsilon^{[I}\partial_\tau\psi^{J]} \ . \tag{1.83}$$

- Finally, if both redefinitions (1.80) and (1.82) are adopted, then we are left with $(0, 2, 2)$ spinor supermultiplet whose components behave as

$$\delta_Q A = i\epsilon_I \partial_\tau \psi^I \, ,$$
$$\delta_Q \psi^I = -\epsilon^I A + 2\epsilon_J A^{IJ} \, ,$$
$$\delta_Q A^{IJ} = -i\epsilon^{[I} \partial_\tau \psi^{J]} \, . \tag{1.84}$$

It is important to stress that we can make redefinitions of bosonic fields via AD maps that involve higher time derivatives. For instance, by applying ∂_τ^2 to the first equation in (1.79) together with the new field introduction

$$\phi = \partial_\tau^{-2} C \, , \tag{1.85}$$

transformations turn out to be free from nonlocal terms if AD for the remaining fields

$$\psi^I = i\partial_\tau^{-1} \xi^I$$
$$\phi^{IJ} = \partial_\tau^{-1} D^{IJ} \tag{1.86}$$

are enforced. Thus, we end with

$$\delta_Q C = -\epsilon_I \partial_\tau \xi^I$$
$$\delta_Q \xi^I = i\epsilon^I C - 2i\epsilon_J \dot{D}^{IJ}$$
$$\delta_Q D^{IJ} = \epsilon^{[I} \xi^{J]} \, . \tag{1.87}$$

From (1.87), one may argue that C is auxiliary while D is physical. The point is which is the meaning of the fields we started from? An invariant action from (1.87) is

$$\mathcal{L} = C^2 + ig\xi\dot{\xi} + \dot{D}_{IJ}\dot{D}^{IJ} \, , \tag{1.88}$$

so that going backward, we can deduce the initial action

$$\mathcal{L} = \ddot{\phi}^2 + ig\dot{\psi}\ddot{\psi} + \ddot{\phi}_{IJ}\dddot{\phi}^{IJ} \, . \tag{1.89}$$

The examples above should convince any reader that Clifford superfields are a starting point to construct a wider class of representation by means of AD maps. Following this idea, one can identify each supermultiplet with a correspondent root label $(a_1, \ldots, a_k)_\pm$ where $a_i \in Z$ are defined according to

$$(\tilde{\phi}, \tilde{\psi}^I, \tilde{\phi}^{IJ}, \ldots)_+ = (\partial_\tau^{-a_0}\phi, \partial_\tau^{a_1}\psi^I, \partial_\tau^{-a_2}\phi^{IJ}, \ldots)_+ \, ,$$
$$(\tilde{\psi}, \tilde{\phi}^I, \tilde{\psi}^{IJ}, \ldots)_{0\_} = (\partial_\tau^{a_0}\psi, \partial_\tau^{-a_1}\phi^I, \partial_\tau^{a_2}\psi^{IJ}, \ldots)_\_ \, , \tag{1.90}$$

and $\pm$ distinction between Clifford superfields of bosonic and fermionic type. For instance, the last supermultiplet (1.87), corresponds to the case $(a_0, a_1, a_2)_\pm = (2, -1, 1)_+$. We name base superfield the one with all zero in the root label $(0, \ldots, 0)_\pm$, underling that in the plus (minus) case, this supermultiplet has to be intended as the one with all bosons (fermion) differentiated in the r.h.s. of variations. They are of particular interest in the supermultiplets whose root labels involve only 0 and 1. All these supermultiplets form what we call root tree.

1.3.2 Reduction

It is shown in Table 1.2 that $N = 8, 7, 6, 5$ irreducible representations have the same dimension. The same happens for the $N = 4, 3$ cases. This fact reflects the possibility to relate those supermultiplets via a reduction procedure. To explain how this method works, consider a form $f_{I_1 \cdots I_K}$ and notice that the indices $I_1, \ldots, I_K$ run on the number of supersymmetries: reducing this number corresponds to diminishing the components contained in the rank k form. The remaining components have to be rearranged into another form. For instance, if we consider a 3-rank form for the $N = 8$ case then the number of components is given by[2] $\binom{8}{3} = 56$ but, reducing to the $N = 7$ case and leaving invariant the rank, we get $\binom{7}{3} = 35$ components. The remaining ones can be rearranged in a 5-rank form. This means that the maximum rank of Clifford superfield expansion is raised until the irreducible representation dimension is reached. However, the right way to look at this rank enhancing is through duality. An enlightening example will be useful. By a proper counting of irreducible representation dimension for the $\mathcal{EGR}(8, 8)$, we are left with $\{\mathcal{U}_L\} \oplus \{\mathcal{M}_L\}$ type Clifford algebraic superfield

$$\phi_{ij} = \phi \delta_{ij} + \phi^{IJ} \left(f_{IJ}\right)_{ij} + \phi^{IJKL} \left(f_{IJKL}\right)_{ij}$$
$$\psi_{\hat{\alpha}i} = \psi^I \left(f_I\right)_{\hat{\alpha}i} + \psi^{IJK} \left(f_{IJK}\right)_{\hat{\alpha}i} , \tag{1.91}$$

where the 4-form has definite duality or, more precisely, the sign in the equation

$$\epsilon^{IJKLMNPQ} f_{MNPQ} = \pm f^{IJKL} , \tag{1.92}$$

has been chosen, halving the number of independent components. To reduce to the $N = 7$ case, we need to eliminate all "8" indices and this can be done by exploiting the duality. For instance, f_{I8} can disappear if transformed into

$$\epsilon^{IJKLMNP8} f_{P8} = \pm f^{IJKLMN} . \tag{1.93}$$

This trick adds the 6-rank to the expansion manifesting the enhancing phenomenon previously discussed. Once the method is understood, it is straightforward to prove that for the $N = 7$ case, the proper superfield expression is

$$\phi_{ij} = \phi \delta_{ij} + \phi^{IJ} \left(f_{IJ}\right)_{ij} + \phi^{IJKL} \left(f_{IJKL}\right)_{ij} + \phi^{IJKLMN} \left(f_{IJKLMN}\right)_{ij}$$
$$\psi_{i\hat{j}} = \psi^I \left(f_I\right)_{i\hat{j}} + \psi^{IJK} \left(f_{IJK}\right)_{i\hat{j}} + \psi^{IJKLM} \left(f_{IJKLM}\right)_{i\hat{j}}$$
$$+ \psi^{IJKLMPQ} \left(f_{IJKMNPQ}\right)_{i\hat{j}} . \tag{1.94}$$

The explicit reduction procedure for $N \leq 8$ can be found in [2] and summarized in Tables 1.3 and 1.4.

---

[2] For the construction of $N = 8$ supersymmetric mechanics, see [9]; the nonlinear chiral multiplet has been used in this connection [10], as well as in related tasks [11].

Table 1.3. $\mathcal{EGR}(4,4)$ and its reduction: algebras representation in terms of forms and division algebra

$\mathcal{EGR}(d,N)$	$\mathcal{AGR}(d,N)$ basis	Division Structure
$\mathcal{EGR}(4,4)$	$\{\mathcal{U}_L\} = \{I, f_{IJ}, \mathcal{E}^{\hat{\mu}}, f_{IJ}\mathcal{E}^{\hat{\mu}}\}$ $\{\mathcal{M}_L\} = \{f_I, f_I\hat{\mathcal{E}}^{\hat{\mu}}\}$	$\mathcal{E}^{\hat{\mu}}, \hat{\mathcal{E}}^{\hat{\mu}}$
$\mathcal{EGR}(4,3)$	$\{\mathcal{U}_L\} = \{I, f_{IJ}, \mathcal{E}^{\hat{\mu}}, f_{IJ}\mathcal{E}^{\hat{\mu}}\}$ $\{\mathcal{M}_L\} = \{f_I, f_I\hat{\mathcal{E}}^{\hat{\mu}}, f_{IJK}, f_{IJK}\hat{\mathcal{E}}^{\hat{\mu}}\}$	$\mathcal{E}^{\hat{\mu}}, \hat{\mathcal{E}}^{\hat{\mu}}$

Here and in the following table, the generators $\mathcal{E}^{\hat{\mu}}$, $\hat{\mathcal{E}}^{\hat{\mu}}$ are respectively the $+$ and $-$ projections of the quaternionic division algebra generators in the Clifford space. The same projection on complex structure originate D, $\hat{D}$.

1.4 Applications

1.4.1 Spinning Particle

Before we begin a detailed analysis of spinning particle system it is important to understand what a spinning particle is. Early models of relativistic particle with spin involving only commuting variables can be divided into the two following classes:

- vectorial models, based on the idea of extending Minkowski space-time by vectorial internal degrees of freedom;
- spinorial models, characterized by the enhancing of configuration space using spinorial commuting variables.

These models lack the following important requirement: after first quantization, they never produce relativistic Dirac equations. Moreover, in the spinorial cases, a tower of all possible spin values appear in the spectrum. Further progress in the development of spinning particle descriptions was achieved by the introduction of anticommuting variables to describe internal degrees of freedom [12]. This idea stems from the classical limit ($h \to 0$) formulation of Fermi systems [13], the so called "pseudoclassical mechanics" referring to the fact that it is not an ordinary mechanical theory because of the presence of Grassmannian variables. By means of pseudoclassical approach, vectorial and spinorial models can be generalized to "spinning particle" and "superparticle" models, respectively. In the first case, the extension to superspace $(x_\mu, \theta_\mu, \theta_5)$

Table 1.4. $\mathcal{EGR}(8,8)$ and its reductions

$\mathcal{EGR}(d,N)$	$\Lambda\mathcal{GR}(d,N)$ Basis	Division Structure
$\mathcal{EGR}(8,8)$	$\{\mathcal{U}_L\} = \{I, f_{IJ}, f_{IJKL}\}$ $\{\mathcal{M}_L\} = \{f_I, f_{IJK}\}$	I
$\mathcal{EGR}(8,7)$	$\{\mathcal{U}_L\} = \{I, f_{IJ}, f_{IJKL}, f_{IJKLMN}\}$ $\{\mathcal{M}_L\} = \{f_I, f_{IJK}, f_{[5]}, f_{[7]}\}$	I
$\mathcal{EGR}(8,6)$	$\{\mathcal{U}_L\} = \{I, f_{I7}, f_{IJ}, f_{IJK7}, f_{IJKL}, f_{[5]7}, f_{[6]}\}$ $\{\mathcal{M}_L\} = \{f_7, f_I, f_{IJ7}, f_{IJK}, , f_{[4]7}, f_{[5]}, f_{[6]7}\}$	$D = f_7$
$\mathcal{EGR}(8,5)$	$\{\mathcal{U}_L\} = \{I, f_{67}, f_{I6}, f_{I7}, f_{IJ}, f_{IJ67}, f_{IJK7},$ $f_{IJKL}, f_{IJKL67}, f_{[5]7}, f_{[5]6}\}$ $\{\mathcal{M}_L\} = \{f_7, f_6, f_I, f_{I67}, f_{IJ7}, f_{IJ6}, f_{IJK},$ $f_{IJk67}, f_{[4]7}, f_{[4]6}, f_{[5]}, f_{[5]67}, \}$	$\mathcal{E}^{\hat{\mu}} = (f_{67}, f_{[5]6}, f_{[5]7})$ $\hat{\mathcal{E}}^{\hat{\mu}} = (f_7, f_6, f_{[5]67})$

Here the subscript $[n]$ is used in place of n anticommuting indices.

is made possible by a pseudovector θ_μ and a pseudoscalar θ_5 [14, 15]. The presence of vector index associated with θ-variables implies the vectorial character of the model.

In the second case, spinorial coordinates are considered, giving rise to ordinary superspace approach whose underlying symmetry is the super-Poincaré group (eventually extended) [13]. The superparticle is nothing but a generalization of relativistic point particle to superspace.

It turns out that after first quantization, the spinning particle model produced Dirac equations and all Grassmann variables are mapped into Clifford algebra generators. Superfields that take values on this kind of quantized superspace are precisely Clifford algebraic superfields described in the previous paragraphs. On the other side, a superspace version of Dirac equation arises from superparticle quantization. Moreover, θ-variables are still present in the quantized version.

To have a more precise idea, we spend a few words discussing the Barducci–Casalbuoni–Lusanna model [14], which is one of the first works on pseudo-classical model. As already mentioned, it is assumed that the configuration space to be described by $(x_\mu, \theta_\mu, \theta_5)$. The Lagrangian of the system

$$\mathcal{L}_{BCL} = -m\sqrt{\left(\dot{x}^\mu - \frac{i}{m}\theta^\mu\dot{\theta}_5\right)\left(\dot{x}_\mu - \frac{i}{m}\theta_\mu\dot{\theta}_5\right)} - \frac{i}{2}\theta_\mu\dot{\theta}^\mu - \frac{i}{2}\theta_5\dot{\theta}_5 \quad (1.95)$$

is invariant under the transformations

$$\delta x_\mu = -\epsilon_\mu a\theta_5 + \epsilon_5 b\theta_\mu ,$$
$$\delta\theta_\mu = \epsilon_\mu ,$$
$$\delta\theta_5 = \epsilon_5 . \quad (1.96)$$

and produces the equations of motion

$$p^2 - m^2 = 0 ,$$
$$p_\mu\theta^\mu - m\theta_5 = 0 , \quad (1.97)$$

after a canonical analysis. These (1.97) are classical limits of Klein–Gordon and Dirac equations, respectively. Moreover, the first quantization maps θ-variables into Clifford algebra generators

$$\theta_\mu \to \gamma_\mu\gamma_5 \quad \text{(pseudovector)} ,$$
$$\theta_5 \to \gamma_5 \quad \text{(pseudoscalar)} , \quad (1.98)$$

so that (1.97) exactly reproduce relativistic quantum behavior of a particle with spin. Even if it is not manifest, it is possible to find a particular direction in the (θ_μ, θ_5) space along which the theory is invariant under the following localized supersymmetry transformation

$$\delta x_\mu = 2\frac{i}{m^2}\epsilon_5(\tau)P_\mu\theta_5(\tau) - \frac{i}{m}\epsilon_5(\tau)\theta_\mu ,$$
$$\delta\theta_\mu = \frac{1}{m}\epsilon_5(\tau)P_\mu ,$$
$$\delta\theta_5 = \epsilon_5(\tau) , \quad (1.99)$$

opening the way to supergravity. Basic concepts on the extension to minimal supergravity-coupled model can be found in [16]. Here the proposed action is a direct generalization of one-dimensional general covariant free particle to include the spin; for the first-order formalism in the massless case we have

$$S = \int d\tau \left\{ P^\mu\dot{X}_\mu - \frac{1}{2}eP^2 - \frac{i}{2}\psi^\mu\dot{\psi}_\mu - \frac{i}{2}\chi\psi^\mu P_\mu \right\} , \quad (1.100)$$

with the local invariances

$$\delta\psi^\mu = \epsilon(\tau)P^\mu, \quad \delta\phi^\mu = i\epsilon(\tau)\psi^\mu, \quad \delta P^\mu = 0 ,$$
$$\delta e = i\epsilon(\tau)\chi, \quad \delta\chi = 2\dot\epsilon(\tau) , \tag{1.101}$$

corresponding to pure supergravity transformations as is shown by calculating the commutators

$$[\delta_{\epsilon_1}, \delta_{\epsilon_2}] X^\mu = \xi\dot X^\mu + i\tilde\epsilon\psi^\mu ,$$
$$[\delta_{\epsilon_1}, \delta_{\epsilon_2}] \psi^\mu = \xi\dot\psi^\mu + \tilde\epsilon P^\mu ,$$
$$[\delta_{\epsilon_1}, \delta_{\epsilon_2}] e = \xi\dot e + \dot\xi e + i\tilde\epsilon\chi ,$$
$$[\delta_{\epsilon_1}, \delta_{\epsilon_2}] \chi = \xi\dot\chi + \dot\xi\chi + 2\dot{\tilde\epsilon} , \tag{1.102}$$

where

$$\xi = 2ie^{-1}\epsilon_2\epsilon_1 ,$$
$$\tilde\epsilon = -\frac{1}{2}\xi\chi . \tag{1.103}$$

In fact, the r.h.s. of (1.102) describes both general coordinate and local supersymmetry transformations.

To produce a mass-shell condition, the massive version of the above model requires the presence of a cosmological term in the action

$$S = -\frac{1}{2} \int d\tau\, e m^2 \tag{1.104}$$

that, in turn, imply the presence of an additional anticommuting field ψ_5, transforming through

$$\delta\psi_5 = m\tilde\epsilon , \tag{1.105}$$

to construct terms that restore the symmetries broken by (1.104). The complete action describing the massive spinning particle version minimally coupled to supergravity multiplet turns out to be

$$S = \int d\tau \left[P^\mu \dot X_\mu - \frac{1}{2}e(P^2 + m^2) - \frac{i}{2}(\psi^\mu\dot\psi_\mu + \psi_5\dot\psi_5) - \frac{i}{2}\chi(\psi^\mu P_\mu + m\psi_5) \right]. \tag{1.106}$$

The second-order formalism for the massless and massive model follow straightforwardly from actions (1.106) and (1.100) eliminating the P fields using their equations of motion.

An advance on this line of research yielded the on-shell N-extension [17]. However, a satisfactory off-shell description with arbitrary N requires the $\mathcal{GR}(d, N)$ approach. In the paragraphs below we describe in detail how this construction is worked out.

Second-Order Formalism for Spinning Particle with Rigid N-Extended Supersymmetry

The basic objects of this model are Clifford algebraic bosonic superfields valued in $\{\mathcal{U}_L\} \oplus \{\mathcal{M}_L\}$ superspace with transformations (1.26). One can easily check that the action

$$S = \int d\tau \{ (\partial_\tau(\phi_1)_i{}^j)(\partial_\tau(\phi_1)_j{}^i) + i(\psi_1)_i{}^{\dot\alpha}\partial_\tau(\psi_1)_{\dot\alpha}^i \} \tag{1.107}$$

is left unchanged by (1.26). The next step consists in separating the physical degrees of freedom in $((\phi_1)_i{}^j, (\psi_1)_{\dot\alpha}^i)$ from nonphysical ones. For the bosonic superfield, valued in $\{\mathcal{U}_L\}$, we separate the trace from the remaining components

$$(\phi_1)_i{}^j = X\delta_i{}^j + \tilde\phi_i{}^j \,,$$
$$X = \phi_i{}^i, \quad \tilde\phi_i{}^i = 0 \,, \tag{1.108}$$

and perform an AD transformation on tilded components

$$\tilde\phi_i{}^j = \partial_\tau^{-1}\mathcal{F}_i{}^j \,, \tag{1.109}$$

to end with the decomposition

$$\phi_i{}^j = X\delta_i{}^j + \partial_\tau^{-1}\mathcal{F}_i{}^j \,, \tag{1.110}$$

constrained by the equation

$$\mathcal{F}_i{}^i = 0. \tag{1.111}$$

The field component X can be interpreted as the spinning particle bosonic coordinate in a background space. Nothing forbids us from considering D supermultiplet of this kind that amounts, to add a D-dimensional background index μ to the superfields

$$(\phi_1)_i{}^j \to (\phi_1^\mu)_i{}^j \,,$$
$$(\psi_1)_{\dot\alpha}^i \to (\psi_1^\mu)_{\dot\alpha}^i. \tag{1.112}$$

In this way the dimension of the background space has no link neither with the number of supersymmetries nor with representation dimension. However, to simplify the notation, background index will be omitted. Transformations involving the fields defined in (1.111) reads

$$\delta_Q X = -\frac{1}{d} i\epsilon^I (R_I)_i{}^{\dot\alpha}(\psi_1)_{\dot\alpha}^i \,,$$
$$\delta_Q \mathcal{F}_i{}^j = {}'i\epsilon^I (R_I)_i{}^{\dot\alpha}\partial_\tau(\psi_1)_{\dot\alpha}^j \,,$$
$$\delta_Q (\psi_1)_{\dot\alpha}^i = \epsilon^I (L_I)_{\dot\alpha}^j \mathcal{F}_j{}^i + \epsilon^I (L_I)_{\dot\alpha}^i \partial_\tau X. \tag{1.113}$$

Even in the fermionic case, we need that only the lowest component in the expansion (1.61) has physical meaning so that the higher level components can be read as auxiliary ones by means of AD map. Fermionic superfield components happen to be distributed in the following manner

$$(\psi_1)_{\dot\alpha}^i = \psi^I (L_I)_{\dot\alpha}^i + \tilde\psi_{\dot\alpha}^i = \psi^I (L_I)_{\dot\alpha}^i + \mu_{\dot\alpha}^i \,, \tag{1.114}$$

where $\psi^I = \frac{1}{d}(R_I)_i{}^{\hat\alpha}(\psi_1)_{\hat\alpha}^i$ and the fermionic superfield $\mu_{\hat\alpha}^i$ obey the constraint equation

$$(R_I)_i{}^{\hat\alpha}\mu_{\hat\alpha}^i = 0 \,. \tag{1.115}$$

After the substitution of the new component fields (1.114), transformations (1.113) became

$$\delta_Q X = -\frac{1}{d}i\epsilon^I (R_I)_i{}^{\hat\alpha}(L_J)_{\hat\alpha}^i \psi^J \,, \tag{1.116}$$

$$\delta_Q \mathcal{F}_i{}^j = i\epsilon^I (R_I)_i{}^{\hat\alpha}(L_J)_{\hat\alpha}^j \partial_\tau \psi^J - i\epsilon^I (R_I)_i{}^{\hat\alpha}\partial_\tau \mu_{\hat\alpha}^j \,, \tag{1.117}$$

$$(L_I)_{\hat\alpha}^i \delta_Q \psi^I + \delta_Q \mu_{\hat\alpha}^i = \epsilon^I (L_I)_{\hat\alpha}^j \mathcal{F}_j{}^i + \epsilon^I (L_I)_{\hat\alpha}^i \partial_\tau X, \tag{1.118}$$

where we used (1.115) to obtain (1.1). Equations (1.1) and (1.2) can be simplified into

$$\delta_Q X = i\epsilon^I \psi_I \,,$$
$$\delta_Q \mathcal{F}_i{}^j = i\epsilon^I (f_{IJ})_i{}^j \partial_\tau \psi^I - i\epsilon^I (R_I)_i{}^{\hat\alpha}\partial_\tau \mu_{\hat\alpha}^j \,, \tag{1.119}$$

if one notice that

$$(R_I)_i{}^{\hat\alpha}(L_J)_{\hat\alpha}^i = -d\delta_{IJ} \,,$$
$$(R_I)_i{}^{\hat\alpha}(L_J)_{\hat\alpha}^j = (f_{IJ})_i{}^j \tag{1.120}$$

while (1.3) need more care. To separate the variation of ψ^I and $\mu_{\hat\alpha}^i$, we multiply by $(R_J)_i{}^{\hat\alpha}$ to eliminate the $\mu_{\hat\alpha}^i$ contribution, thanks to (1.115). As a result we get

$$- d\delta_Q \psi_J = \epsilon^I (f_{JI})_i{}^j \mathcal{F}_j{}^i - d\epsilon_J \partial_\tau X$$
$$\Downarrow$$
$$\delta_Q \psi_I = \epsilon_I \partial_\tau X - \frac{1}{d}\epsilon^J (f_{IJ})_i{}^j \mathcal{F}_j{}^i. \tag{1.121}$$

Substituting back (1.121) into (1.3) we finally have

$$\delta_Q \mu_{\hat\alpha}^i = -(L_I)_{\hat\alpha}^i \left[\epsilon^I \partial_\tau X - \frac{1}{d}\epsilon^J (f^I)_i{}^j \mathcal{F}_j{}^i\right] + \epsilon^I (L_I)_{\hat\alpha}^j \mathcal{F}_j{}^i + \epsilon^I (L_I)_{\hat\alpha}^i \partial_\tau X$$

$$= \left[\epsilon^I (\hat{f}_I)_{\hat\alpha}^j + \frac{1}{d}\epsilon^J (\hat{f}_I)_{\hat\alpha}^i (f^I_J)_i{}^j\right] \mathcal{F}_j{}^i \,. \tag{1.122}$$

The supermultiplet $(X, \mathcal{F}_i{}^j, \psi_I, \mu_{\hat\alpha}^i)$ together with transformations (1.119), (1.121), and (1.122), is called "universal spinning particle multiplet" (USPM). Acting with the maps (1.111) and (1.114) on the action (1.107) we obtain the USPM invariant action

$$S = \int d\tau \{d(\partial_\tau X \partial_\tau X - i\psi_I \partial_\tau \psi_I) + \mathcal{F}_i{}^j \mathcal{F}_i{}^j + i\mu_i{}^{\hat\alpha}\partial_\tau \mu_{\hat\alpha}^i\} \tag{1.123}$$

that represent the second-order approach to the spinning particle problem with global supersymmetry. A remarkable difference between the AD presented in Subsect. 1.3.1 and the AD used to derive USPM resides in the fact that in the latter case we map $\tilde{\phi}_i{}^j$ and $\tilde{\psi}_{\hat{\alpha}}{}^i$ which are Clifford algebraic superfield while in the previous we work at the component level. Finally, it is important to keep in mind that the superfields $\tilde{\phi}_i{}^j$ and $\tilde{\psi}_{\hat{\alpha}}{}^i$ take values on the normal part of the enveloping algebra which is equivalent to say that they can be expanded on the basis (1.60).

First-Order Formalism for Spinning Particle with Rigid N-Extended Supersymmetry

To formulate a first-order formalism, one more fermionic supermultiplet is required. This time the superfields $((\phi_2)_i{}^j, (\psi_2)_i{}^{\hat{\alpha}})$, valued in $\{\mathcal{U}_L\} \oplus \{\mathcal{M}_R\}$ superspace, transform according to

$$\delta_Q(\phi_2)_i{}^j = -i\epsilon^I (L_I)_{\hat{\alpha}}^j \partial_\tau (\psi_2)_i{}^{\hat{\alpha}}$$
$$\delta_Q(\psi_2)_i{}^{\hat{\alpha}} = \epsilon^I (R_I)_j{}^{\hat{\alpha}} (\phi_2)_i{}^j. \tag{1.124}$$

The expansions needed turns out to be

$$(\phi_2)_i{}^j = P\delta_i{}^j + \mathcal{G}_i{}^j, \quad \mathcal{G}_i{}^i = 0$$
$$(\psi_2)_i{}^{\hat{\alpha}} = \bar{\psi}^I (R_I)_i{}^{\hat{\alpha}} + \mathcal{X}_i{}^{\hat{\alpha}}, \quad (L_I)_{\hat{\alpha}}^i \mathcal{X}_i{}^{\hat{\alpha}} = 0 \tag{1.125}$$

that bring us to the trasformations

$$\delta_Q P = i\epsilon^I \bar{\psi}_I ,$$
$$\delta_Q \mathcal{G}_i{}^j = -i\partial_\tau \epsilon_J (\hat{f}_{IJ})_i{}^j \bar{\psi}_I - i\epsilon^K (L_K)_{\hat{\alpha}}^j \mathcal{X}_i{}^{\hat{\alpha}} ,$$
$$\delta_Q \bar{\psi}_I = \epsilon_I P + d^{-1}\epsilon^J (\hat{f}_{JI})_j{}^i \mathcal{G}_i{}^j ,$$
$$\delta_Q \mathcal{X}_i{}^{\hat{\alpha}} = -d^{-1}\epsilon^J (\hat{f}^I)_i{}^{\hat{\alpha}} (\hat{f}_{JI})_j{}^i \mathcal{G}_i{}^j + \epsilon^I (\hat{f}_I)_j{}^{\hat{\alpha}} \mathcal{G}_i{}^j . \tag{1.126}$$

Here, the scalar supermultiplet $((\phi_1)_i{}^j, (\psi_1)_{\hat{\alpha}}^i)$ has to be treated in the following different way: the off-trace superfield $\mu_{\hat{\alpha}}^i$ undergoes an AD

$$\mu_{\hat{\alpha}}^i \rightarrow \partial_\tau^{-1} \Lambda_{\hat{\alpha}}^i , \tag{1.127}$$

that slightly changes the variation (1.119), (1.121) and (1.122) into

$$\delta_Q X = i\epsilon^I \psi_I ,$$
$$\delta_Q \mathcal{F}_i{}^j = i\epsilon^I (f_{IJ})_i{}^j \partial_\tau \psi^I - i\epsilon^I (R_I)_i{}^{\hat{\alpha}} \Lambda_{\hat{\alpha}}^j ,$$
$$\delta_Q \psi_I = \epsilon_I \partial_\tau X + \frac{1}{d}\epsilon^J (f_{IJ})_i{}^j \mathcal{F}_j{}^i ,$$
$$\delta_Q \Lambda_{\hat{\alpha}}^i = \left[\epsilon^I (L_I)_{\hat{\alpha}}^j - \frac{1}{d}\epsilon^J (\hat{f}_I)_{\hat{\alpha}}^i (f^I{}_J)_i{}^j \right] \partial_\tau \mathcal{F}_j{}^i . \tag{1.128}$$

The action can be thought as the sum of two separated pieces

$$S_{P^2} = \frac{1}{d} \int d\tau \{ (\phi_2)_i{}^j (\phi_2)_j{}^i + i (\psi_2)_{\hat{\alpha}}^i \partial_\tau (\psi_2)_i{}^{\hat{\alpha}} \} ,$$

$$S_{PV} = \frac{1}{d} \int d\tau \{ (\phi_2)_i{}^j \partial_\tau (\phi_1)_j{}^i + i (\psi_2)_i{}^{\hat{\alpha}} \partial_\tau (\psi_1)_{\hat{\alpha}}^i \} , \qquad (1.129)$$

so that, in analogy with the free particle description where the Lagrangian has the form

$$\mathcal{L} = PV - \frac{1}{2} P^2 \qquad (1.130)$$

we consider

$$S = S_{PV} - \frac{1}{2} S_{P^2} , \qquad (1.131)$$

as the correct first-order free spinning particle model. By eliminating the fermionic supermultiplet superfields by their equations of motion, we fall into the second-order description. It is clear that the fermionic supermultiplet is nothing but the conjugated of USPM. After (1.125) substitution the proposed action (1.131) assume the final aspect

$$S_{sp} = \int d\tau \left\{ P \partial_\tau X + \frac{1}{d} \mathcal{G}_i{}^j \mathcal{F}_j{}^i - i \bar{\psi}^I \partial_\tau \psi_I + \frac{i}{d} \mathcal{X}_i{}^{\hat{\alpha}} \Lambda_{\hat{\alpha}}^i \right.$$

$$\left. - \frac{1}{2} P^2 - \frac{1}{2d} \mathcal{G}_i{}^j \mathcal{G}_j{}^i - \frac{i}{2} \bar{\psi}^I \partial_\tau \bar{\psi}_I - \frac{i}{2d} \mathcal{X}_i{}^{\hat{\alpha}} \partial_\tau \mathcal{X}_{\hat{\alpha}}^i \right\} , \qquad (1.132)$$

where the auxiliary superfields P^2, $\mathcal{F}_i{}^j$, $\Lambda_{\hat{\alpha}}^i$, $\mathcal{G}_i{}^j$, and $\mathcal{X}_i{}^{\hat{\alpha}}$ are manifest.

Massive Theory

The massive theory is obtained by adding to the previous first- and second-order actions the appropriate terms where it figures an additional supermultiplet $(\hat{\psi}_i{}^{\hat{\alpha}}, \hat{G}_i{}^j)$, which is fermionic in nature

$$\delta_Q \hat{\psi}_i{}^{\hat{\alpha}} = \epsilon^I (R_I)_j{}^{\hat{\alpha}} \hat{G}_i{}^j ,$$

$$\delta_Q \hat{G}_i{}^j = i \epsilon^I (L_I)_{\hat{\alpha}}^j \partial_\tau \hat{\psi}_i{}^{\hat{\alpha}} , \qquad (1.133)$$

and is inserted through the action

$$S_M = \int d\tau [i \hat{\psi}_{\hat{\alpha}}^i \partial_\tau \hat{\psi}_i{}^{\hat{\alpha}} + \hat{G}_j{}^i \hat{G}_i{}^j + M \hat{G}_i{}^i] . \qquad (1.134)$$

Here the bosonic auxiliary trace $\mathcal{G}_i{}^i$ plays a different role with respect to the other off-trace component because it is responsible, by its equation of motion, for setting the mass equal to M. It can be easily recognized the resemblance between the Sherk–Shwartz method [18] and the above way to proceed if we interpret the mass multiplet as a $(D+1)$th Minkowski momentum component without coordinate analogue. We underline that the "mass multiplet" (1.133) is crucial if we want to insert the mass and preserve the preexisting symmetries, as it happens for the ψ_5 field (1.105).

First- and Second-Order Formalism for Spinning Particle Coupled to Minimal N-Extended Supergravity

For completeness, we include the coupling of the above models to minimal one-dimensional supergravity. The supergravity multiplet escape from $\mathcal{GR}(d,N)$ embedding because its off-shell fields content (e, χ_I),

$$\delta_Q e = -4ie^2 \epsilon_I \chi_I ,$$
$$\delta_Q \chi^I = -\partial_\tau \epsilon_I , \tag{1.135}$$

consists of one real boson and N real fermions. General coordinate variations are

$$\delta_{GC} e = \dot{e}\xi - e\dot{\xi} ,$$
$$\delta_{GC} \chi^I = \partial_\tau (\chi^I \xi) . \tag{1.136}$$

The gauging of supersymmetry requires the introduction of connections by means of generators $\mathcal{A}_{IJ}$ valued on an arbitrary Lie algebra. Local supersymmetric variations come from (1.119), (1.121), and (1.122) by replacing

$$\partial_\tau \rightarrow \mathcal{D} = e\partial_\tau + e\chi^I Q_I + \frac{1}{2} w^{JK} \mathcal{A}_{JK} , \tag{1.137}$$

while the gravitino one can be written

$$\delta_Q \chi^I = - \left[\delta_J^I \partial_\tau - \frac{1}{2} e^{-1} w^{KL} (f_{KL})_J^{\ I} \right] \epsilon^J . \tag{1.138}$$

One can explicitly check that the local supersymmetric invariant action for the above-mentioned transformations, is

$$S = \int d\tau \left\{ e^{-1} \left[P\mathcal{D}_\tau X + \frac{1}{d} \mathcal{G}_i^{\ j} \mathcal{F}_j^{\ i} - i\bar{\psi}^I \mathcal{D}_\tau \psi_I + \frac{i}{d} \mathcal{X}_i^{\ \hat{\alpha}} \Lambda_{\hat{\alpha}}^{\ i} \right. \right.$$
$$\left. - \frac{1}{2} P^2 - \frac{1}{2d} \mathcal{G}_i^{\ j} \mathcal{G}_j^{\ i} - \frac{i}{2} \bar{\psi}^I \mathcal{D}_\tau \bar{\psi}_I - \frac{i}{2d} \mathcal{X}_i^{\ \hat{\alpha}} \mathcal{D}_\tau \mathcal{X}_{\hat{\alpha}}^{\ i} \right]$$
$$- i\chi_I \left[\bar{\psi}_I P + \frac{1}{d} (f_{IJ})_j^{\ i} \mathcal{G}_i^{\ j} \bar{\psi}_J + \frac{1}{d} (L_I)_{\hat{\alpha}}^{\ i} \mathcal{G}_i^{\ j} \mathcal{X}_j^{\ \hat{\alpha}} \right.$$
$$\left. \left. \psi_I P + \frac{1}{d} (f_{IJ})_j^{\ i} \mathcal{G}_i^{\ j} \psi_J - \frac{1}{d} (L_I)_{\hat{\alpha}}^{\ i} \mathcal{F}_i^{\ j} \mathcal{X}_j^{\ \hat{\alpha}} \right] \right\} , $$
$$\tag{1.139}$$

providing the first-order massless model for spinning particle minimally coupled to N-extended supergravity on the worldline. Equation of motion associated to P field reads

$$P = \mathcal{D}_\tau X - i\chi_I \bar{\psi}_I - i\chi_I \psi_I , \tag{1.140}$$

that, substituted in (1.139), give us the second-order formulation

$$S = \int d\tau \Big\{ \frac{1}{2} \mathcal{D}_\tau X \mathcal{D}_\tau X + \frac{1}{d} \mathcal{G}_i{}^j \mathcal{F}_j{}^i - i\bar{\psi}^I \mathcal{D}_\tau \psi_I + \frac{i}{d} \mathcal{X}_i{}^{\hat{\alpha}} \Lambda_{\hat{\alpha}}^i$$

$$- \frac{1}{2d} \mathcal{G}_i{}^j \mathcal{G}_j{}^i - \frac{i}{2} \bar{\psi}^I \mathcal{D}_\tau \bar{\psi}_I - \frac{i}{2d} \mathcal{X}_i{}^{\hat{\alpha}} \mathcal{D}_\tau \mathcal{X}_{\hat{\alpha}}^i$$

$$- \frac{i}{d} \chi_I \Big[(f_{IJ})_j{}^i \mathcal{G}_i{}^j \bar{\psi}_J + (L_I)_{\hat{\alpha}}^i \mathcal{G}_i{}^j \mathcal{X}_j{}^{\hat{\alpha}}$$

$$\Big(f_{IJ} \Big)_j{}^i \mathcal{G}_i{}^j \psi_J - (L_I)_{\hat{\alpha}}^i \mathcal{F}_i{}^j \mathcal{X}_j{}^{\hat{\alpha}} + \chi\chi \text{terms} \Big] \Big\}. \qquad (1.141)$$

Finally, the massive theory is obtained following the ideas of the previous paragraph. The massive supermultiplet (1.133) is coupled to supergravity supermultiplet by the extension of (1.134) to the local supersymmetric case

$$S_M = \int d\tau \{ i e^{-1} \hat{\psi}_{\hat{\alpha}}^i \partial_\tau \hat{\psi}_i{}^{\hat{\alpha}} + e^{-1} \hat{G}_j{}^i \hat{G}_i{}^j + e^{-1} M \hat{G}_i{}^i$$

$$+ i \chi^I (L_I)_{\hat{\alpha}}^i \hat{\psi}_j^{\hat{\alpha}} \hat{G}_i{}^j - i M \chi^I (R_I)_i{}^{\hat{\alpha}} \hat{\psi}_{\hat{\alpha}}^i \}, \qquad (1.142)$$

that is exactly what we need to add to the action (1.139) to achieve a completely off-shell massive first-order description. We close this review by noting that important issues regarding zero-modes of the models discussed above have yet to be resolved. So, we do not regard this as a completed subject yet.

1.4.2 $N = 8$ Unusual Representations

There exists also some "unusual" representations in this approach to one-dimensional supersymmetric quantum mechanics. As an illustration of these, the discussion will now treat such a case for $\mathcal{GR}(8,8)$. It may be verified that a suitable representation is provided by the 8×8 matrices

$$
\begin{aligned}
L_1 &= i\sigma^3 \otimes \sigma^2 \otimes \sigma^1 \ , & L_5 &= i\sigma^1 \otimes \sigma^1 \otimes \sigma^2 \ , \\
L_2 &= i\sigma^3 \otimes \sigma^2 \otimes \sigma^3 \ , & L_6 &= i\sigma^1 \otimes \sigma^3 \otimes \sigma^2 \ , \\
L_3 &= i\sigma^3 \otimes I_2 \otimes \sigma^2 \ , & L_7 &= i\sigma^1 \otimes \sigma^2 \otimes I_2 \ , \\
L_4 &= -i\sigma^2 \otimes I_2 \otimes I_2 \ , & L_8 &= I_2 \otimes I_2 \otimes I_2 \ .
\end{aligned}
\qquad (1.143)
$$

An octet of scalar fields A_I and spinor fields Ψ_I may be introduced. The supersymmetry variation of these are defined by

$$\delta_Q A_J = i\epsilon^I (L_J)_{IK} \Psi_K \ , \qquad \delta_Q \Psi_K = -\epsilon^I (R_N)_{KI} \Big(\partial_\tau A_N \Big), \qquad (1.144)$$

where I, J, K, etc. now take on the values 1, 2,...,8. Proper closure of the supersymmetry algebra requires in addition to (1.6) also the fact that

$$(R_N)_{KJ}(L_N)_{IM} + (R_N)_{KI}(L_N)_{JM} = -2\delta_{IJ}\delta_{KM} \ , \qquad (1.145)$$

which may be verified for the representation in (1.143). This is the fact that identifies the representation in (1.144) as being an "unusual" representation.

Table 1.5. The representation in (1.144) has been given the name "ultra-multiplets"

p	Degeneracy
8	1
7	2
6	1
5	2
4	4
3	2
2	1
1	2
0	1

The sum of the degeneracies adds to 16.

The representation in (1.144) through the action of various AD maps generates many closely related representations. In fact, it can be seen that for any integer p (with $0 \leq p \leq 8$) there exist (p, 8, 8 $-p$) representations!

Table 1.5 shows that there are, for example, two distinct supermultiplets that have seven propagating bosons. To gain some insight into how this profusion of supermultiplets comes into being, it is convenient to note that the matrices in (1.143) can be arranged according to the identifications

$$\alpha_{\hat{A}} = \begin{bmatrix} L_1 \\ L_2 \\ L_3 \end{bmatrix}, \qquad \beta_{\hat{A}} = \begin{bmatrix} L_5 \\ L_6 \\ L_7 \end{bmatrix},$$

$$\Theta = L_4, \qquad L_8 = \delta, \tag{1.146}$$

where the quantities L_I is split into triplets of matrices $\alpha_{\hat{A}}$ and $\beta_{\hat{A}}$, as well as the single matrix Θ and the identity matrix δ.

With respect to this same decomposition the eight bosonic fields may be written as

$$A_I = \{\mathcal{P}_{\hat{A}}, \ \mathcal{A}_{\hat{A}}, \ \mathcal{A}, \ \mathcal{P}\}. \tag{1.147}$$

Now the two distinct cases where $p = 2$ occur from the respective AD maps

$$\{\mathcal{P}_{\hat{A}}, \ \mathcal{A}_{\hat{A}}, \ \mathcal{A}, \ \mathcal{P}\} \ \rightarrow \ \{\mathcal{P}_{\hat{A}}, \ \mathcal{A}_{\hat{A}}, \ \partial_\tau^{-1}\mathcal{A}, \ \mathcal{P}\} \tag{1.148}$$

and

$$\{\mathcal{P}_{\hat{A}}, \ \mathcal{A}_{\hat{A}}, \ \mathcal{A}, \ \mathcal{P}\} \ \rightarrow \ \{\mathcal{P}_{\hat{A}}, \ \mathcal{A}_{\hat{A}}, \ \mathcal{A}, \ \partial_\tau^{-1}\mathcal{P}\}. \tag{1.149}$$

1.5 Graphical Supersymmetric Representation Technique: Adinkras

The root labels defined in (1.90) seem to be good candidates to classify linear representation of supersymmetry. However, a more careful analysis reveals that there is not a one to one correspondence between admissible transformations and labels. For instance the $N = 1$ scalar supermultiplet can be identified by both $(0,0)_+$ and $(0,1)_-$ root labels.

To exploit fully the power of the developed formalism, we need to introduce a more fundamental technique that from one side eliminate all the ambiguities and from another side reveal new structures. A useful way to encode all the informations contained in a supermultiplet, is provided by a graphical formulation where each graph is christen Adinkra in honour of Asante populations of Ghana, West Africa, accustomed to express concepts that defy usual words, by symbols. This approach was pioneered in [19].

Basic pictures used to represent supersymmetry are circles (nodes), white for bosons and black for fermions component fields, connected by arrows that are chosen in such a way to point the higher component field which is assumed to be the one that does not appear differentiated in the r.h.s. of transformation properties. The general rule to follow in constructing variations from Adinkras is

$$\delta_Q f_i = \pm i^b \partial_\tau^a f_j , \tag{1.150}$$

where f_i, f_j are two adjacent component fields, $b = 1$ ($b = 0$) if f_j is a fermion (boson) and $a = 1$ ($a = 0$) if fj is the lower (higher) component field. The sign has to be the same for both the nodes connected. Its relevance became clear only for $N > 1$ as will be discussed in the Subsect. 1.5.1. In the following we introduce a general procedure to classify root tree supermultiplets, that works for arbitrary N.

1.5.1 $N = 1$ Supermultiplets

It is straightforward to recognize the $N = 1$ scalar supermultiplet labeled by $(0,0)_+$

$$\begin{aligned} \delta_Q \phi &= i\epsilon\psi, \\ \delta_Q \psi &= \epsilon\partial_\tau\phi, \end{aligned} \qquad \Rightarrow \tag{1.151}$$

and the first level dualized supermultiplet $(1,0)_+$

$$\begin{aligned} \delta_Q \phi &= i\epsilon\partial_\tau\psi, \\ \delta_Q \psi &= \epsilon\phi. \end{aligned} \qquad \Rightarrow \tag{1.152}$$

that corresponds to the spinorial supermultiplet. The order of the nodes is conventionally chosen to keep contact with component fields order of the bosonic root labels (i.e., marked with a plus sign). Alternatively, starting from (1.151), we can dualize the second level, falling in the (1.152) option. The last possibility is to dualize both levels, but again we go back to the scalar case. Now that we have run out all the bosonic root label possibilities, we can outline the following sequence of congruences

$$(0,0)_+ \simeq (1,1)_+ \,,$$
$$(1,0)_+ \simeq (0,1)_+. \tag{1.153}$$

In this framework the AD (1.74) is seen to be implemented by a simple change in the orientation of the arrow. We refer to this simple sequence, made of all the inequivalent root tree supermultiplets of the bosonic type, as the "base sequence."

Besides the AD we have another kind of duality, namely the Klein flip (KF) [20], which corresponds to the exchanging of bosons and fermions. If we apply the KF to the previous Adinkras it happens that we get what we call the "mirror sequence"

$$\delta_Q \psi = \epsilon \phi,$$
$$\delta_Q \phi = i\epsilon \partial_\tau \psi, \tag{1.154}$$

$$\delta_Q \psi = \epsilon \partial_\tau \phi,$$
$$\delta_Q \phi = i\epsilon \psi. \tag{1.155}$$

Here the KF is responsible for a changing of the supermultiplets nature from bosonic to fermionic. Accordingly, to maintain the order of fermionic root labels, we put a fermionic node on the upper position. As in the base sequence, even in the mirror one, we have congruences between root labels, precisely $(0,0)_- \simeq (1,1)_-$ are referred to (1.154) while $(1,0)_- \simeq (0,1)_-$ to (1.155). The power of Adinkras became manifest when we try to find which supermultiplet of the base sequence is equivalent to the supermultiplets in the mirror one. It is straightforward to see that up to 180° rotations, only two Adinkras are inequivalent

$$(0,0)_+ \simeq (1,1)_+ \simeq (0,1)_- \simeq (1,0)_- \,,$$
$$(0,0)_- \simeq (1,1)_- \simeq (0,1)_+ \simeq (1,0)_+ \,. \tag{1.156}$$

All the above results about $N = 1$ root tree supermultiplet, can be reassumed in a compact way in Fig. 1.1, where boxed nodes refer to auxiliary fields. Actually there exists a way to define auxiliary fields by means of Adinkras without appealing to the dynamics. In the following we will denote as auxiliary all the fields whose associated bosonic (fermionic) nodes are sink (source), namely all the arrows point to (comes out from) the node.

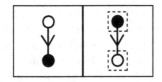

Fig. 1.1. $N = 1$ root tree elements

1.5.2 $N = 2$ Supermultiplets

Even in the $N = 2$ case we start from the scalar supermultiplet (1.13) whose root label is $(0, 0, 0)_+$. Choosing the representation

$$L_1 = R_1 = i\sigma_2, \quad L_2 = -R_2 = I_2 , \tag{1.157}$$

the resulting explicit transformation properties are

$$
\begin{aligned}
\delta_Q \phi_1 &= -i\epsilon^1 \psi_2 + i\epsilon^2 \psi_1 , \\
\delta_Q \phi_2 &= i\epsilon^1 \psi_1 + i\epsilon^2 \psi_2 , \\
\delta_Q \psi_1 &= \epsilon^1 \dot{\phi}_2 + \epsilon^2 \dot{\phi}_1 , \\
\delta_Q \psi_2 &= -\epsilon^1 \dot{\phi}_1 + \epsilon^2 \dot{\phi}_2 .
\end{aligned}
\tag{1.158}
$$

Accordingly, the Adinkra associated with (1.158) can be drawn as

$$\tag{1.159}$$

The filled arrow is inserted to take into account that appears a minus sign in the (1.150) involving ϕ_1 and ψ_2. The $N = 2$ case furnishes new features that will be present in all higher supersymmetric extensions. One of them is the sum rule that can be stated as follows: multiplying the signs chosen in the (1.150) for a closed path in the Adinkra, we should get a minus sign. Clearly, the graph (1.159) satisfy this condition. Moreover, it is possible to flip the sign of a field associated to a node. The net effect on the Adinkra is a shift of the red arrow or the appearing of two more red arrows confirming that after this kind of flip the sum rule still holds. Another evident property is that parallel arrows correspond to the same supersymmetry. It is easy to foresee that the N-extended Adinkras live in an N-dimensional space so that graphical difficulties will arise for $N \geq 4$. However, suitable techniques will be developed below to treat higher dimensional cases.

The AD can be generalized to arbitrary N-extended cases by saying that its application to a field is equivalent to reversing *all* the arrows connected to the corresponding node. However, as will be cleared in the next paragraphs, if we want to move inside the root tree, we have to implement AD level by level. This means that in (1.158), the AD is necessarily implemented on both ψ_1 and ψ_2 fermionic fields.

For arbitrary value of N, the proper way to manage the signs is to consider the scalar supermultiplet Adinkra associated to (1.78), as the starting point to construct all the other root tree supermultiplets implementing AD, Klein flip, and sign flipping of component fields. Since the scalar supermultiplet has well defined signs by construction, the resulting Adinkras will be consistent with the underlying theory. This allows us to forget about the red arrows and consider equivalent all the graphs that differ from each other by a sign redefinition of a component field. Once the problem of signs is understood, let us go back to the classification problem. Following the line of the $N = 1$ case, from the scalar Adinkra (1.159) we derive the base sequence whose inequivalent graphs, with root labels on the right, are

$$\rightarrow \qquad (0,0,0)_+ \simeq (1,1,1)_+ \qquad (1.160)$$

$$\rightarrow \qquad (0,0,1)_+ \simeq (1,0,0)_+ \qquad (1.161)$$

$$\rightarrow \qquad (0,1,0)_+ \qquad (1.162)$$

The KF applied to the above Adinkras, provides the mirror sequence

$$\rightarrow \qquad (0,0,0)_- \simeq (1,1,1)_- \qquad (1.163)$$

$$\rightarrow \qquad (0,0,1)_- \simeq (1,0,0)_- \qquad (1.164)$$

$$\rightarrow \qquad (0,1,0)_- . \qquad (1.165)$$

To classify the $N = 2$ root tree supermultiplets, the last step is the matching of the base sequence with the mirror sequence to recognize topologically equivalent graphs. It turns out that only four of them originate different dual supermultiplets. In fact, Adinkra (1.160) is nothing but Adinkra (1.165) rotated by 90°. The same relation holds between the Adinkras (1.162) and (1.163). To make the remaining Adinkra relationships clear, we can arrange them in the following way

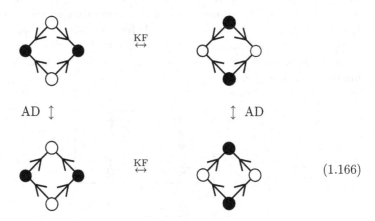

$$(1.166)$$

so that left column is connected by the klein flip to the right column while the upper row is the automorphic dual of the lower one.

1.5.3 Adinkras Folding

As observed in the previous section, Adinkras associated to $N \geq 3$ extended supermultiplets may become problematic to draw and consequently to classify. Fortunately, there exists a very simple way to reduce the dimensionality of the graphs preserving the topological structure memory. The process consists in moving the nodes and arrows into each other in a proper way. In doing this two basic rules have to be satisfied:

1. only nodes of the same type can be overlapped,

2. we can make arrows lay upon each other only if they are oriented in the same way.

In the first rule, when we talk about nodes of the same type, we refer not only to the bosonic or fermionic nature but even to physical or auxiliary dynamical behavior. To clarify how this process works, let us graphically examine the simplest example by folding the Adinkra (1.160)

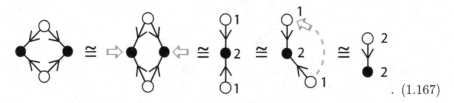

$$. \quad (1.167)$$

On the left of each node it is reported its multiplicity. Thus from a two-dimensional Adinkra we end up to a one-dimensional one, increasing the multiplicity of the nodes. We emphasize that in the example above, we have a sequence of two different folding. After the first one we are left with a partially folded Adinkra while in the end we obtain a fully folded one. It is important for the following developments, to have in mind that we have various levels of folding for the same Adinkra. A remarkable property of the root tree elements is that they can be always folded into a linear chain. Applying this technique to the arrangement scheme (1.166), the $N = 2$ root tree Adinkras can be organized in Fig. 1.2.

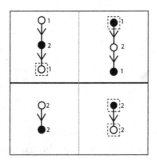

Fig. 1.2. Fully folded $N = 2$ root tree elements

1.5.4 Escheric Supermultiplets

In this section we want to give some hints about how to describe supermultiplets that are not in the root tree. We anticipated that implementing AD on singular nodes may bring us outside the root tree sequence. Let us examine this aspect in some detail. Starting from variations (1.158) we dualize

$$\phi_1 = \partial_\tau^{-1} A , \tag{1.168}$$

to obtain

$$\delta_Q A = -i\epsilon^1 \dot{\psi}_2 + i\epsilon^2 \dot{\psi}_1 ,$$
$$\delta_Q \phi_2 = i\epsilon^1 \psi_1 + i\epsilon^2 \psi_2 ,$$
$$\delta_Q \psi_1 = \epsilon^1 \dot{\phi}_2 + \epsilon^2 A ,$$
$$\delta_Q \psi_2 = -\epsilon^1 A + \epsilon^2 \dot{\phi}_2 , \tag{1.169}$$

associated to Adinkra (1.161). Then let the AD map act on the left fermionic node

$$\psi_2 = i\partial_\tau \xi , \tag{1.170}$$

to end into

$$\delta_Q A = -i\epsilon^1 \ddot{\xi} + i\epsilon^2 \dot{\psi}_1 ,$$
$$\delta_Q \phi_2 = i\epsilon^1 \psi_1 - \epsilon^2 \ddot{\xi} ,$$
$$\delta_Q \psi_1 = \epsilon^1 \dot{\phi}_2 + \epsilon^2 A ,$$
$$\delta_Q \xi = i\epsilon^1 \int^{\tilde{t}} d\tilde{t} A - i\epsilon^2 \phi_2 , \tag{1.171}$$

whose corresponding Adinkra symbol is

$$\tag{1.172}$$

where the new modified arrow is used to describe the appearance of the anti-derivative in the r.h.s. of ψ_2 variation. Let us notice that the usual ordering of the nodes in the Adinkra (1.172) makes no sense because each node is upper than the previous and lower then the next one. This situation was one of the main theme of some drawings of the graphic artist Maurits Cornelis Escher (see, for instance, the lithograph "Ascending and Descending"). For this reason we will refer to these kind of supermultiplets as escheric. One of the main features of Adinkra (1.172) is that it cannot be folded into a lower dimensional graph. This forces us to introduce a new important concept which is the rank of an Adinkra, namely the dimensions spanned by the fully folded graph diminished by one. The case (1.172) provides an $N = 2$ example of a rank one Adinkra while the root trees are always composed of rank zero Adinkras.

A similar result can be found even in the $N = 1$ case. It is possible to go outside the root tree enforcing the duality

$$\phi \rightarrow \partial_\tau^{-2} \tilde{\phi} , \tag{1.173}$$

on transformation properties (1.151), in order to obtain

$$\delta_Q \tilde{\phi} = i\epsilon \partial_\tau^2 \psi \ ,$$

$$\delta_Q \psi = \epsilon \int^{\tilde{t}} d\tilde{t} \ \tilde{\phi}(\tilde{t}) \ , \tag{1.174}$$

and the new $N = 1$ escheric symbol

$$\tag{1.175}$$

with equivalent root labels $(2,0)_+ \cong (0,2)_+ \cong (1,0)_- \cong (0,1)_-$.

The integral in the r.h.s. of the above transformation properties assume a particularly interesting meaning whenever the integrated boson lives in a compact manifold. If this is the case then the integral term counts the number of wrappings of the considered bosonic field.

The above discussion about escheric supermultiplet is somehow linked via AD maps to supersymmetric multiplets presented so far. However, the contact with the previous approach is completely lost by considering the Adinkra

$$\tag{1.176}$$

and associating to this graph the variations exploiting the general method of (1.150). It turns out that transformation properties referred to (1.176) are

$$\delta_Q \phi_1 = -i\epsilon^1 \psi_2 + i\epsilon^2 \psi_1 \ ,$$
$$\delta_Q \phi_2 = i\epsilon^1 \psi_1 + i\epsilon^2 \psi_2 \ ,$$
$$\delta_Q \psi_1 = \epsilon^1 \dot{\phi}_2 + \epsilon^2 \phi_1 \ ,$$
$$\delta_Q \psi_2 = -\epsilon^1 \dot{\phi}_1 + \epsilon^2 \phi_2 \ , \tag{1.177}$$

where each field is associated to each site in the same way of Adinkra (1.159) and a proper minus sign has been inserted in order to accomplish the sum rule. The first thing to figure out in order to understand what kind of supersymmetric properties are hidden under this new supermultiplet, is the commutator between two variations. It is straightforward to prove that after the fields and parameters complexification

$$\phi = \phi_1 + i\phi_2, \qquad \psi = \psi_1 + i\psi_2, \qquad \epsilon = \epsilon^1 + i\epsilon^2 \ , \tag{1.178}$$

we have[3]

$$[\delta_Q(\epsilon_1), \delta_Q(\epsilon_2)] = -2i\bar{\epsilon}_{[1}\epsilon_{2]}\partial_\tau - \frac{i}{2}(\epsilon_1\epsilon_2 - \bar{\epsilon}_1\bar{\epsilon}_2)\,\delta_Y \qquad (1.179)$$

where δ_Y acts on the fields in the following way

$$\begin{aligned} \delta_Y(\phi, \psi) &= (\partial_\tau^2 - 1)(\phi, \psi)\,, \\ \delta_Y(\bar{\phi}, \bar{\psi}) &= -(\partial_\tau^2 - 1)(\bar{\phi}, \bar{\psi}). \end{aligned} \qquad (1.180)$$

Expressing the variations in terms of supersymmetric charges Q, $\bar{Q}$ and central charge Y

$$\begin{aligned} \delta_Q(\epsilon) &= \epsilon Q + \bar{\epsilon}\bar{Q} \\ \delta_Y &= \frac{Y}{4}\,, \end{aligned} \qquad (1.181)$$

we obtain the central extended algebra

$$\{Q, \bar{Q}\} = H, \quad Q^2 = iY, \quad [H, Y] = 0\,, \qquad (1.182)$$

where Y plays the role of a purely imaginary central charge. Although real central extension of similar algebras has been studied [21], the purely imaginary case still lacks a completely clear interpretation.

The escheric Adinkras make clear how this graphical approach can offer the possibility to describe theories that lie outside the formalism developed in the previous sections and eventually can make arise to new nontrivial features.

1.5.5 Through Higher N

In principle, the techniques of the previous paragraphs are suitable even for $N \geq 3$. Thus, for $N = 3$ case, the scalar supermultiplet field transformations can be written down using the representation for $\mathcal{GR}(4, 3)$ given by

$$L_1 = R_1 = i\sigma_1 \otimes \sigma_2 = \begin{pmatrix} 0 & 0 & 0 & 1 \\ 0 & 0 & 1 & 0 \\ 0 & -1 & 0 & 0 \\ -1 & 0 & 0 & 0 \end{pmatrix}\,,$$

$$L_2 = R_2 = i\sigma_2 \otimes I_2 = \begin{pmatrix} 0 & 1 & 0 & 0 \\ -1 & 0 & 0 & 0 \\ 0 & 0 & 0 & 1 \\ 0 & 0 & -1 & 0 \end{pmatrix}\,,$$

$$L_3 = R_3 = -i\sigma_3 \otimes \sigma_2 = \begin{pmatrix} 0 & 0 & -1 & 0 \\ 0 & 0 & 0 & 1 \\ 1 & 0 & 0 & 0 \\ 0 & -1 & 0 & 0 \end{pmatrix}\,, \qquad (1.183)$$

---

[3] The lower indices of the supersymmetry parameters are referred to different supersymmetries while the upper one are associated to the two real charges of the same supersymmetry.

so that explicitly we have

$$\delta\phi_1 = -i\epsilon^1\psi_4 - i\epsilon^2\psi_2 + i\epsilon^3\psi_3 \,,$$
$$\delta\phi_2 = -i\epsilon^1\psi_3 + i\epsilon^2\psi_1 - i\epsilon^3\psi_4 \,,$$
$$\delta\phi_3 = i\epsilon^1\psi_2 - i\epsilon^2\psi_4 - i\epsilon^3\psi_1 \,,$$
$$\delta\phi_4 = i\epsilon^1\psi_1 + i\epsilon^2\psi_3 + i\epsilon^3\psi_2 \,,$$
$$\delta\psi_1 = \epsilon^1\dot\phi_4 + \epsilon^2\dot\phi_2 - \epsilon^3\dot\phi_3 \,,$$
$$\delta\psi_2 = \epsilon^1\dot\phi_3 - \epsilon^2\dot\phi_1 + \epsilon^3\dot\phi_4 \,,$$
$$\delta\psi_3 = -\epsilon^1\dot\phi_2 + \epsilon^2\dot\phi_4 + \epsilon^3\dot\phi_1 \,,$$
$$\delta\psi_4 = -\epsilon^1\dot\phi_1 - \epsilon^2\dot\phi_3 - \epsilon^3\dot\phi_2 \,, \qquad (1.184)$$

that, translated in terms of graph, are equivalent to

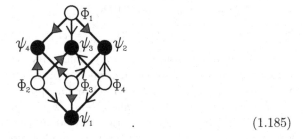

$$(1.185)$$

Next, we dualize via klein flip to obtain

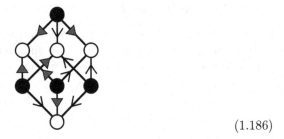

$$(1.186)$$

that, together with (1.185), are respectively the starting point to construct all the elements of the base and mirror sequences using all possible levelwise AD. Alternatively, one can derive all the base sequence from (1.185) and then performs a klein flip on each base element in order to deduce the mirror Adinkras. Finally, we fold all the topologically inequivalent Adinkras organizing them into Fig. 1.3.

As expected, all the fully folded root tree Adinkras are one dimensional. Moreover, one can verify that each closed path without arrows within, satisfy the sum rule. The $N = 3$ case furnishes the possibility to generalize the sum rule by saying that if a is the number of arrows that are circuited by the path then the sign of the sum rule turns out to be $(-1)^{a+1}$. It is of some importance to notice that by completely folding the Adinkras, the levels of the nodes can

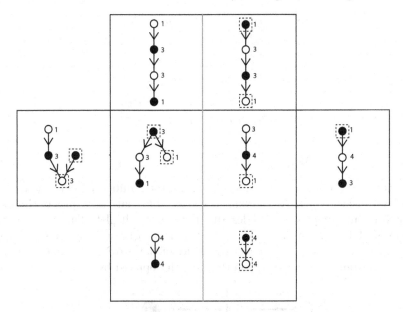

Fig. 1.3. Fully folded $N = 3$ root tree elements

be upsetted. Nevertheless, if we consider the fully unfolded one-dimensional Adinkras obtained by folding the $N = 3$ graphs, then we can still implement AD level by level in order to get all the root tree. In other words the depth of the ADs (i.e., the minimum number of dimensions reached by the folded Adinkra when the AD is applied) used to deduce the root tree is one.

If we assume that each supersymmetry corresponds to an orthogonal direction, as stated above, then the nodes are placed on the vertices of an N-dimensional hypercube. Consequently, at $N = 4$ we find $2^N|_{N=4} = 16$ nodes among which eight are bosonic and eight are fermionic. This is in contrast with the irreducible representation dimension that is $4 + 4 = 8$ as reported in Table 1.2. The problem to face is how to reduce consistently the dimension of the representation that arise from the $N = 4$ Adinkras in order to obtain the irreducible representation described in Subsect. 1.2.3. The following two methods are effective to solve this problem: in the first one we identify consistently the nodes to obtain the proper transformation properties, while in the second one we recognize irreducible sub-Adinkras making rise to gauge degrees of freedom. Let us consider the first method (the other one will be analyzed in the next paragraph) constructing the $N = 4$ scalar supermultiplet as example. A possible choice of $\mathcal{GR}(\triangle, \triangle)$ generators turns out to be composed of six generators of the $N = 3$ case (1.183) plus the two generators

$$L_4 = -R_4 = -I_2 \otimes I_2 \tag{1.187}$$

It is straightforward to figure out the following variations:

$$\delta\phi_1 = -i\epsilon^1\psi_4 - i\epsilon^2\psi_2 + i\epsilon^3\psi_3 - i\epsilon^4\psi_1 \,,$$
$$\delta\phi_2 = -i\epsilon^1\psi_3 + i\epsilon^2\psi_1 - i\epsilon^3\psi_4 - i\epsilon^4\psi_2 \,,$$
$$\delta\phi_3 = i\epsilon^1\psi_2 - i\epsilon^2\psi_4 - i\epsilon^3\psi_1 - i\epsilon^4\psi_3 \,,$$
$$\delta\phi_4 = i\epsilon^1\psi_1 + i\epsilon^2\psi_3 + i\epsilon^3\psi_2 - i\epsilon^4\psi_4 \,,$$
$$\delta\psi_1 = \epsilon^1\dot\phi_4 + \epsilon^2\dot\phi_2 - \epsilon^3\dot\phi_3 - \epsilon^4\dot\phi_1 \,,$$
$$\delta\psi_2 = \epsilon^1\dot\phi_3 - \epsilon^2\dot\phi_1 + \epsilon^3\dot\phi_4 - \epsilon^4\dot\phi_2 \,,$$
$$\delta\psi_3 = -\epsilon^1\dot\phi_2 + \epsilon^2\dot\phi_4 + \epsilon^3\dot\phi_1 - \epsilon^4\dot\phi_3 \,,$$
$$\delta\psi_4 = -\epsilon^1\dot\phi_1 - \epsilon^2\dot\phi_3 - \epsilon^3\dot\phi_2 - \epsilon^4\dot\phi_4 \,, \tag{1.188}$$

associated to the scalar supermultiplet. Since the number of fields are doubled by the translation into Adinkras, it is conceivable that two copies of the same Adinkra could fit properly to describe the supermultiplet. The $N = 3$ scalar Adinkra (1.185) is suitable to encode the first three supersymmetries while the extra supersymmetry connect the nodes of the two $N = 3$ scalar Adinkras copies. Graphically, the situation can be well depicted by

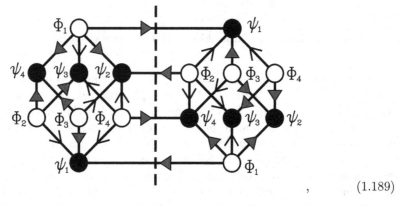

$$, \tag{1.189}$$

where we omitted the arrows from ϕ_3 to ϕ_3 (the arrows between ϕ_2 and ψ_2, and between ϕ_2 and ψ_2, are not repeated in the external nodes). We can see that the fourth supersymmetry connects opposite nodes of the Adinkra (1.185) so that we can render the drawing (1.189) in the following more compact way:

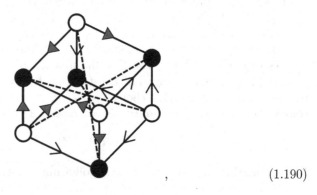

$$, \tag{1.190}$$

where we agree that the dashed diagonal arrows describe the same supersymmetry even if they are not parallel. Let us notice that the second $N = 3$ Adinkra in the (1.189) picture, is the mirrored copy of the first. In fact, the dashed line in the middle represents the mirror plane inserted to underline this property. The subtlety in this construction is hidden in the way to connect the two $N = 3$ Adinkras using the fourth supersymmetry. In fact, four consistent choices of sign flips make rise to as many inequivalent supermultiplets that behave to the same conjugacy class. These four scalar supermultiplets were considered in [20]. To describe them it is useful to fold the Adinkra (1.189) in the following way:

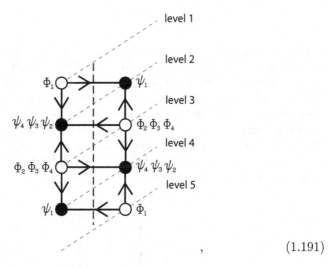

$$,\qquad\qquad (1.191)$$

where the diagonal dashed lines stand for the levels of the $N = 4$ supermultiplets. It is clear that the right side with respect of the mirror plan is redundant since it can be deduced from the left one. Therefore, it is sufficient to draw only the left side of the graph (1.191) in order to allow us to add the sign flips that identify each scalar supermultiplet as it follows

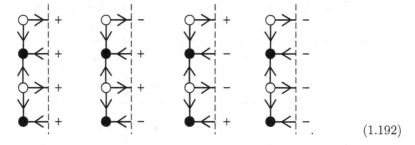

$$(1.192)$$

This quaternionic structure can be neglected if we assume that each $N = 4$ Adinkra in the root tree stands for a conjugacy class. If we adhere to this point of view, then it is a good exercise for the reader to derive all the fully folded root tree elements of the $N = 4$ case using the techniques described so far. The best way to proceed is to reduce the Adinkra (1.189) to its most unfolded one-dimensional version which is

$$(1.193)$$

obtained by identifying the nodes along the levels represented by dashed diagonal lines in the graph (1.191). We underline that the nontrivial structure of the levels is not manifest in the drawing (1.189) but it becomes evident once we fold it into the linear graph (1.193). One can check that the root tree elements can be obtained dualizing along these levels and the resulting graphs can be arranged in Fig. 1.4. The reader is also encouraged to try to implement the AD not respecting the suggested levels. For instance, if we consider the levels of each $N = 3$ sub-Adinkra cube to apply AD, then it is easy to see that escheric loops may come out.

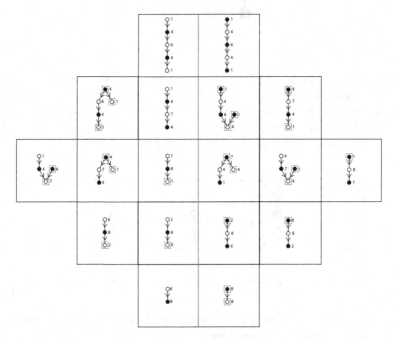

Fig. 1.4. Fully folded $N = 4$ root tree elements

1.5.6 Gauge Invariance

Before starting with the discussion of the gauge aspects of Adinkras, we need to describe explicitly the $N = 4$ chiral supermultiplet. To this end, let us apply a third level AD to the Adinkra (1.189) and fold it in the following way:

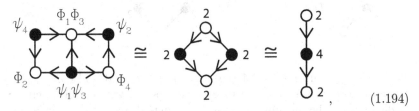

$$(1.194)$$

where we disregarded the second $N = 3$ cube and the fourth supersymmetry arrows. The simplest root label associated to this supermultiplet is $(00100)_+$ and it corresponds to the shadow of the $N = 1$, $d = 4$ chiral supermultiplet. For this reason we refer to it as the $N = 4$ chiral supermultiplet. Analogously, the shadow of $N = 1$, $d = 4$ vector supermultiplet can be constructed dualizing the first, fourth, and fifth level of the scalar supermultiplet associated to the Adinkra (1.189). By doing this we are left with

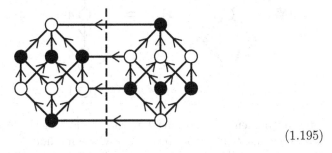

$$(1.195)$$

which is foldable to the following form

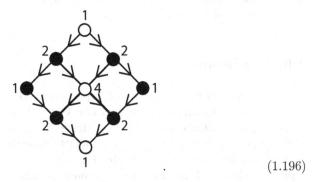

$$(1.196)$$

In this structure, there is embedded the chiral supermultiplet as an Adinkra. It is possible to remove it from the top of the vector Adinkra in order to obtain two irreducible representations

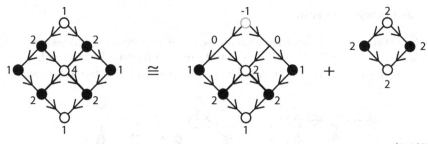

$$(1.197)$$

We see that a subtraction of the nodes is performed and consequently, the topmost node of the vector Adinkra assumes a negative multiplicity. Such a node acquires the meaning of a residual gauge degree of freedom. By moving the gauge node along the initial structure of Adinkra (1.195), we fix it on the nearest remained node, as shown in the figure

$$(1.198)$$

This phenomenon is nothing but the shadow of the $N = 4$, $d = 1$ Wess-Zumino gauge fixing procedure. Clearly, the method described above offers an alternative possibility to reduce the reducible supermultiplet coming out from an Adinkra symbol into two reducible representations via the introduction of gauge degrees of freedom.

1.6 Conclusions

By a geometrical interpretation of supersymmetric mechanics, we reviewed a classification scheme that exploits real Clifford algebras which are in one-to-one correspondence with the geometrical framework of Garden Algebras. For supersymmetric mechanics we explicitly described the link between the number of supersymmetries and the dimension and geometry of their faithful representations. All methods used to construct the explicit representation of such algebras are reviewed in detail. Particular emphasis has been dedicated to the duality relations among different supermultiplets at fixed number of supersymmetries using Clifford algebraic superfields. The formalism developed turned out to be necessary, as well as effective, when applied to the spinning particle problem, providing, quite straightforwardly, first- and second-order supersymmetric actions both in the case of global and local N-extended supersymmetry. Another new application example has been provided by an $N = 8$ unusual representation, suggesting how to derive many related representation via automorphic duality.

The second part of these lectures concerned the translation of all the results obtained so far into a simple graphical language whose symbols are called "Adinkras." In particular, we encoded all properties of each supermultiplet into an Adinkra graph in order to classify and better clarify the duality relations between supermultiplets. Using a folding procedure to reduce the dimensions of the Adinkras, we succeeded in classifying, up to $N = 4$, a large class of supermultiplets (root trees) using linear graphs. Moreover, it has been demonstrated that this graphical technique offers the possibility to derive new supermultiplets through dualities, possibly with the appearance of central charges or topological charges.

Even though the attempt to formalize a method to relate Adinkras to supermultiplets has been carried out successfully in these lectures, many aspects still need a proper investigation on mathematical footing. A step forward in this direction has been presented in a recent work [22]. However, the $N \geq 4$ cases still present many unresolved classification subtleties mainly due to the nontrivial topology structure of the Adinkras. Another line of research that may be followed deals with the implications of the duality relations between supermultiplets on higher dimensional field theories. The oxidation procedure is a nice tool that can be used to proceed in this way. Recently, exploiting the automorphic duality, it has been shown [23] that it is possible to relate not only the $N = 4$ root tree supermultiplets, but even the associated interacting sigma models[4]. Anyway, if we work outside the root tree, it is still not completely clear what kind of theories can be constructed with such supermultiplets. Especially, it should be interesting to better understand how to introduce central charges through dualities. It is our belief that the techniques reviewed here will provide new insight toward the solution of this open problem.

Acknowledgments

E.O. would like to thank the University of Maryland and S.J. Gates Jr. for the warm hospitality during the development of this work. Furthermore, E.O. would like to express his gratitude to Lubna Rana for helpful discussions. The research of S.B. is partially supported by the European Commu-

---

[4] Notice that in [23] linear and nonlinear chiral supermultiplets were obtained by the reduction of the linear supermultiplet with four bosonic and four fermionic degrees of freedom [8, 24].

nity's Marie Curie Research Training Network under contract MRTN-CT-2004-005104 Forces of Universe, and by INTAS-00-00254 grant. The research of S.J.G. is partially supported by the National Science Foundation Grant PHY-0354401.

References

1. S.J. Gates Jr, L. Rana: Phys. Lett. B **369**, 262 (1996) [arXiv:hep-th/9510151]
2. S.J. Gates Jr, W.D. Linch, J. Phillips: hep-th/0211034
3. S. Okubo: J. Math. Phys. **32**(7) 1657 (1991); S. Okubo: J. Math. Phys. **32**(7) 1669 (1991)
4. A. Pashnev, F. Toppan: J. Math. Phys. **42**, 5257 (2001) arXiv:hep-th/0010135
5. S.J. Gates Jr, L. Rana: Phys. Lett. B **352**, 50 (1995) [arXiv:hep-th/9504025]
6. S. de Buyl, M. Henneaux, B. Julia, L. Paulot: [arXiv:hep-th/0312251]
7. A. Keurentjes: Class. Quant. Grav. **20**, S525–S532 (2003) [arXiv:hep-th/0212050]
8. E. Ivanov, S. Krivonos, O. Lechtenfeld: Class. Quant. Grav. **21**, 1031–1050 (2004) [arXiv:hep-th/0310299]
9. S. Bellucci, E. Ivanov, S. Krivonos, O. Lechtenfeld: Nucl. Phys. B **699**, 226 (2004) [arXiv:hep-th/0406015]; S. Bellucci, E. Ivanov, S. Krivonos, O. Lechtenfeld: Nucl. Phys. B **684**, 321 (2004) [arXiv:hep-th/0312322]; S. Bellucci, S. Krivonos, A. Nersessian: Phys. Lett. B **605**, 181 (2005) [arXiv:hep-th/0410029]; S. Bellucci, S. Krivonos, A. Nersessian, A. Shcherbakov: [arXiv:hep-th/0410073]; S. Bellucci, S. Krivonos, A. Shcherbakov: Phys. Lett. B **612**, 283 (2005) [arXiv:hep-th/0502245]
10. S. Bellucci, A. Beylin, S. Krivonos, A. Shcherbakov: Phys. Lett. B **633** (2006) 382–388 [arXiv:hep-th/0511054]
11. S. Bellucci, A. Beylin, S. Krivonos, A. Nersessian, E. Orazi: Phys. Lett. B **616**, 228 (2005) [arXiv:hep-th/0503244]
12. J. Martin, Proc. R. Soc. Lond. A **251**, 536 (1959)
13. R. Casalbuoni: Nuovo Cim. A **33**, 389 (1976)
14. A. Barducci, R. Casalbuoni, L. Lusanna: Nuovo. Cim. **35**, 307 (1977)
15. F.A. Berezin, M.S. Marinov: Ann. Phys. **104**, 336 (1977)
16. L. Brink, S. Deser, B. Zumino, P. Di Vecchia, P. Howe: Phys. Lett. B **64**, 435 (1976); B.P. Di Vecchia, P. Howe: Nucl. Phys. B **118** 76 (1977)
17. V.D. Gershun and V.I. Tkach, JETP Lett. **29**, 288 (1979); ibid., Pisma Zh. Eksp. Teor. Fiz. **29**, 320 (1979); P. Howe, S. Penati, M. Pernici, P. Townsend: Phys. Lett. B **215**, 555 (1988); ibid. Class. Quant. Grav. **6**, 1125 (1989); S.M. Kuzenko and Z.V. Yarevskaya, Mod. Phys. Lett. A **11**, 1653 (1996) [arXiv:hep-th/9512115].
18. J. Sherk, H. Schwarz: Nucl. Phys. B **153**, 61 (1979)
19. M. Faux, S.J. Gates Jr.: Phys. Rev. D **71**, 065002 (2005) [hep-th/0408004]
20. S.J. Gates, S.V. Ketov: Phys. Lett. B **418**, 119–124 (1998) [arXiv:hep-th/9504077]
21. M. Faux, D. Spector: Phys. Rev. D **70**, 085014 (2004) [hep-th/0311095]
22. C.F. Doran, M.G. Faux, S.J. Gates Jr, T. Hubsch, K.M. Iga, G.D. Landweber: math-ph/0512016

23. S. Bellucci, S. Krivonos, A. Marrani, E. Orazi: Phys. Rev. D **73** (2006) 025011 [hep-th/0511249]
24. E. Ivanov, S. Krivonos, O. Lechtenfeld: JHEP **0303**, 014 (2003); S. Bellucci, S. Krivonos, A. Sutulin: Phys. Lett. B **605**, 406 (2005); S. Bellucci, E. Ivanov, A. Sutulin: Nucl. Phys. B **722**, 297 (2005)

2

Supersymmetric Mechanics in Superspace

S. Bellucci[1] and S. Krivonos[2]

[1] INFN-Laboratori Nazionali di Frascati, C.P. 13, 00044 Frascati, Italy
[2] Bogoliubov Laboratory of Theoretical Physics, Joint Institute for Nuclear
   Research, 141980, Dubna, Moscow region, Russia

2.1 Introduction

These lectures were given at Laboratori Nazionali di Frascati in the month of
March 2005. The main idea was to provide our young colleagues, who joined us
in our attempts to understand the structure of N-extended supersymmetric
one-dimensional systems, with short descriptions of the methods and techniques we use. This was reflected in the choice of material and in the style of
presentation. We base our treatment mainly on the superfield point of view.
Moreover, we prefer to deal with $N = 4$ and $N = 8$ superfields. At present,
there exists an extensive literature on the component approach to extended
supersymmetric theories in $d = 1$ while the manifestly supersymmetric formulation in terms of properly constrained superfields is much less known.
Nevertheless, we believe that just such formulations are preferable.

To make these lectures more or less self-consistent, we started from the
simplest examples of one-dimensional supersymmetric theories and paid a lot
of attention to the peculiarities of $d = 1$ supersymmetry. From time to time
we presented the calculations in a very detailed way. In other cases we omitted
the details and gave only the final answers. In any case these lectures cannot
be considered as a textbook in any respect. They can be considered as our
personal point of view on the one-dimensional superfield theories and on the
methods and techniques we believe to be important.

Especially, all this concerns Sect. 2.3, where we discuss the nonlinear realization method. We did not present any proofs in this section. Instead we
focused on the details of calculations.

2.2 Supersymmetry in $d = 1$

The extended supersymmetry in one dimension has plenty of peculiarities
which make it quite different from its higher-dimensional analogs. Indeed,
even the basic statement of any supersymmetric theory in $d > 1$ – the equality

S. Bellucci and S. Krivonos: *Supersymmetric Mechanics in Superspace*, Lect. Notes Phys. **698**,
49–96 (2006)
DOI 10.1007/3-540-33314-2_2

of bosonic and fermionic degrees of freedom – is not valid in $d = 1$. As a result, there are many new supermultiplets in one dimension which have no higher-dimensional ancestors. On the other hand, the constraints describing on-shell supermultiplets, being reduced to $d = 1$, define off-shell multiplets! Therefore, it makes sense to start with the basic properties of one-dimensional supermultiplets and give a sort of vocabulary with all linear finite-dimensional $N = 1, 2, 4$ and $N = 8$ supermultiplets. This is the goal of the present section.

2.2.1 Super-Poincaré Algebra in $d = 1$

In one dimension there is no Lorentz group and therefore all bosonic and fermionic fields have no space-time indices. The simplest free action for one bosonic field ϕ and one fermionic field ψ reads

$$S = \gamma \int dt \left[\dot{\phi}^2 - \frac{i}{2} \psi \dot{\psi} \right] . \tag{2.1}$$

In what follows it will be useful to treat the scalar field as dimensionless and assign dimension $\mathrm{cm}^{-1/2}$ to fermions. Therefore, all our actions will contain the parameter γ with the dimension $[\gamma] = \mathrm{cm}$.

The action (2.1) provides the first example of a supersymmetric invariant action. Indeed, it is a rather simple exercise to check its invariance with respect to the following transformations:

$$\delta\phi = -i\epsilon\psi, \qquad \delta\psi = -\epsilon\dot{\phi} . \tag{2.2}$$

As usual, the infinitesimal parameter ϵ anticommutes with fermionic fields and with itself. What is really important about transformations (2.2) is their commutator

$$\delta_2\delta_1\phi = \delta_2 \left(-\epsilon_1\psi \right) = i\epsilon_1\epsilon_2\dot{\phi} ,$$
$$\delta_1\delta_2\phi = i\epsilon_2\epsilon_1\dot{\phi} \; \Rightarrow \; [\delta_2, \delta_1]\,\phi = 2i\epsilon_1\epsilon_2\dot{\phi} . \tag{2.3}$$

Thus, from (2.3) we may see the main property of supersymmetry transformations: they commute on translations. In our simplest one-dimensional framework this is the time translation. This property has the followin form in terms of the supersymmetry generator Q:

$$\{Q, Q\} = -2P . \tag{2.4}$$

The anticommutator (2.4), together with

$$[Q, P] = 0 \tag{2.5}$$

describe $N = 1$ super-Poincaré algebra in $d = 1$. It is rather easy to guess the structure of N-extended super-Poincaré algebra: it includes N *real* super-charges $Q^A, A = 1, \ldots, N$ with the following anti commutators:

$$\{Q^A, Q^B\} = -2\delta^{AB} P, \quad [Q^A, P] = 0 . \tag{2.6}$$

Let us stress that the reality of the supercharges is very important, as well as having the same sign in the r.h.s. of $\{Q^A, Q^B\}$ for all Q^A. From time to time one can see wrong statements in the literature about the number of supersymmetries in the theories when authors forget about these absolutely needed properties.

From (2.2) we see that the minimal $N = 1$ supermultiplet includes one bosonic and one fermionic field. A natural question arises: how many components we need, in order to realize the N-extended superalgebra (2.6)? The answer has been found in a paper by Gates and Rana [1]. Their idea is to mimic the transformations (2.2) for all N supertranslations as follows:

$$\delta\phi_i = -i\epsilon^A (L_A)_i^{\hat{i}} \psi_{\hat{i}}, \qquad \delta\psi_{\hat{i}} = -\epsilon^A (R_A)_{\hat{i}}^{i} \dot{\phi}_i . \tag{2.7}$$

Here the indices $i = 1, \ldots, d_b$ and $\hat{i} = 1, \ldots, d_f$ count the numbers of bosonic and fermionic components, while $(L_A)_i^{\hat{i}}$ and $(R_A)_{\hat{i}}^{i}$ are N arbitrary, for the time being, matrices. The additional conditions one should impose on the transformations (2.7) are

- they should form the N-extended superalgebra (2.6)
- they should leave invariant the free action constructed from the involved fields.

These conditions result in some equations on the matrices L_A and R_A which has been solved in [1]. These results for the most interesting cases are summarized in Table 2.1. Here, $d = d_b = d_f$ is the number of bosonic/fermionic

Table 2.1. Minimal supermultiplets in N-extended supersymmetry

N	1	2	3	4	5	6	7	8	9	10	12	16
d	1	2	4	4	8	8	8	8	16	32	64	128

components. From Table 2.1 we see that there are four special cases with $N = 1, 2, 4, 8$ when d coincides with N. Just these cases we will discuss in the present lectures. When $N > 8$ the minimal dimension of the supermultiplets rapidly increases and the analysis of the corresponding theories becomes very complicated. For many reasons, the most interesting case seems to be the $\mathcal{N} = 8$ supersymmetric mechanics. Being the highest $\mathcal{N}$ case of *minimal* $\mathcal{N}$-extended supersymmetric mechanics admitting realization on $\mathcal{N}$ bosons (physical and auxiliary) and $\mathcal{N}$ fermions, the systems with eight supercharges are the highest $\mathcal{N}$ ones, among the extended supersymmetric systems, which still possess a nontrivial geometry in the bosonic sector [2]. When the number of supercharges exceeds 8, the target spaces are restricted to be symmetric spaces. Moreover, $\mathcal{N} = 8$ supersymmetric mechanics should be related via a proper dimensional reduction with four-dimensional $\mathcal{N} = 2$ supersymmetric field theories. So, one may hope that some interesting properties of the latter will survive after reduction.

2.2.2 Auxiliary Fields

In the previous subsection, we considered the realization of N-extended supersymmetry on the bosonic fields of the same dimensions. Indeed, in (2.7) all fields appear on the same footing. It is a rather special case which occurs only in $d = 1$. In higher dimensions the appearance of the auxiliary fields is inevitable. They appear also in one dimension. Moreover, in $d = 1$, one may convert any physical field to auxiliary and vice versa [1]. To clarify this very important property of one-dimensional theories, let us consider the simplest example of $N = 2, d = 1$ supermultiplets.

The standard definition of $N = 2, d = 1$ super-Poincaré algebra follows from (2.6):

$$\{Q^1, Q^1\} = \{Q^2, Q^2\} = -2P, \{Q^1, Q^2\} = 0 \ . \tag{2.8}$$

It is very convenient to redefine the supercharges as follows:

$$Q \equiv \frac{1}{\sqrt{2}} (Q_1 + iQ_2) \, , \ \overline{Q} \equiv \frac{1}{\sqrt{2}} (Q_1 - iQ_2) \ ,$$

$$\{Q, \overline{Q}\} = -2P, \ Q^2 = \overline{Q}^2 = 0 \ . \tag{2.9}$$

From Table 2.1 we know that the minimal $N = 2$ supermultiplet contains two bosonic and two fermionic components. The supersymmetry transformations may be easily written as

$$\begin{cases} \delta\phi = -i\bar{\epsilon}\psi \, , \\ \delta\psi = -2\epsilon\dot{\phi} \, , \end{cases} \qquad \begin{cases} \delta\bar{\phi} = -i\epsilon\bar{\psi} \, , \\ \delta\bar{\psi} = -2\bar{\epsilon}\dot{\bar{\phi}} \, . \end{cases} \tag{2.10}$$

Here, ϕ and ψ are complex bosonic and fermionic components. The relevant free action has a very simple form

$$S = \gamma \int dt \left(\dot{\phi}\dot{\bar{\phi}} - \frac{i}{2}\psi\dot{\bar{\psi}} \right) \ . \tag{2.11}$$

Now, we introduce new bosonic variables V and A

$$V = \phi + \bar{\phi}, \qquad A = i\left(\dot{\phi} - \dot{\bar{\phi}} \right) \ . \tag{2.12}$$

What is really impressive is that, despite the definition of A in terms of time derivatives of the initial bosonic fields, the supersymmetry transformations can be written in terms of $\{V, \psi, \bar{\psi}, A\}$ only

$$\delta V = -i\left(\bar\epsilon\psi + \epsilon\bar\psi\right), \quad \delta A = \bar\epsilon\dot\psi - \epsilon\dot{\bar\psi},$$

$$\delta\psi = -\epsilon\left(\dot V - iA\right), \quad \delta\bar\psi = -\bar\epsilon\left(\dot V + iA\right). \tag{2.13}$$

Thus, the fields $V, \psi, \bar\psi, A$ form an $N = 2$ supermultiplet, but the dimension of the component A is now cm^{-1}. If we rewrite the action (2.11) in terms of new components

$$S = \gamma \int dt \left(\frac{1}{4}\dot V\dot V - \frac{i}{2}\psi\dot{\bar\psi} + \frac{1}{4}A^2\right), \tag{2.14}$$

one may see that the field A appears in the action without derivatives and their equation of motion is purely algebraic

$$A = 0. \tag{2.15}$$

In principle, we can exclude this component from the Lagrangian (2.12) using (2.15). As a result we will have the Lagrangian written in terms of $V, \psi, \bar\psi$ only. But, without A, supersymmetry will close only up to equations of motion. Indeed, the variation of (2.15) with respect to (2.13) will enforce equations of motion for fermions

$$\dot\psi = \dot{\bar\psi} = 0.$$

The next variation of these equations gives

$$\ddot V = 0.$$

Such components are called auxiliary fields. In what follows we will use the notation $(\mathbf{n}, \mathbf{N}, \mathbf{N} - \mathbf{n})$ to describe a supermultiplet with n physical bosons, N fermions, and $N - n$ auxiliary bosons. Thus, the transformations (2.13) describe a transition from the multiplet $(\mathbf{2}, \mathbf{2}, \mathbf{0})$ to $(\mathbf{1}, \mathbf{2}, \mathbf{1})$. One may continue this process and so pass from the multiplet $(\mathbf{1}, \mathbf{2}, \mathbf{1})$ to the $(\mathbf{0}, \mathbf{2}, \mathbf{2})$ one, by introducing the new components $B = \dot\phi, \bar B = \dot{\bar\phi}$. The existence of such a multiplet containing no physical bosons at all is completely impossible in higher dimensions. Let us note that the inverse procedure is also possible [1]. Therefore, the field contents of *linear, finite-dimensional* off-shell multiplets of $N = 2, 4, 8, d = 1$ supersymmetry read

$$N = 2: \ (\mathbf{2}, \mathbf{2}, \mathbf{0}), \ (\mathbf{1}, \mathbf{2}, \mathbf{1}) \ (\mathbf{0}, \mathbf{2}, \mathbf{2})$$

$$N = 4: \ (\mathbf{4}, \mathbf{4}, \mathbf{0}), \ (\mathbf{3}, \mathbf{4}, \mathbf{1}), \ (\mathbf{2}, \mathbf{4}, \mathbf{2}), \ (\mathbf{1}, \mathbf{4}, \mathbf{3}), \ (\mathbf{0}, \mathbf{4}, \mathbf{4})$$

$$N = 8: \ (\mathbf{8}, \mathbf{8}, \mathbf{0}), \ \ldots, \ (\mathbf{0}, \mathbf{8}, \mathbf{8}) \tag{2.16}$$

Finally, one should stress that both restrictions – linearity and finiteness – are important. There are nonlinear supermultiplets [3–5], but it is not always possible to change their number of physical/auxiliary components. In the case of $N = 8$ supersymmetry one may define infinite-dimensional supermultiplets, but for them the interchanging of the physical and auxiliary components is impossible. We notice, in passing, that recently for the construction of $N = 8$ supersymmetric mechanics [5–7] the nonlinear chiral multiplet has been used [8].

2.2.3 Superfields

Now, we will turn to the main subject of these lectures – Superfields in N-extended $d = 1$ Superspace. One may ask – Why do we need superfields? There are a lot of motivations, but here we present only two of them. First of all, it follows from the previous subsections that only a few supermultiplets from the whole "zoo" of them presented in (2.33), contain no auxiliary fields. Therefore, for the rest of the cases, working in terms of physical components we will deal with on-shell supersymmetry. This makes life very uncomfortable – even checking the supersymmetry invariance of the action becomes a rather complicated task, while in terms of superfields everything is manifestly invariant. Secondly, it is a rather hard problem to write the interaction terms in the component approach. Of course, the superfield approach has its own problems. One of the most serious, when dealing with extended supersymmetry, is to find the irreducibility constraints which decrease the number of components in the superfields. Nevertheless, the formulation of the theory in a manifestly supersymmetric form seems preferable, not only because of its intrinsic beauty, but also since it provides an efficient technique, in particular, in quantum calculations.

The key idea of manifestly invariant formulations of supersymmetric theories is using superspace, where supersymmetry is realized geometrically by coordinate transformations. Let us start with $N = 2$ supersymmetry. The natural definition of $N = 2$ superspace $\mathbb{R}^{(1|2)}$ involves time t and two anticommuting coordinates $\theta, \bar{\theta}$

$$\mathbb{R}^{(1|2)} = (t, \theta, \bar{\theta}) \ . \tag{2.17}$$

In this superspace $N = 2$ super-Poincaré algebra (2.9) can be easily realized

$$\delta\theta = \epsilon, \quad \delta\bar{\theta} = \bar{\epsilon}, \quad \delta t = -i\left(\epsilon\bar{\theta} + \bar{\epsilon}\theta\right) \ . \tag{2.18}$$

The $N = 2$ superfields $\Phi(t, \theta, \bar{\theta})$ are defined as functions on this superspace. The simplest superfield is the scalar one, which transforms under (2.15) as follows:

$$\Phi'(t', \theta', \bar{\theta}') = \Phi(t, \theta, \bar{\theta}) \ . \tag{2.19}$$

From (2.19) one may easily find the variation of the superfield in passive form

$$\delta\Phi \equiv \Phi'(t, \theta, \bar{\theta}) - \Phi(t, \theta, \bar{\theta}) = -\epsilon\left(\frac{\partial}{\partial\theta} - i\bar{\theta}\frac{\partial}{\partial t}\right)\Phi - \bar{\epsilon}\left(\frac{\partial}{\partial\bar{\theta}} - i\theta\frac{\partial}{\partial t}\right)\Phi$$
$$\equiv -\epsilon Q\Phi - \bar{\epsilon}\bar{Q}\Phi \ . \tag{2.20}$$

Thus, we get the realization of supercharges $Q, \bar{Q}$ in superspace

$$Q = \frac{\partial}{\partial\theta} - i\bar{\theta}\frac{\partial}{\partial t}, \quad \bar{Q} = \frac{\partial}{\partial\bar{\theta}} - i\theta\frac{\partial}{\partial t}, \quad \{Q, \bar{Q}\} = -2i\frac{\partial}{\partial t} \ . \tag{2.21}$$

To construct covariant objects in superspace, we have to define covariant derivatives and covariant differentials of the coordinates. Under the transformations (2.18), $d\theta$ and $d\bar{\theta}$ are invariant, but dt is not. It is not too hard to find the proper covariantization of dt

$$dt \quad \rightarrow \quad \Delta t = dt - i\, d\bar{\theta}\theta - i\, d\theta\bar{\theta}\,. \tag{2.22}$$

Indeed, one can check that $\delta\Delta t = 0$ under (2.18). Having at hands the covariant differentials, one may define the covariant derivatives

$$\left(dt\frac{\partial}{\partial t} + d\theta\frac{\partial}{\partial\theta} + d\bar{\theta}\frac{\partial}{\partial\bar{\theta}}\right)\Phi \equiv \left(\Delta t\nabla_t + d\theta\, D + d\bar{\theta}\,\overline{D}\right)\Phi\,,$$

$$\nabla_t = \frac{\partial}{\partial t},\ D = \frac{\partial}{\partial\theta} + i\bar{\theta}\frac{\partial}{\partial t},\ \overline{D} = \frac{\partial}{\partial\bar{\theta}} + i\theta\frac{\partial}{\partial t},\ \{D,\overline{D}\} = 2i\partial_t\,. \tag{2.23}$$

As important properties of the covariant derivatives let us note that they anticommute with the supercharges (2.28).

The superfield Φ contains the ordinary bosonic and fermionic fields as coefficients in its $\theta,\bar{\theta}$ expansion. A convenient covariant way to define these components is to define them as follows:

$$V = \Phi|,\quad \psi = i\overline{D}\Phi|,\quad \bar{\psi} = -iD\Phi|,\quad A = \frac{1}{2}\left[D,\overline{D}\right]\Phi|\,, \tag{2.24}$$

where $|$ means the restriction to $\theta = \bar{\theta} = 0$. One may check that the transformations of the components (2.24), which follow from (2.20), coincide with (2.13). Thus, the general bosonic $N = 2$ superfield Φ describes the $(\mathbf{1},\mathbf{2},\mathbf{1})$ supermultiplet. The last thing we need to know, in order to construct the superfield action is the rule for integration over Grassmann coordinates $\theta,\bar{\theta}$. By definition, this integration is equivalent to a differentiation

$$\int dt\, d\theta\, d\bar{\theta}\mathcal{L} \equiv \int dt D\overline{D}\mathcal{L} = \frac{1}{2}\int dt\left[D,\overline{D}\right]\mathcal{L}\,. \tag{2.25}$$

Now, we are ready to write the free action for the $N = 2$ $(\mathbf{1},\mathbf{2},\mathbf{1})$ supermultiplet

$$S = \gamma\int dt\, d\theta\, d\bar{\theta}\, D\Phi\overline{D}\Phi\,. \tag{2.26}$$

It is a simple exercise to check that, after integration over $d\theta$, $d\bar{\theta}$ and passing to the components (2.24), the action (2.26) coincides with (2.14).

The obvious question now is how to describe in superfields the supermultiplet $(\mathbf{2},\mathbf{2},\mathbf{0})$. The latter contains two physical bosons, therefore the proper superfield should be a complex one. But without any additional conditions the complex $N = 2$ superfield $\phi,\bar{\phi}$ describes a $(\mathbf{2},\mathbf{4},\mathbf{2})$ supermultiplet. The solution is to impose the so-called chirality constraints on the superfield

$$D\phi = 0,\qquad \overline{D}\bar{\phi} = 0\,. \tag{2.27}$$

It is rather easy to find the independent components of the chiral superfield (2.48)

$$\tilde{\phi} = \phi|, \quad \bar{\psi} = i\overline{D}\phi|, \quad \tilde{\bar{\phi}} = \bar{\phi}|, \quad \psi = iD\bar{\phi}| . \tag{2.28}$$

Let us note that, due to the constraints (2.48), the auxiliary components are expressed through time derivatives of the physical ones

$$A = \frac{1}{2}\left[D, \overline{D}\right]\phi| = \frac{1}{2}D\overline{D}\phi| = i\dot{\phi} . \tag{2.29}$$

The free action has a very simple form

$$S = \gamma \int dt \, d\theta \, d\bar{\theta} \left(\phi\dot{\bar{\phi}} - \dot{\phi}\bar{\phi}\right) . \tag{2.30}$$

As the immediate result of the superfield formulation, one may write the actions of $N = 2$ σ-models for both supermultiplets

$$S_\sigma = \gamma \int dt \, d\theta \, d\bar{\theta} \, F_1(\Phi)D\Phi\overline{D}\Phi ,$$

$$S_\sigma = \gamma \int dt \, d\theta \, d\bar{\theta} \, F_2(\phi, \bar{\phi}) \left(\phi\dot{\bar{\phi}} - \dot{\phi}\bar{\phi}\right) , \tag{2.31}$$

where $F_1(\Phi)$ and $F_2(\phi, \bar{\phi})$ are two arbitrary functions defining the metric in the target space.

Another interesting example is provided by the action of $N = 2$ superconformal mechanics [9]

$$S_{\text{Conf}} = \gamma \int dt \, d\theta \, d\bar{\theta} \left[D\Phi\overline{D}\Phi + 2m \log \Phi\right] . \tag{2.32}$$

The last $N = 2$ supermultiplet in Table 2.1, with content $(\mathbf{0}, \mathbf{2}, \mathbf{2})$, may be described by the chiral fermionic superfield Ψ:

$$D\Psi = 0, \qquad \overline{D}\bar{\Psi} = 0 . \tag{2.33}$$

Thus, we described in superfields all $N = 2$ supermultiplets. But really interesting features appear in the $N = 4$ supersymmetric theories which we will consider in the next subsection.

2.2.4 $N = 4$ Supermultiplets

The $N = 4, d = 1$ superspace $\mathbb{R}^{(1|4)}$ is parameterized by the coordinates

$$\mathbb{R}^{(1|4)} = \left(t, \theta_i, \bar{\theta}^j\right) , \qquad (\theta_i)^\dagger = \bar{\theta}^i , \qquad i, j = 1, 2 . \tag{2.34}$$

The covariant derivatives may be defined in a full analogy with the $N = 2$ case as

$$D^i = \frac{\partial}{\partial\theta_i} + i\bar\theta^i\frac{\partial}{\partial t}, \qquad \overline{D}_j = \frac{\partial}{\partial\bar\theta_j} + i\theta_j\frac{\partial}{\partial t}, \qquad \{D^i, \overline{D}_j\} = 2i\delta^i_j\partial_t . \quad (2.35)$$

Such a representation of the algebra of $N = 4, d = 1$ spinor covariant derivatives manifests an automorphism $SU(2)$ symmetry (from the full $SO(4)$ automorphism symmetry of $N = 4, d = 1$ superspace) realized on the doublet indices i, j. The transformations from the coset $SO(4)/SU(2)$ rotate D^i and $\overline{D}^j$ through each other.

Now, we are going to describe in superfields all possible $N = 4$ supermultiplets from Table 2.1.

$N = 4$, $d = 1$ "Hypermultiplet": $(4, 4, 0)$

We shall start with the most general case when the supermultiplet contains four physical bosonic components. To describe this supermultiplet we have to introduce four $N = 4$ superfields $q^i, \bar{q}_j$. These superfields should be properly constrained to reduce 32 bosonic and 32 fermionic components, which are present in the unconstrained $q^i, \bar{q}_j$, to 4 bosonic and 4 fermionic ones. One may show that the needed constraints read

$$D^{(i}q^{j)} = 0, \qquad \overline{D}^{(i}q^{j)} = 0, \qquad D^{(i}\bar{q}^{j)} = 0, \qquad \overline{D}^{(i}\bar{q}^{j)} = 0. \quad (2.36)$$

This $N = 4, d = 1$ multiplet was considered in [3, 4, 10–12] and also was recently studied in $N = 4, d = 1$ harmonic superspace [13]. It resembles the $N = 2, d = 4$ hypermultiplet. However, in contrast to the $d = 4$ case, the constraints (2.125) define an off-shell multiplet in $d = 1$.

The constraints (2.36) leave in the $N = 4$ superfield q^i just four spinor components

$$D_i q^i, \quad \overline{D}_i q^i, \quad D_i \bar{q}^i, \quad \overline{D}_i \bar{q}^i, \quad (2.37)$$

while all higher components in the θ-expansion are expressed as time-derivatives of the lowest ones. This can be immediately seen from the following consequences of (2.36):

$$D^i\overline{D}_j q^j = 4i\dot{q}^i, \qquad D^i D_j q^j = 0.$$

The general sigma-model action for the supermultiplet $(4, 4, 0)$ reads[1]

$$S_\sigma = \int dt\, d^4\theta K(q, \bar{q}), \quad (2.38)$$

where $K(q, \bar{q})$ is an arbitrary function on q^i and q_j. When expressed in components, the action (2.38) has the following form:

$$S_\sigma = \int dt \left[\frac{\partial^2 K(q, \bar{q})}{\partial q^i \partial \bar{q}_i} \dot{q}^j \dot{\bar{q}}_j + \text{fermions} \right]. \quad (2.39)$$

---

[1] The standard convention for integration in $N = 4, d = 1$ superspace is $\int dt\, d^4\theta = \frac{1}{16}\int dt D^i D_i \overline{D}^j \overline{D}_j$

Another interesting example is the superconformally invariant superfield action [4]

$$S_{\text{Conf}} = -\int dt\, d^4\theta\, \frac{\ln(q^i \bar{q}_i)}{q^j \bar{q}_j}\,. \tag{2.40}$$

A more detailed discussion of possible actions for the q^i multiplet can be found in [13]. In particular, there exists a superpotential-type off-shell invariant which, however, does not give rise in components to any scalar potential for the physical bosons. Instead, it produces a Wess–Zumino type term of the first order in the time derivative. It can be interpreted as a coupling to a four-dimensional background abelian gauge field. The superpotential just mentioned admits a concise manifestly supersymmetric superfield formulation, as an integral over an analytic subspace of $N = 4, d = 1$ harmonic superspace [13].

Notice that the q^i supermultiplet can be considered as a fundamental one, since all other $N = 4$ supermultiplets can be obtained from q^i by reduction. We will consider how such reduction works in the last section of these lectures.

$N = 4$, $d = 1$ "Tensor" Multiplet: $(\mathbf{3}, \mathbf{4}, \mathbf{1})$

The "tensor" multiplet includes three $N = 4$ bosonic superfields which can be combined in an $N = 4$ real isovector superfield V^{ij} ($V^{ij} = V^{ji}$ and $\overline{V^{ik}} = \epsilon_{ii'}\epsilon_{kk'}V^{i'k'}$). The irreducibility constraints may be written in the manifestly $SU(2)$-symmetric form

$$D^{(i}V^{jk)} = 0\,, \qquad \overline{D}^{(i}V^{jk)} = 0\,. \tag{2.41}$$

The constraints (2.41) could be obtained by a direct dimensional reduction from the constraints defining the $N = 2, d = 4$ tensor multiplet [14], in which one suppresses the $SL(2, C)$ spinor indices of the $d = 4$ spinor derivatives, thus keeping only the doublet indices of the R-symmetry $SU(2)$ group. This is the reason why we can call it $N = 4, d = 1$ "tensor" multiplet. Of course, in one dimension no differential (notoph-type) constraints arise on the components of the superfield V^{ij}. The constraints (2.41) leave in V^{ik} the following independent superfield projections:

$$V^{ik}\,, \quad D^i V^{kl} = -\frac{1}{3}(\epsilon^{ik}\chi^l + \epsilon^{il}\chi^k)\,, \quad \bar{D}^i V^{kl} = \frac{1}{3}(\epsilon^{ik}\bar{\chi}^l + \epsilon^{il}\bar{\chi}^k)\,, \quad D^i \bar{D}^k V_{ik}\,, \tag{2.42}$$

where

$$\chi^k \equiv D^i V_i^k\,, \qquad \bar{\chi}_k = \overline{\chi^k} = \bar{D}_i V_k^i\,. \tag{2.43}$$

Thus, its off-shell component field content is just $(\mathbf{3}, \mathbf{4}, \mathbf{1})$. The $N = 4, d = 1$ superfield V^{ik} subjected to the conditions (2.41) was introduced in [15] and, later on, rediscovered in [16–18].

As in the case of the superfield q^i, the general sigma-model action for the supermultiplet $(\mathbf{3}, \mathbf{4}, \mathbf{1})$ may be easily written as

$$S_\sigma = \int dt\, d^4\theta K(V) \,, \tag{2.44}$$

where $K(V)$ is an arbitrary function on V.

As the last remark in this subsection, let us note that the "tensor" multiplet can be constructed in terms of the "hypermultiplet." Indeed, let us represent V^{ij} as the following composite superfield:

$$\widetilde{V}^{11} = -i\sqrt{2}\,q^1\bar{q}^1 \,, \quad \widetilde{V}^{22} = -i\sqrt{2}\,q^2\bar{q}^2 \,, \quad \widetilde{V}^{12} = -\frac{i}{\sqrt{2}}\left(q^1\bar{q}^2 + q^2\bar{q}^1\right) \,. \tag{2.45}$$

One can check that, as a consequence of the "hypermultiplet" constraints (2.36), the composite superfield $\widetilde{V}^{ij}$ automatically obeys (2.41). This is just the relation established in [13].

The expressions (2.45) supply a rather special solution to the "tensor" multiplet constraints. In particular, they express the auxiliary field of $\widetilde{V}^{ij}$ through the time derivative of the physical components of q^i, which contains no auxiliary fields.[2] As a consequence, the superpotential of V^{ik} is a particular case of the q^i superpotential, which produces no genuine scalar potential for physical bosons and gives rise for them only to a Wess–Zumino type term of the first order in the time derivative.

$N = 4$, $d = 1$ Chiral Multiplet: $(2, 4, 2)$

The chiral $N = 4$ supermultiplet is the simplest one. It can be described, in full analogy with the $N = 2$ case, by a complex superfield ϕ subjected to the constraints [20, 21]

$$D^i\phi = 0 \,, \qquad \overline{D}_j\bar{\phi} = 0 \,. \tag{2.46}$$

The sigma-model type action for this multiplet

$$S_\sigma = \int d^4\theta K(\phi, \bar{\phi}) \tag{2.47}$$

may be immediately extended to include the potential terms

$$S_{\text{pot}} = \int d^2\bar{\theta} F(\phi) + \int d^2\theta \bar{F}(\bar{\phi}) \,. \tag{2.48}$$

In more details, such a supermultiplet and the corresponding actions have been considered in [22].

---

[2] This is a nonlinear version of the phenomenon which holds in general for $d = 1$ supersymmetry and was discovered at the linearized level in [1, 19]

The "Old Tensor" Multiplet: $(1, 4, 3)$

The last possibility corresponds to the single bosonic superfield u. In this case no linear constraints appear, since four fermionic components are expressed through four spinor derivatives of u. As it was shown in [22], one should impose some additional, second order in spinor derivatives, irreducibility constraints on u

$$D^i D_i \, e^u = \overline{D}_i \overline{D}^i \, e^u = \left[D^i, \overline{D}_i \right] e^u = 0 \,, \tag{2.49}$$

in order to pick up in u the minimal off-shell field content $(\mathbf{1,4,3})$. Once again, the detailed discussion of this case can be found in [22].

In fact, we could rederive the multiplet u from the tensor multiplet V^{ik} discussed in the previous subsection. Indeed, one can construct the composite superfield

$$e^{\tilde{u}} = \frac{1}{\sqrt{V^2}} \,, \tag{2.50}$$

which obeys just the constraints (2.49) as a consequence of (2.41). The relation (2.50) is of the same type as the previously explored substitution (2.45) and expresses two out of the three auxiliary fields of $\tilde{u}$ via physical bosonic fields of V^{ik} and time derivatives thereof. The superconformal invariant action for the superfield u can be also constructed [22]

$$S_{\mathrm{Conf}} = \int \mathrm{d}t \, \mathrm{d}^4\theta \, e^u \, u \,. \tag{2.51}$$

By this, we complete the superfield description of all linear $N = 4, d = 1$ supermultiplets. But the story about $N = 4, d = 1$ supermultiplets is not finished. There are nonlinear supermultiplets which make everything much more interesting. Let us discuss here only one example – the nonlinear chiral multiplet [4, 23].

$N = 4, \, d = 1$ Nonlinear Chiral Multiplet: $(2, 4, 2)$

The idea of nonlinear chiral supermultiplets comes about as follows. If two bosonic superfields $\mathcal{Z}$ and $\overline{\mathcal{Z}}$ parameterize the two-dimensional sphere $SU(2)/U(1)$ instead of flat space, then they transform under $SU(2)/U(1)$ transformations with the parameters $a, \bar{a}$ as

$$\delta \mathcal{Z} = a + \bar{a}\mathcal{Z}^2 \,, \qquad \delta \overline{\mathcal{Z}} = \bar{a} + a\overline{\mathcal{Z}}^2 \,. \tag{2.52}$$

With respect to the same group $SU(2)$, the $N = 4$ covariant derivatives form a doublet

$$\delta D_i = -a\overline{D}_i \,, \qquad \delta \overline{D}_i = \bar{a}D_i \,. \tag{2.53}$$

One may immediately check that the ordinary chirality conditions

$$D_i \mathcal{Z} = 0, \qquad \overline{D}_i \overline{\mathcal{Z}} = 0$$

are not invariant with respect to (2.52), (2.53) and they should be replaced, if we wish to keep the $SU(2)$ symmetry. It is rather easy to guess the proper $SU(2)$ invariant constraints

$$D_i \mathcal{Z} = -\alpha \mathcal{Z} \overline{D}_i \mathcal{Z} , \qquad \overline{D}_i \overline{\mathcal{Z}} = \alpha \overline{\mathcal{Z}} D_i \overline{\mathcal{Z}} , \qquad \alpha = \text{const} . \qquad (2.54)$$

So, using the constraints (2.54), we restore the $SU(2)$ invariance, but the price for this is just the nonlinearity of the constraints. Let us stress that $N = 4, d = 1$ supersymmetry is the minimal one where the constraints (2.54) may appear, because the covariant derivatives (and the supercharges) form a doublet of $SU(2)$ which cannot be real.

The $N = 4, d = 1$ nonlinear chiral supermultiplet involves one complex scalar bosonic superfield $\mathcal{Z}$ obeying the constraints (2.54). If the real parameter $\alpha \neq 0$, it is always possible to pass to $\alpha = 1$ by a redefinition of the superfields $\mathcal{Z}, \overline{\mathcal{Z}}$. So, one has only two essential values $\alpha = 1$ and $\alpha = 0$. The latter case corresponds to the standard $N = 4, d = 1$ chiral supermultiplet. Now, one can write the most general $N = 4$ supersymmetric Lagrangian in $N = 4$ superspace

$$S = \int dt\, d^2\theta\, d^2\bar{\theta}\, K(\mathcal{Z}, \overline{\mathcal{Z}}) + \int dt\, d^2\bar{\theta}\, F(\mathcal{Z}) + \int dt\, d^2\theta\, \overline{F}(\overline{\mathcal{Z}}) . \qquad (2.55)$$

Here, $K(\mathcal{Z}, \overline{\mathcal{Z}})$ is an arbitrary function of the superfields $\mathcal{Z}$ and $\overline{\mathcal{Z}}$, while $F(\mathcal{Z})$ and $\overline{F}(\overline{\mathcal{Z}})$ are arbitrary holomorphic functions depending only on $\mathcal{Z}$ and $\overline{\mathcal{Z}}$, respectively. Let us stress that our superfields $\mathcal{Z}$ and $\overline{\mathcal{Z}}$ obey the nonlinear variant of chirality conditions (2.54), but nevertheless the last two terms in the action S (2.55) are still invariant with respect to the full $N = 4$ supersymmetry. Indeed, the supersymmetry transformations of the integrand of, for example, the second term in (2.55) read

$$\delta F(\mathcal{Z}) = -\epsilon^i D_i F(\mathcal{Z}) + 2i\epsilon^i \bar{\theta}_i \dot{F}(\mathcal{Z}) - \bar{\epsilon}_i \overline{D}^i F(\mathcal{Z}) + 2i\bar{\epsilon}_i \theta^i \dot{F}(\mathcal{Z}) . \qquad (2.56)$$

Using the constraints (2.54) the first term in the r.h.s. of (2.56) may be rewritten as

$$- \epsilon^i D_i F = -\epsilon^i F_{\mathcal{Z}} D_i \mathcal{Z} = \alpha \epsilon^i F_{\mathcal{Z}} \mathcal{Z} \overline{D}_i \mathcal{Z} \equiv \alpha \epsilon^i \overline{D}_i \int d\mathcal{Z}\, F_{\mathcal{Z}} \mathcal{Z} . \qquad (2.57)$$

Thus, all terms in (2.56) either are full time derivatives or disappear after integration over $d^2\bar{\theta}$.

The irreducible component content of $\mathcal{Z}$, implied by (2.54), does not depend on α and can be defined as

$$z = \mathcal{Z}| , \quad \bar{z} = \overline{\mathcal{Z}}| , \quad A = -i\overline{D}^i D_i \mathcal{Z}| ,$$
$$\bar{A} = -i D^i D_i \overline{\mathcal{Z}}| , \quad \psi^i = \overline{D}^i \mathcal{Z}| , \quad \bar{\psi}^i = -D^i \overline{\mathcal{Z}}| , \qquad (2.58)$$

where $|$ means restricting expressions to $\theta_i = \bar{\theta}^j = 0$. All higher-dimensional components are expressed as time derivatives of the irreducible ones. Thus, the $N = 4$ superfield $\mathcal{Z}$ constrained by (2.54) has the same field content as the linear chiral supermultiplet.

After integrating in (2.55) over the Grassmann variables and eliminating the auxiliary fields $A, \bar{A}$ by their equations of motion, we get the action in terms of physical components

$$S = \int dt \left\{ g\dot{z}\dot{\bar{z}} - i\alpha \frac{\dot{z}\bar{z}}{1+\alpha^2 z\bar{z}} F_z + i\alpha \frac{\dot{\bar{z}}z}{1+\alpha^2 z\bar{z}} \overline{F}_{\bar{z}} - \frac{F_z \overline{F}_{\bar{z}}}{g(1+\alpha^2 z\bar{z})^2} + \text{fermions} \right\} , \tag{2.59}$$

where

$$g(z,\bar{z}) = \frac{\partial^2 K(z,\bar{z})}{\partial z \partial \bar{z}}, \quad F_z = \frac{dF(z)}{dz}, \quad \overline{F}_{\bar{z}} = \frac{d\overline{F}(\bar{z})}{d\bar{z}}. \tag{2.60}$$

From the bosonic part of the action (2.59) one may conclude that the system contains a *nonzero magnetic field* with the potential

$$\mathcal{A}_0 = i\alpha \frac{F_z \bar{z} \, dz}{1+\alpha^2 z\bar{z}} - i\alpha \frac{\overline{F}_{\bar{z}} z \, d\bar{z}}{1+\alpha^2 z\bar{z}} , \quad d\mathcal{A}_0 = i\alpha \frac{F_z + \overline{F}_{\bar{z}}}{(1+\alpha^2 z\bar{z})^2} \, dz \wedge d\bar{z} . \tag{2.61}$$

The strength of this magnetic field is given by the expression

$$B = \alpha \frac{(F_z + \overline{F}_{\bar{z}})}{(1+\alpha^2 z\bar{z})^2 g} . \tag{2.62}$$

As for the fermionic part of the kinetic term, it can be represented as follows:

$$\mathcal{S}_{\text{KinF}} = \frac{i}{4} \int dt (1+\alpha^2 z\bar{z}) g \left(\psi \frac{D\bar{\psi}}{dt} - \bar{\psi} \frac{D\psi}{dt} \right) , \tag{2.63}$$

where

$$D\psi = d\psi + \Gamma \psi \, dz + T^+ \bar{\psi} \, dz , \quad D\bar{\psi} = d\bar{\psi} + \bar{\Gamma} \bar{\psi} \, d\bar{z} + T^- \psi \, d\bar{z} , \tag{2.64}$$

and

$$\Gamma = \partial_z \log \left((1+\alpha^2 z\bar{z}) g \right) , \quad T^\pm = \pm \frac{\alpha}{1+\alpha^2 z\bar{z}} . \tag{2.65}$$

Clearly enough, Γ, $\bar{\Gamma}$, $T^\pm$ define the components of the connection defining the configuration superspace. The components Γ and $\bar{\Gamma}$ could be identified with the components of the symmetric connection on the base space equipped with the metric $(1+\alpha^2 z\bar{z}) g \, dz \, d\bar{z}$, while the rest does not have a similar interpretation.

Thus, we conclude that the main differences between the $N = 4$ supersymmetric mechanics with nonlinear chiral supermultiplet and the standard one are the coupling of the fermionic degrees of freedom to the background, via the deformed connection, the possibility to introduce a magnetic field, and the deformation of the bosonic potential.

So far, we presented the results without explanations about how they were found. It appears to be desirable to put the construction and study of

N-extended supersymmetric models on a systematic basis by working out the appropriate superfield techniques. Such a framework exists and is based on a superfield nonlinear realization of the $d = 1$ superconformal group. It was pioneered in [22] and recently advanced in [4, 6, 24]. Its basic merits are, firstly, that in most cases it automatically yields the irreducibility conditions for $d = 1$ superfields and, secondly, that it directly specifies the superconformal transformation properties of these superfields. The physical bosons and fermions, together with the $d = 1$ superspace coordinates, prove to be coset parameters associated with the appropriate generators of the superconformal group. Thus, the differences in the field content of the various supermultiplets are attributed to different choices of the coset supermanifold inside the given superconformal group.

2.3 Nonlinear Realizations

In the previous section, we considered the $N = 4, d = 1$ linear supermultiplets and constructed some actions. But the most important questions concerning the irreducibility constraints and the transformation properties of the superfields were given as an input. In this section, we are going to demonstrate that most of the constraints and all transformation properties can be obtained automatically as results of using the nonlinear realization approach. Our consideration will be mostly illustrative – we will skip presenting any proofs. Instead, we will pay attention to the ideas and technical features of this approach.

2.3.1 Realizations in the Coset Space

The key statement of the nonlinear realization approach may be formulated as follows.

Theorem 1.   *If a group G acts transitively[3] on some space, and the subgroup H preserves a given point of this space, then this action of the group G may be realized by left multiplications on the coset G/H, while the coordinates which parameterize the coset G/H are just the coordinates of the space.*

As the simplest example let us consider the four-dimensional Poincaré group $\{P_\mu, M_{\mu\nu}\}$. It is clear that the transformations which preserve some point are just rotations around this points. In other words, $H = \{M_{\mu\nu}\}$. Therefore, due to our Theorem, a natural realization of the Poincaré group can be achieved in the coset G/H. This coset contains only the translations P_μ, and it is natural to parameterize this coset as

---

[3] This means that the transformations from G relate any two arbitrary points in the space

$$G/H = e^{ix^\mu P_\mu} \, . \tag{2.66}$$

It is evident that the Poincaré group may be realized on the four coordinates $\{x^\mu\}$. What is important here is that we do not need to know how our coordinates transform under the Poincaré group. Instead, we can deduce these transformations from the representation (2.66).

Among different cosets there are special ones which are called orthonormal. They may be described as follows. Let us consider the group G with generators $\{X_i, Y_\alpha\}$ which obey the following relations:

$$\begin{aligned}
[Y_\alpha, Y_\beta] &= iC^\gamma_{\alpha\beta} Y_\gamma \, , \\
[Y_\alpha, X_i] &= iC^j_{\alpha i} X_j + iC^\beta_{\alpha i} Y_\beta \, , \\
[X_i, X_j] &= iC^k_{ij} X_k + iC^\alpha_{ij} Y_\alpha \, ,
\end{aligned} \tag{2.67}$$

where C are structure constants. We see that the generators Y_α form the subgroup H. The coset G/H is called orthonormal if $C^\beta_{\alpha i} = 0$. In other words, this property means that the generators X_i transform under some representation of the stability subgroup H. In what follows we will consider only such a coset. A more restrictive class of cosets – the symmetric spaces – corresponds to the additional constraints $C^k_{ij} = 0$.

A detailed consideration of the cosets and their geometric properties may be found in [25].

A very important class is made by the cosets which contain space-time symmetry generators as well as the generators of internal symmetries. To deal with such cosets we must

- introduce the coordinates for space-time translations (or/and supertranslations);
- introduce the parameters for the rest of the generators in the coset. These additional parameters are treated as *fields* which depend on the space-time coordinates.

The fields (superfields) which appear as parameters of the coset will have inhomogeneous transformation properties. They are known as Goldstone fields. Their appearance is very important: they are definitely needed to construct an action, which is invariant with respect to transformations from the coset G/H. Let us repeat this point: in the nonlinear realization approach only the H-symmetry is manifest. The invariance under G/H transformations is achieved through the interaction of matter fields with Goldstone ones.

Finally, let us note that the number of essential Goldstone fields does not always coincide with the number of coset generators. As we will see later, some of the Goldstone fields often can be expressed through other Goldstone fields. This is the so called Inverse Higgs phenomenon [26].

Now, it is time to demonstrate how all this works on the simplest examples.

2.3.2 Realizations: Examples and Technique

$N = 2$, $d = 1$ Super Poincaré

Let us start with the simplest example of the $N = 2, d = 1$ super-Poincaré algebra that contains two supertranslations $Q, \bar{Q}$ which anticommute on the time-translation P (2.9)

$$\{Q, \bar{Q}\} = -2P .$$

In this case the stability subgroup is trivial and all generators are in the coset. Therefore, one should introduce coordinates for all generators:

$$g = G/H = e^{itP} \, e^{\theta Q + \bar{\theta}\bar{Q}} . \tag{2.68}$$

These coordinates $\{t, \theta, \bar{\theta}\}$ span $N = 2, d = 1$ superspace. Now, we are going to find the realization of $N = 2$ superalgebra (2.9) in this superspace. The first step is to find the realization of the translation P. So, we act on the element (2.68) from the left by g_0

$$g_0 = e^{iaP} : \; g_0 \cdot g = e^{iaP} \cdot e^{itP} \, e^{\theta Q + \bar{\theta}\bar{Q}} = e^{i(t+a)P} \, e^{\theta Q + \bar{\theta}\bar{Q}} \equiv e^{it'P} e^{\theta'Q + \bar{\theta}'\bar{Q}} . \tag{2.69}$$

Thus, we get the standard transformations of the coordinates

$$P: \quad \begin{cases} \delta t = a \\ \delta\theta = \delta\bar{\theta} = 0 . \end{cases} \tag{2.70}$$

Something more interesting happens for the supertranslations

$$g_1 = e^{\epsilon Q + \bar{\epsilon}\bar{Q}} : \; g_1 \cdot g = e^{\epsilon Q + \bar{\epsilon}\bar{Q}} \cdot e^{itP} \, e^{\theta Q + \bar{\theta}\bar{Q}} = e^{itP} \, e^{\epsilon Q + \bar{\epsilon}\bar{Q}} \, e^{\theta Q + \bar{\theta}\bar{Q}} . \tag{2.71}$$

Now, we need to bring the product of the exponents in (2.71) to the standard form

$$e^{it'P} \, e^{\theta'Q + \bar{\theta}'\bar{Q}} .$$

To do this, one should use the Campbell–Hausdorff formulae

$$e^A \cdot e^B = \exp\left(A + B + \frac{1}{2}[A, B] + \frac{1}{12}[A, [A, B]] - \frac{1}{12}[[A, B], B] + \cdots\right) \tag{2.72}$$

Thus, we will get

$$e^{\epsilon Q + \bar{\epsilon}\bar{Q}} \, e^{\theta Q + \bar{\theta}\bar{Q}} = e^{(\theta+\epsilon)Q + (\bar{\theta}+\bar{\epsilon})\bar{Q} + (\epsilon\bar{\theta} + \bar{\epsilon}\theta)P} . \tag{2.73}$$

So, the supertranslation is realized as follows:

$$Q, \bar{Q}: \quad \begin{cases} \delta t = -i(\epsilon\bar{\theta} + \bar{\epsilon}\theta) \\ \delta\theta = \epsilon, \; \delta\bar{\theta} = \bar{\epsilon} . \end{cases} \tag{2.74}$$

This is just the transformation (2.18) we used before. Of course, in this rather simple case we could find the answer without problems. But the lesson is that the same procedure works always, for any (super)group and any coset. Let us consider a more involved example.

$d = 1$ Conformal Group

The conformal algebra in $d = 1$ contains three generators: translation P, dilatation D and conformal boost K obeying the following relations:

$$i\,[P, K] = -2D, \quad i\,[D, P] = P, \quad i\,[D, K] = -2K \ . \tag{2.75}$$

The stability subgroup is again trivial and our coset may be parameterized as

$$g = e^{itP}\, e^{iz(t)K}\, e^{iu(t)D} \ . \tag{2.76}$$

Let us stress that we want to obtain a one-dimensional realization. Therefore, we can introduce only one coordinate – time t. But we have three generators in the coset. The unique solution[4] is to consider the two other coordinates as functions of time. Thus, we have to introduce two fields $z(t)$ and $u(t)$ which are just Goldstone fields.

Let us find the realization of the conformal group in our coset (2.76).

The translation is realized trivially, as in (2.70), so we will start from the dilatation

$$g_0 = e^{iaD} : \quad g_0 \cdot g = e^{iaD} \cdot e^{itP}\, e^{iz(t)K}\, e^{iu(t)D} \ . \tag{2.77}$$

Now, we have a problem – how to commute the first exponent in (2.77) with the remaining ones? The Campbell–Hausdorff formulae (2.72) do not help us too much, because the series does not terminate. A useful trick is to represent r.h.s. of (2.77) in the form

$$\underbrace{e^{iaD}\, e^{itP}\, e^{-iaD}}_{1}\ \underbrace{e^{iaD}\, e^{iz(t)K}\, e^{-iaD}}_{2}\ e^{iaD}\, e^{iu(t)D} \ . \tag{2.78}$$

To evaluate (2.78) we will use the following formulas due to Bruno Zumino [27]:

$$e^A B e^{-A} \equiv e^A \wedge B, \qquad e^A e^B e^{-A} \equiv e^{e^A \wedge B} \ , \tag{2.79}$$

where

$$A^n \wedge B \equiv \underbrace{[A, [A, [\dots\ , B]]\dots]}_{n} \tag{2.80}$$

Using (2.79) we can immediately find

$$D: \quad \begin{cases} \delta t = at \\ \delta u = a, \quad \delta z = -az \ . \end{cases} \tag{2.81}$$

We see that the field $u(t)$ is shifted by the constant parameter a under dilatation. Such a Goldstone field is called a *dilaton*.

---

[4] Of course, we may put some of the generators K and D, or even both of them, in the stability subgroup H. But in this case all matter fields should realize a representation of $H = \{K, D\}$ which never happens

Finally, we should find the transformations of the coordinates under conformal boost K:

$$g_1 = \mathrm{e}^{ibK} : \quad g_1 \cdot g = \underbrace{\mathrm{e}^{ibK}\, \mathrm{e}^{itP}}_{1}\, \mathrm{e}^{iz(t)K}\, \mathrm{e}^{iu(t)D}. \tag{2.82}$$

Here, in order to commute the first two terms we will use the following trick: let us represent the first two exponents in the r.h.s. of (2.82) as follows:

$$\mathrm{e}^{ibK}\, \mathrm{e}^{itP} = \mathrm{e}^{itP} \cdot \tilde{g} \ \Rightarrow \ \tilde{g} = \mathrm{e}^{-itP}\, \mathrm{e}^{ibK}\, \mathrm{e}^{itP}\ . \tag{2.83}$$

So, we again can use (2.79) to calculate[5] $\tilde{g}$

$$\tilde{g} = \mathrm{e}^{ibK + 2ibtD + ibt^2 P} \approx \mathrm{e}^{ibK}\, \mathrm{e}^{2ibtD}\, \mathrm{e}^{ibt^2 P}\ .$$

Thus, we have

$$K: \quad \begin{cases} \delta t = bt^2 \\ \delta u = 2bt, \quad \delta z = b - 2btz \end{cases} \tag{2.84}$$

Now, we know how to find the transformation properties of the coordinates and Goldstone (super)fields for any cosets. The next important question is how to construct the invariant and/or covariant objects.

2.3.3 Cartan's Forms

The Cartan's forms for the coset $g = G/H$ are defined as follows:

$$g^{-1}\, \mathrm{d}g = i\omega^i X_i + i\omega^\alpha Y_\alpha\ , \tag{2.85}$$

where the generators $\{X_i, Y_\alpha\}$ obey to (2.67).

By the definition (2.85) the Cartan's forms are invariant with respect to left multiplication of the coset element g. Let us represent the result of the left multiplication of the coset element g as follows:

$$g_0 \cdot g = \tilde{g} \cdot h\ . \tag{2.86}$$

Here, $\tilde{g}$ belongs to the coset, while the element h lies in the stability subgroup H. Now we have

$$i\omega^i X_i + i\omega^\alpha Y_\alpha = h^{-1}\ \underbrace{\left(\tilde{g}^{-1}\, \mathrm{d}\tilde{g}\right)}_{i\tilde{\omega}^i X_i + i\tilde{\omega}^\alpha Y_\alpha}\ h + h^{-1}\, \mathrm{d}h\ . \tag{2.87}$$

If the coset $g = G/H$ is othonormal, then

$$\tilde{\omega}^i X_i = h \cdot \omega^i X_i \cdot h^{-1}\ ,$$
$$\tilde{\omega}^\alpha Y_\alpha = h \cdot \omega^\alpha Y_\alpha \cdot h^{-1} + i\, \mathrm{d}h \cdot h^{-1}\ . \tag{2.88}$$

---

[5] We are interested in the infinitesimal transformations and omit all terms which are higher then linear in the parameters

Thus, we see that the forms ω^i which belong to the coset transform homogeneously, while the forms ω^α on the stability subgroup transform like connections and can be used to construct covariant derivatives.

Finally, one should note that, in the exponential parameterization we are using through these lectures, the evaluation of the Cartan's forms is based on the following identity [27]:

$$e^{-A} \, \mathrm{d}e^A = \frac{1 - e^{-A}}{A} \wedge \mathrm{d}A \, . \tag{2.89}$$

Now it is time for some examples.

$N = 2$, $d = 1$ Super Poincaré

Choosing the parameterization of the group element as in (2.68) one may immediately find the Cartan's forms

$$\omega_P = \mathrm{d}t - i \left(\mathrm{d}\bar\theta\theta + \mathrm{d}\theta\bar\theta \right), \qquad \omega_Q = \mathrm{d}\theta, \ \bar\omega_Q = \mathrm{d}\bar\theta \, . \tag{2.90}$$

Thus, the Cartan's forms (2.90) just coincide with the covariant differentials (2.41) we guessed before.

$d = 1$ Conformal Group

The Cartan's forms for the coset (2.76) may be easily calculated using (2.89)

$$\omega_P = e^{-u}, \quad \omega_D = \mathrm{d}u - 2z \, \mathrm{d}t, \quad \omega_K = e^u \left[\mathrm{d}z + z^2 \, \mathrm{d}t \right] \, . \tag{2.91}$$

What is the most interesting here is the structure of the form ω_D. Indeed, from the previous consideration we know that ω_D being coset forms, transforms homogeneously. Therefore, the following condition:

$$\omega_D = 0 \ \Rightarrow \ z = \frac{1}{2}\dot u \tag{2.92}$$

is invariant with respect to the whole conformal group! This means that the Goldstone field $z(t)$ is unessential and can be expressed in terms of the dilaton $u(t)$. This is the simplest variant of the inverse Higgs phenomenon [26]. With the help of the Cartan' forms (2.92) one could construct the simplest invariant action

$$S = -\int \left(\omega_K + m^2 \omega_P \right) = \int \mathrm{d}t \left[\frac{1}{4} e^u \dot u^2 - m^2 e^{-u} \right]$$

$$= \int \mathrm{d}t \left[\dot\rho^2 - \frac{m^2}{\rho^2} \right], \quad \rho \equiv e^{\frac{u}{2}} \, , \tag{2.93}$$

with m being a parameter of the dimension of mass. The action (2.93) is just the conformal mechanics action [28].

It is rather interesting that we could go a little bit further.

The basis (2.75) in the conformal algebra $so(1,2)$ can naturally be called "conformal," as it implies the standard $d = 1$ conformal transformations for the time t. Now, we pass to another basis in the same algebra

$$\hat{K} = mK - \frac{1}{m}P \, , \qquad \hat{D} = mD \, . \tag{2.94}$$

This choice will be referred to as the "AdS basis" for a reason to be clear soon.

The conformal algebra (2.75) in the AdS basis (2.94) reads

$$i[P, \hat{D}] = -mP \, , \quad i[\hat{K}, \hat{D}] = 2P + m\hat{K} \, , \quad i[P, \hat{K}] = -2\hat{D} \, . \tag{2.95}$$

An element of $SO(1,2)$ in the AdS basis is defined to be

$$g = e^{iyP} \, e^{i\phi(y)\hat{D}} \, e^{i\Omega(y)\hat{K}} \, . \tag{2.96}$$

Now, we are in a position to explain the motivation for the nomenclature "AdS basis." The generator $\hat{K}$ (2.94) can be shown to correspond to an $SO(1,1)$ subgroup of $SO(1,2)$. Thus, the parameters y and $\phi(y)$ in (2.96) parameterize the coset $SO(1,2)/SO(1,1)$, i.e., AdS_2. The parameterization (2.96) of AdS_2 is a particular case of the so-called "solvable subgroup parameterization" of the AdS spaces. The $d = 4$ analog of this parameterization is the parameterization of the AdS_5 space in such a way that its coordinates are still parameters associated with the 4-translation and dilatation generators P_m, D of $SO(2,4)$, while it is the subgroup $SO(1,4)$ with the algebra $\propto \{P_m - K_m, so(1,3)\}$ which is chosen as the stability subgroup.

The difference in the geometric meanings of the coordinate pairs $(t, u(t))$ and $(y, \phi(y))$ is manifested in their different transformation properties under the same $d = 1$ conformal transformations. Left shifts of the $SO(1,2)$ group element in the parameterization (2.96) induce the following transformations:

$$\delta y = a(y) + \frac{1}{m^2} \, c \, e^{2m\phi} \, , \quad \delta\phi = \frac{1}{m} \, \dot{a}(y) = \frac{1}{m} \, (b + 2c \, y) \, , \quad \delta\Omega = \frac{1}{m} \, c \, e^{m\phi} \, . \tag{2.97}$$

We observe the modification of the special conformal transformation of y by a field-dependent term.

The relevant left-invariant Cartan forms are given by the following expressions:

$$\hat{\omega}_D = \frac{1 + \Lambda^2}{1 - \Lambda^2} \, d\phi - 2\frac{\Lambda}{1 - \Lambda^2} \, e^{-m\phi} \, dy \, ,$$

$$\hat{\omega}_P = \frac{1 + \Lambda^2}{1 - \Lambda^2} \, e^{-m\phi} \, dy - 2\frac{\Lambda}{1 - \Lambda^2} \, d\phi \, ,$$

$$\hat{\omega}_K = m\frac{\Lambda}{1 - \Lambda^2} \left(\Lambda e^{-m\phi} \, dy - d\phi \right) + \frac{d\Lambda}{1 - \Lambda^2} \, , \tag{2.98}$$

where

$$\Lambda = \tanh \Omega . \tag{2.99}$$

As in the previous realization, the field $\Lambda(y)$ can be eliminated by imposing the inverse Higgs constraint

$$\hat{\omega}_D = 0 \ \Rightarrow \ \partial_y \phi = 2 e^{-m\phi} \frac{\Lambda}{1 + \Lambda^2} , \tag{2.100}$$

whence Λ is expressed in terms of ϕ

$$\Lambda = \partial_y \phi \, e^{m\phi} \frac{1}{1 + \sqrt{1 - e^{2m\phi} (\partial_y \phi)^2}} . \tag{2.101}$$

The $SO(1,2)$ invariant distance on AdS_2 can be defined, prior to imposing any constraints, as

$$ds^2 = -\hat{\omega}_P^2 + \hat{\omega}_D^2 = -e^{-2m\phi} \, dy^2 + d\phi^2 . \tag{2.102}$$

Making the redefinition

$$U = e^{-m\phi} ,$$

it can be cast into the standard Bertotti–Robinson metrics form

$$ds^2 = -U^2 \, dy^2 + (1/m^2) U^{-2} \, dU^2 , \tag{2.103}$$

with $1/m$ as the inverse AdS_2 radius,

$$\frac{1}{m} = R . \tag{2.104}$$

The invariant action can now be constructed from the new Cartan forms (2.98) which, after substituting the inverse Higgs expression for Λ, (2.101), read

$$\hat{\omega}_P = e^{-m\phi} \sqrt{1 - e^{2m\phi} (\partial_y \phi)^2} \, dy ,$$

$$\hat{\omega}_K = -\frac{m}{2} e^{-m\phi} \left(1 - \sqrt{1 - e^{2m\phi} (\partial_y \phi)^2} \right) dy + \text{Tot. deriv.} \times dy. \tag{2.105}$$

The invariant action reads

$$S = - \int \left(\tilde{\mu} \, \hat{\omega}_P - q \, e^{-m\phi} \right) = - \int dy \, e^{-m\phi} \left(\tilde{\mu} \sqrt{1 - e^{2m\phi} (\partial_y \phi)^2} - q \right) . \tag{2.106}$$

After the above field redefinitions it is recognized as the radial-motion part of the "new" conformal mechanics action [29]. Notice that the second term in (2.106) is invariant under (2.97), up to a total derivative in the integrand. The action can be rewritten in a manifestly invariant form (with a tensor Lagrangian) by using the explicit expression for $\hat{\omega}_K$ in (2.105)

$$S = \int [(q - \tilde{\mu})\,\hat{\omega}_P - (2/m)q\,\hat{\omega}_K] \;. \tag{2.107}$$

Now, we are approaching the major point. We see that the "old" and "new" conformal mechanics models are associated with two different nonlinear realizations of the same $d = 1$ conformal group $SO(1,2)$ corresponding, respectively, to the two different choices (2.75) and (2.96) of the parameterization of the group element. The invariant actions in both cases can be written as integrals of linear combinations of the left-invariant Cartan forms. But the latter *cannot* depend on the choice of parameterization. Then the actions (2.93) and (2.106) should in fact coincide with each other, up to a redefinition of the free parameters of the actions. Thus, two conformal mechanics models are *equivalent* modulo redefinitions of the involved time coordinate and field. This statement should be contrasted with the previous view of the "old" conformal mechanics model as a "nonrelativistic" approximation of the "new" one.

2.3.4 Nonlinear Realizations and Supersymmetry

One of the interesting applications of the nonlinear realizations technique is that of establishing irreducibility constraints for superfields.

In the present section, we focus on the case of $N = 4, d = 1$ supersymmetry (with 4 real supercharges) and propose to derive its various irreducible off-shell superfields from different nonlinear realizations of the most general $N = 4, d = 1$ superconformal group $D(2,1;\alpha)$. An advantage of this approach is that it simultaneously specifies the superconformal transformation properties of the superfields, though the latter can equally be used for constructing nonconformal supersymmetric models as well. As the essence of these techniques, any given irreducible $N = 4, d = 1$ superfield comes out as a Goldstone superfield parameterizing, together with the $N = 4, d = 1$ superspace coordinates, some supercoset of $D(2,1;\alpha)$. The method was already employed in the paper [13] where the off-shell multiplet $(\mathbf{3}, \mathbf{4}, \mathbf{1})$ was rederived from the nonlinear realization of $D(2,1;\alpha)$ in the coset with an $SL(2,R) \times [SU(2)/U(1)]$ bosonic part (the second $SU(2) \subset D(2,1;\alpha)$ was placed into the stability subgroup).

Here, we consider nonlinear realizations of the same conformal supergroup $D(2,1;\alpha)$ in its other coset superspaces. In this way we reproduce the $(\mathbf{4}, \mathbf{4}, \mathbf{0})$ multiplet and also derive two new nonlinear off-shell multiplets. The $(\mathbf{4}, \mathbf{4}, \mathbf{0})$ multiplet is represented by superfields parameterizing a supercoset with the bosonic part being $SL(2,R) \times SU(2)$, where the dilaton and the three parameters of $SU(2)$ are identified with the four physical bosonic fields. One of the new Goldstone multiplets is a $d = 1$ analog of the so-called nonlinear multiplet of $N = 2, d = 4$ supersymmetry. It has the same off-shell contents $(\mathbf{3}, \mathbf{4}, \mathbf{1})$ as the multiplet employed in [13], but it obeys a different constraint and enjoys different superconformal transformation properties. It corresponds to the specific nonlinear realization of $D(2,1;\alpha)$, where the dilatation generator and one of the two $SU(2)$ subgroups are placed into the stability subgroup. One more new multiplet of a similar type is obtained by placing into the stability subgroup, along with the dilatation and three $SU(2)$ generators, also the $U(1)$ generator from the second $SU(2) \subset D(2,1;\alpha)$. It has the same field content as a chiral $N = 4, d = 1$ multiplet, i.e., $(\mathbf{2,4,2})$. Hence, it may be termed as the nonlinear chiral supermultiplet. It is exceptional, in the sense that no analogs for it are known in $N = 2, d = 4$ superspace.

Supergroup $D(2,1;\alpha)$ and Its Nonlinear Realizations

We use the standard definition of the superalgebra $D(2,1;\alpha)$ with the notations of [13]. It contains nine bosonic generators which form a direct sum of $sl(2)$ with generators P, D, K and two $su(2)$ subalgebras with generators $V, \overline{V}, V_3$ and $T, \overline{T}, T_3$, respectively

$$i\,[D,P] = P\,, \quad i\,[D,K] = -K\,, \quad i\,[P,K] = -2D\,, \quad i\,[V_3,V] = -V\,,$$
$$i\,[V_3,\overline{V}] = \overline{V}\,, \quad i\,[V,\overline{V}] = 2V_3\,, \quad i\,[T_3,T] = -T\,, \quad i\,[T_3,\overline{T}] = \overline{T}\,,$$
$$i\,[T,\overline{T}] = 2T_3. \tag{2.108}$$

The eight fermionic generators $Q^i, \overline{Q}_i, S^i, \overline{S}_i$ are in the fundamental representations of all bosonic subalgebras (in our notation only one $su(2)$ is manifest, viz. the one with generators $V, \overline{V}, V_3$)

$$i\,[D,Q^i] = \frac{1}{2}Q^i, \quad i\,[D,S^i] = -\frac{1}{2}S^i, \quad i\,[P,S^i] = -Q^i, \quad i\,[K,Q^i] = S^i,$$
$$i\,[V_3,Q^1] = \frac{1}{2}Q^1, \quad i\,[V_3,Q^2] = -\frac{1}{2}Q^2, \quad i\,[V,Q^1] = Q^2, \quad i\,[V,\overline{Q}_2] = -\overline{Q}_1,$$
$$i\,[V_3,S^1] = \frac{1}{2}S^1, \quad i\,[V_3,S^2] = -\frac{1}{2}S^2, \quad i\,[V,S^1] = S^2, \quad i\,[V,\overline{S}_2] = -\overline{S}_1,$$
$$i\,[T_3,Q^i] = \frac{1}{2}Q^i, \quad i\,[T_3,S^i] = \frac{1}{2}S^i, \quad i\,[T,Q^i] = \overline{Q}^i, \quad i\,[T,S^i] = \overline{S}^i \tag{2.109}$$

(and c.c.). The splitting of the fermionic generators into the Q and S sets is natural and useful, because $Q^i, \overline{Q}_k$ together with P form $N = 4, d = 1$ super-Poincaré subalgebra, while $S^i, \overline{S}_k$ generate superconformal translations

$$\{Q^i, \overline{Q}_j\} = -2\delta^i_j P, \qquad \{S^i, \overline{S}_j\} = -2\delta^i_j K\,. \tag{2.110}$$

The nontrivial dependence of the superalgebra $D(2,1;\alpha)$ on the parameter α manifests itself only in the cross-anticommutators of the Poincaré and conformal supercharges

$$\{Q^i, S^j\} = -2(1+\alpha)\epsilon^{ij}\overline{T}, \quad \{Q^1, \overline{S}_2\} = 2\alpha\overline{V}, \quad \{Q^1, \overline{S}_1\}$$
$$= -2D - 2\alpha V_3 + 2(1+\alpha)T_3\,,$$
$$\{Q^2, \overline{S}_1\} = -2\alpha V, \quad \{Q^2, \overline{S}_2\} = -2D + 2\alpha V_3 + 2(1+\alpha)T_3 \tag{2.111}$$

(and c.c.). The generators P, D, K are chosen to be hermitian, and the remaining ones obey the following conjugation rules:

$$(T)^\dagger = \overline{T}, \quad (T_3)^\dagger = -T_3, \quad (V)^\dagger = \overline{V},$$
$$(V_3)^\dagger = -V_3, \quad \overline{(Q^i)} = \overline{Q}_i, \quad \overline{(S^i)} = \overline{S}_i. \tag{2.112}$$

The parameter α is an arbitrary real number. For $\alpha = 0$ and $\alpha = -1$ one of the $su(2)$ algebras decouples and the superalgebra $su(1,1|2) \oplus su(2)$ is recovered. The superalgebra $D(2,1;1)$ is isomorphic to $osp(4^*|2)$.

We will be interested in diverse nonlinear realizations of the superconformal group $D(2,1;\alpha)$ in its coset superspaces. As a starting point we shall consider the following parameterization of the supercoset:

$$g = e^{itP} e^{\theta_i Q^i + \bar\theta^i \overline{Q}_i} e^{\psi_i S^i + \bar\psi^i \overline{S}_i} e^{izK} e^{iuD} e^{i\varphi V + i\bar\varphi \overline{V}} e^{\phi V_3}. \tag{2.113}$$

The coordinates $t, \theta_i, \bar\theta^i$ parameterize the $N = 4, d = 1$ superspace. All other supercoset parameters are Goldstone $N = 4$ superfields. The group $SU(2) \propto (V, \overline{V}, V_3)$ linearly acts on the doublet indices i of spinor coordinates and Goldstone fermionic superfields, while the bosonic Goldstone superfields $\varphi, \bar\varphi, \phi$ parameterize this $SU(2)$. Another $SU(2)$, as a whole, is placed in the stability subgroup and acts only on fermionic Goldstone superfields and θ's, mixing them with their conjugates. With our choice of the $SU(2)$ coset, we are led to assume that $\alpha \neq 0$.

The left-covariant Cartan one-form Ω with values in the superalgebra $D(2,1;\alpha)$ is defined by the standard relation

$$g^{-1} d g = \Omega. \tag{2.114}$$

In what follows we shall need the explicit structure of several important one-forms in the expansion of Ω over the generators,

$$\omega_D = i \, du - 2 \left(\bar\psi^i \, d\theta_i + \psi_i \, d\bar\theta^i \right) - 2iz \, d\tilde t,$$

$$\omega_V = \frac{e^{-i\phi}}{1 + \Lambda\overline\Lambda} \left[i \, d\Lambda + \hat\omega_V + \Lambda^2 \hat{\bar\omega}_V - \Lambda \hat\omega_{V_3} \right],$$

$$\bar\omega_V = \frac{e^{i\phi}}{1 + \Lambda\overline\Lambda} \left[i \, d\overline\Lambda + \hat{\bar\omega}_V + \overline\Lambda^2 \hat\omega_V + \overline\Lambda \hat\omega_{V_3} \right],$$

$$\omega_{V_3} = d\phi + \frac{1}{1 + \Lambda\overline\Lambda} \left[i \left(d\Lambda\overline\Lambda - \Lambda d\overline\Lambda \right) + \left(1 - \Lambda\overline\Lambda \right) \hat\omega_{V_3} - 2 \left(\Lambda \hat{\bar\omega}_V - \overline\Lambda \hat\omega_V \right) \right]. \tag{2.115}$$

Here,

$$\hat\omega_V = 2\alpha \left[\psi_2 \, d\bar\theta^1 - \bar\psi^1 \left(d\theta_2 - \psi_2 \, d\tilde t \right) \right],$$

$$\hat{\bar\omega}_V = 2\alpha \left[\bar\psi^2 \, d\theta_1 - \psi_1 \left(d\bar\theta^2 - \bar\psi^2 \, d\tilde t \right) \right],$$

$$\hat\omega_{V_3} = 2\alpha \left[\psi_1 \, d\bar\theta^1 - \bar\psi^1 \, d\theta_1 - \psi_2 \, d\bar\theta^2 + \bar\psi^2 \, d\theta_2 + \left(\bar\psi^1 \psi_1 - \bar\psi^2 \psi_2 \right) d\tilde t \right], \tag{2.116}$$

$$d\tilde{t} \equiv dt + i\left(\theta_i\,d\bar{\theta}^i + \bar{\theta}^i\,d\theta_i\right)\,, \tag{2.117}$$

and

$$\Lambda = \frac{\tan\sqrt{\varphi\bar{\varphi}}}{\sqrt{\varphi\bar{\varphi}}}\varphi \qquad \bar{\Lambda} = \frac{\tan\sqrt{\varphi\bar{\varphi}}}{\sqrt{\varphi\bar{\varphi}}}\bar{\varphi}\,. \tag{2.118}$$

The semicovariant (fully covariant only under Poincaré supersymmetry) spinor derivatives are defined by

$$D^i = \frac{\partial}{\partial\theta_i} + i\bar{\theta}^i\partial_t\,, \quad \overline{D}_i = \frac{\partial}{\partial\bar{\theta}^i} + i\theta_i\partial_t\,, \quad \{D^i, \overline{D}_j\} = 2i\delta^i_j\partial_t\,. \tag{2.119}$$

Let us remind that the transformation properties of the $N = 4$ superspace coordinates and the basic Goldstone superfields under the transformations of the supergroup $D(2,1;\alpha)$ could be easily found, as we did in the previous sections. Here, we give the explicit expressions only for the variations of our superspace coordinates and superfields, with respect to two $SU(2)$ subgroup. They are generated by the left action of the group element

$$g_0 = e^{iaV + i\bar{a}\overline{V}}\,e^{ibT + i\bar{b}\overline{T}} \tag{2.120}$$

and read

$$\delta\theta_1 = \bar{b}\bar{\theta}^2 - \bar{a}\theta_2\,, \quad \delta\theta_2 = -\bar{b}\bar{\theta}^1 + a\theta_1\,,$$
$$\delta\Lambda = a + \bar{a}\Lambda^2\,, \quad \delta\bar{\Lambda} = \bar{a} + a\bar{\Lambda}^2\,, \quad \delta\phi = i\left(a\bar{\Lambda} - \bar{a}\Lambda\right)\,. \tag{2.121}$$

$N = 4$, $d = 1$ "Hypermultiplet"

The basic idea of our method is to impose the appropriate $D(2,1;\alpha)$ covariant constraints on the Cartan forms (2.114), (2.115), so as to end up with some minimal $N = 4, d = 1$ superfield set carrying an irreducible off-shell multiplet of $N = 4, d = 1$ supersymmetry. Due to the covariance of the constraints, the ultimate Goldstone superfields will support the corresponding nonlinear realization of the superconformal group $D(2,1;\alpha)$.

Let us elaborate on this in some detail. It was the desire to keep $N = 4, d = 1$ Poincaré supersymmetry unbroken that led us to associate the Grassmann coordinates $\theta_i, \bar{\theta}^i$ with the Poincaré supercharges in (2.113) and the fermionic Goldstone superfields $\psi_i, \bar{\psi}^i$ with the remaining four supercharges which generate conformal supersymmetry. The minimal number of physical fermions in an irreducible $N = 4, d = 1$ supermultiplet is four, and it nicely matches with the number of fermionic Goldstone superfields in (2.113), the first components of which can so be naturally identified with the fermionic fields of the ultimate Goldstone supermultiplet. On the other hand, we can vary the number of bosonic Goldstone superfields in (2.113): by putting some of them equal to zero we can enlarge the stability subgroup by the corresponding generators and so switch to another coset with a smaller set of parameters. Thus, for different choices of the stability subalgebra, the coset (2.113) will contain a different number of bosonic superfields, but always the same number of fermionic superfields $\psi_i, \bar{\psi}^i$. Yet, the corresponding sets of bosonic and fermionic Goldstone superfields contain too many field components, and it is natural to impose on them the appropriate covariant constraints, in order to reduce the number of components, as much as possible. For preserving off-shell $N = 4$ supersymmetry these constraints must be purely kinematical, i.e., they must not imply any dynamical restriction, such as equations of motion.

Some of the constraints just mentioned above should express the Goldstone fermionic superfields in terms of spinor derivatives of the bosonic ones. On the other hand, as soon as the first components of the fermionic superfields $\psi_i, \bar{\psi}^k$ are required to be the only physical fermions, we are led to impose much the stronger condition that *all* spinor derivatives of *all* bosonic superfields be properly expressed in terms of ψ_i, $\bar{\psi}^i$. Remarkably, the latter conditions will prove to be just the irreducibility constraints picking up irreducible $N = 4$ supermultiplets.

Here, we will consider in details only the most general case when the coset (2.113) contains all four bosonic superfields $u, \varphi, \bar{\varphi}, \phi$. Looking at the structure of the Cartan 1-forms (2.115), it is easy to find that the covariant constraints which express all spinor covariant derivatives of bosonic superfields in terms of the Goldstone fermions amount to setting equal to zero the spinor projections of these 1-forms (these conditions are a particular case of the inverse Higgs effect [26]). Thus, in the case at hand we impose the following constraints:

$$\omega_D = \omega_V \mid = \bar{\omega}_V \mid = \omega_{V_3} \mid = 0 , \tag{2.122}$$

where $\mid$ means restriction to spinor projections. These constraints are manifestly covariant under the whole supergroup $D(2, 1; \alpha)$. They allow one to express the Goldstone spinor superfields as the spinor derivatives of the residual bosonic Goldstone superfields $u, \Lambda, \bar{\Lambda}, \phi$ and imply some irreducibility constraints for the latter

$$\begin{aligned}
D^1\Lambda &= -2i\alpha\Lambda \left(\bar{\psi}^1 + \Lambda\bar{\psi}^2 \right) , \quad D^1\bar{\Lambda} = -2i\alpha \left(\bar{\psi}^2 - \bar{\Lambda}\bar{\psi}^1 \right) , \\
D^1\phi &= -2\alpha \left(\bar{\psi}^1 + \Lambda\bar{\psi}^2 \right) , \quad D^2\Lambda = 2i\alpha \left(\bar{\psi}^1 + \Lambda\bar{\psi}^2 \right) , \\
D^2\bar{\Lambda} &= -2i\alpha\bar{\Lambda} \left(\bar{\psi}^2 - \bar{\Lambda}\bar{\psi}^1 \right) , \quad D^2\phi = 2\alpha \left(\bar{\psi}^2 - \bar{\Lambda}\bar{\psi}^1 \right) , \\
D^1u &= 2i\bar{\psi}^1, \quad D^2u = 2i\bar{\psi}^2, \quad \dot{u} = 2z
\end{aligned} \tag{2.123}$$

(and c.c.). The irreducibility conditions, in this and other cases which we shall consider further, arise due to the property that the Goldstone fermionic superfields are simultaneously expressed by (2.123) in terms of spinor derivatives of different bosonic superfields. Then, eliminating these spinor superfields, we end up with the relations between the spinor derivatives of bosonic Goldstone superfields. To make the use of these constraints the most feasible, it is advantageous to pass to the new variables

$$q^1 = \frac{e^{\frac{1}{2}(\alpha u - i\phi)}}{\sqrt{1 + \Lambda\bar{\Lambda}}}\Lambda, \quad q^2 = -\frac{e^{\frac{1}{2}(\alpha u - i\phi)}}{\sqrt{1 + \Lambda\bar{\Lambda}}}, \quad \bar{q}_1 = \frac{e^{\frac{1}{2}(\alpha u + i\phi)}}{\sqrt{1 + \Lambda\bar{\Lambda}}}\bar{\Lambda}, \quad \bar{q}_2 = -\frac{e^{\frac{1}{2}(\alpha u + i\phi)}}{\sqrt{1 + \Lambda\bar{\Lambda}}}.$$

(2.124)

In terms of these variables the irreducibility constraints acquire the manifestly $SU(2)$ covariant form

$$D^{(i}q^{j)} = 0, \qquad \overline{D}^{(i}q^{j)} = 0. \tag{2.125}$$

This is just the $N = 4, d = 1$ hypermultiplet we considered in the previous section.

The rest of the supermultiplets from Table 2.1 may be obtained similarly. For example, the tensor $N = 4, d = 1$ supermultiplet corresponds to the coset with the V_3 generator in the stability subgroup and so on. The detailed discussion of all cases may be found in [4].

2.4 $N = 8$ Supersymmetry

Most of the models explored in the previous sections possess $N \leq 4, d = 1$ supersymmetries. Much less is known about higher-N systems. Some of them were addressed many years ago in the seminal paper [15], within an on-shell Hamiltonian approach. Some others (with $N = 8$) received attention later [30, 31].

As we stressed many times, the natural formalism for dealing with supersymmetric models is the off-shell superfield approach. Thus, for the construction of new SQM models with extended $d = 1$ supersymmetry, one needs, first of all, the complete list of the corresponding off-shell $d = 1$ supermultiplets and the superfields which encompass these multiplets. One of the peculiarities of $d = 1$ supersymmetry is that some of its off-shell multiplets cannot be obtained via a direct dimensional reduction from the multiplets of higher-d supersymmetries with the same number of spinorial charges. Another peculiarity is that some on-shell multiplets of the latter have *off-shell* $d = 1$ counterparts.

In the previous section we considered nonlinear realizations of the finite-dimensional $N = 4$ superconformal group in $d = 1$. We showed that the irreducible superfields representing one or another off-shell $N = 4, d = 1$ supermultiplet come out as Goldstone superfields parameterizing one or another coset manifold of the superconformal group. The superfield irreducibility constraints naturally emerge as a part of manifestly covariant inverse Higgs [26] conditions on the relevant Cartan superforms.

This method is advantageous in that it automatically specifies the superconformal properties of the involved supermultiplets, which are of importance. The application of the nonlinear realization approach to the case of $N = 8, d = 1$ supersymmetry was initiated in [6]. There, nonlinear realizations of the $N = 8, d = 1$ superconformal group $OSp(4^*|4)$ in its two different cosets were considered, and it was shown that two interesting $N = 8, d = 1$

multiplets, with off-shell field contents (**3, 8, 5**) and (**5, 8, 3**), naturally come out as the corresponding Goldstone multiplets. These supermultiplets admit a few inequivalent splittings into pairs of irreducible off-shell $N = 4, d = 1$ multiplets, such that different $N = 4$ superconformal subgroups of $OSp(4^\star|4)$, viz., $SU(1,1|2)$ and $OSp(4^\star|2)$, are manifest for different splittings. Respectively, the off-shell component action of the given $N = 8$ multiplet in general admits several different representations in terms of $N = 4, d = 1$ superfields.

Now we are going to present a superfield description of all other linear off-shell $N = 8, d = 1$ supermultiplets with eight fermions, in both $N = 8$ and $N = 4$ superspaces.

When deriving an exhaustive list of off-shell $N = 8$ supermultiplets and the relevant constrained $N = 8, d = 1$ superfields, we could proceed in the same way as in the case of $N = 4$ supermultiplets, i.e., by considering nonlinear realizations of all known $N = 8$ superconformal groups in their various cosets. However, this task is more complicated, as compared to the $N = 4$ case, in view of the existence of many inequivalent $N = 8$ superconformal groups ($OSp(4^\star|4)$, $OSp(8|2)$, $F(4)$, and $SU(1,1|4)$, see e.g. [2]), with numerous coset manifolds.

To avoid these complications, we take advantage of two fortunate circumstances. Firstly, as we already know, the field contents of *linear* off-shell multiplets of $N = 8, d = 1$ supersymmetry with eight physical fermions range from (**8, 8, 0**) to (**0, 8, 8**), with the intermediate multiplets corresponding to all possible splittings of eight bosonic fields into physical and auxiliary ones. Thus, we are aware of the full list of such multiplets, independently of the issue of their interpretation as the Goldstone ones parameterizing the proper superconformal cosets.

The second circumstance allowing us to advance without resorting to the nonlinear realizations techniques is the aforesaid existence of various splittings of $N = 8$ multiplets into pairs of irreducible $N = 4$ supermultiplets. We know how to represent the latter in terms of constrained $N = 4$ superfields, so it proves to be a matter of simple algebra to guess the form of the four extra supersymmetries mixing the $N = 4$ superfields inside each pair and extending the manifest $N = 4$ supersymmetry to $N = 8$. After fixing such pairs, it is again rather easy to embed them into appropriately constrained $N = 8, d = 1$ superfields.

2.4.1 $N = 8$, $d = 1$ Superspace

The maximal automorphism group of $N = 8, d = 1$ super-Poincaré algebra (without central charges) is $SO(8)$ and so eight real Grassmann coordinates of $N = 8, d = 1$ superspace $\mathbb{R}^{(1|8)}$ can be arranged into one of three eight-dimensional real irreps of $SO(8)$. The constraints defining the irreducible $N = 8$ supermultiplets in general break this $SO(8)$ symmetry. So, it is preferable to split the eight coordinates into two real quartets

$$\mathbb{R}^{(1|8)} = (t, \theta_{ia}, \vartheta_{\alpha A}), \quad \overline{(\theta_{ia})} = \theta^{ia}, \quad \overline{(\vartheta_{\alpha A})} = \vartheta^{\alpha A}, \quad i, a, \alpha, A = 1, 2,$$
(2.126)

in terms of which only four commuting automorphism $SU(2)$ groups will be explicit. The further symmetry breaking can be understood as the identification of some of these $SU(2)$, while extra symmetries, if they exist, mix different $SU(2)$ indices. The corresponding covariant derivatives are defined by

$$D^{ia} = \frac{\partial}{\partial \theta_{ia}} + i\theta^{ia}\partial_t, \quad \nabla^{\alpha A} = \frac{\partial}{\partial \vartheta_{\alpha A}} + i\vartheta^{\alpha A}\partial_t.$$
(2.127)

By construction, they obey the algebra

$$\{D^{ia}, D^{jb}\} = 2i\epsilon^{ij}\epsilon^{ab}\partial_t, \quad \{\nabla^{\alpha A}, \nabla^{\beta B}\} = 2i\epsilon^{\alpha\beta}\epsilon^{AB}\partial_t.$$
(2.128)

2.4.2 $N = 8$, $d = 1$ Supermultiplets

As we already mentioned, our real strategy of deducing a superfield description of the $N = 8, d = 1$ supermultiplets consisted in selecting an appropriate pair of constrained $N = 4, d = 1$ superfields and then guessing the constrained $N = 8$ superfield. Now, just to make the presentation more coherent, we turn the argument around and start with postulating the $N = 8, d = 1$ constraints. The $N = 4$ superfield formulations will be deduced from the $N = 8$ ones.

Supermultiplet (0, 8, 8)

The off-shell $N = 8, d = 1$ supermultiplet (**0, 8, 8**) is carried out by two real fermionic $N = 8$ superfields Ψ^{aA}, $\Xi^{i\alpha}$ subjected to the following constraints:

$$D^{(ia}\Xi^{j)}_\alpha = 0, \quad D^{i(a}\Psi^{b)}_A = 0, \quad \nabla^{(\alpha A}\Xi^{\beta)}_i = 0, \quad \nabla^{\alpha(A}\Psi^{B)}_a = 0, \quad (2.129)$$

$$\nabla^{\alpha A}\Psi^a_A = D^{ia}\Xi^\alpha_i, \quad \nabla^{\alpha A}\Xi^i_\alpha = -D^{ia}\Psi^A_a.$$
(2.130)

To understand the structure of this supermultiplet in terms of $N = 4$ superfields we proceed as follows. As a first step, let us single out the $N = 4$ subspace in the $N = 8$ superspace $\mathbb{R}^{(1|8)}$ as the set of coordinates

$$\mathbb{R}^{(1|4)} = (t, \theta_{ia}) \subset \mathbb{R}^{(1|8)},$$
(2.131)

and expand the $N = 8$ superfields over the extra Grassmann coordinate $\vartheta_{\alpha A}$. Then we observe that the constraints (2.130) imply that the spinor derivatives of all involved superfields with respect to $\vartheta_{\alpha A}$ can be expressed in terms of spinor derivatives with respect to θ_{ia}. This means that the only essential $N = 4$ superfield components of Ψ^{aA} and $\Xi^{i\alpha}$ in their ϑ-expansion are the first ones

$$\psi^{aA} \equiv \Psi^{aA}|_{\vartheta=0}, \quad \xi^{i\alpha} \equiv \Xi^{i\alpha}|_{\vartheta=0}.$$
(2.132)

These fermionic $N = 4$ superfields are subjected, in virtue of (2.129) and (2.130), to the irreducibility constraints in $N = 4$ superspace

$$D^{a(i}\xi^{j)\alpha} = 0, \qquad D^{i(a}\psi^{b)A} = 0 . \qquad (2.133)$$

As it follows from Sect. 2.2, these superfields are just two fermionic $N = 4$ hypermultiplets, each carrying $(\mathbf{0}, \mathbf{4}, \mathbf{4})$ independent component fields. So, being combined together, they accommodate the whole off-shell component content of the $N = 8$ multiplet $(\mathbf{0}, \mathbf{8}, \mathbf{8})$, which proves that the $N = 8$ constraints (2.129), (2.130) are the true choice.

Thus, from the $N = 4$ superspace perspective, the $N = 8$ supermultiplet $(\mathbf{0}, \mathbf{8}, \mathbf{8})$ amounts to the sum of two $N = 4, d = 1$ fermionic hypermultiplets with the off-shell component content $(\mathbf{0}, \mathbf{4}, \mathbf{4}) \oplus (\mathbf{0}, \mathbf{4}, \mathbf{4})$.

The transformations of the implicit $N = 4$ Poincaré supersymmetry, completing the manifest one to the full $N = 8$ supersymmetry, have the following form in terms of the $N = 4$ superfields defined above:

$$\delta\psi^{aA} = \frac{1}{2}\eta^{A\alpha}D^{ia}\xi_{i\alpha}, \qquad \delta\xi^{i\alpha} = -\frac{1}{2}\eta^{\alpha}_A D^i_a \psi^{aA} . \qquad (2.134)$$

The invariant free action can be written as

$$S = \int dt\, d^4\theta \left[\theta^{ia}\theta^b_i \psi^A_a \psi_{bA} + \theta^{ia}\theta^j_a \xi^{\alpha}_i \xi_{j\alpha} \right] . \qquad (2.135)$$

Because of the presence of explicit theta's in the action (2.135), the latter is not manifestly invariant even with respect to the manifest $N = 4$ supersymmetry. Nevertheless, one can check that (2.135) is invariant under this supersymmetry, which is realized on the superfields as

$$\delta^*\psi^{aA} = -\varepsilon_{jb}Q^{jb}\psi^{aA}, \qquad \delta^*\xi^{i\alpha} = -\varepsilon_{jb}Q^{jb}\xi^{i\alpha} , \qquad (2.136)$$

where

$$Q^{ia} = \frac{\partial}{\partial\theta_{ia}} - i\theta^{ia}\partial_t , \qquad (2.137)$$

ε_{ia} is the supertranslation parameter and $*$ denotes the "active" variation (taken at a fixed point of the $N = 4$ superspace).

Supermultiplet $(\mathbf{1}, \mathbf{8}, \mathbf{7})$

This supermultiplet can be described by a single scalar $N = 8$ superfield $\mathcal{U}$ which obeys the following irreducibility conditions:

$$D^{ia}D^j_a\mathcal{U} = -\nabla^{\alpha j}\nabla^i_\alpha\mathcal{U} , \qquad (2.138)$$

$$\nabla^{(\alpha i}\nabla^{\beta)j}\mathcal{U} = 0 , \qquad D^{i(a}D^{jb)}\mathcal{U} = 0 . \qquad (2.139)$$

Let us note that the constraints (2.138) reduce the manifest R-symmetry to $[SU(2)]^3$, due to the identification of the indices i and A of the covariant derivatives D^{ia} and $\nabla^{\alpha A}$.

This supermultiplet possesses a unique decomposition into the pair of $N = 4$ supermultiplets as $(\mathbf{1}, \mathbf{8}, \mathbf{7}) = (\mathbf{1}, \mathbf{4}, \mathbf{3}) \oplus (\mathbf{0}, \mathbf{4}, \mathbf{4})$. The corresponding $N = 4$ superfield projections can be defined as

$$u = \mathcal{U}|_{\vartheta=0} , \qquad \psi^{i\alpha} = \nabla^{\alpha i}\mathcal{U}|_{\vartheta=0} , \tag{2.140}$$

and they obey the standard constraints

$$D^{(ia}\psi^{j)\alpha} = 0 , \qquad D^{i(a}D^{jb)}u = 0 . \tag{2.141}$$

The second constraint directly follows from (2.139), while the first one is implied by the relation

$$\frac{\partial}{\partial t}D^{(i}_a\nabla^{j)}_\alpha\mathcal{U} = 0 , \tag{2.142}$$

which can be proven by applying the differential operator $D^{kb}\nabla^{\beta l}$ to the $N = 8$ superfield constraint (2.138) and making use of the algebra of covariant derivatives.

The additional implicit $N = 4$ supersymmetry is realized on these $N = 4$ superfields as follows:

$$\delta u = -\eta_{i\alpha}\psi^{i\alpha}, \qquad \delta\psi^{i\alpha} = -\frac{1}{2}\eta^\alpha_j D^{ia}D^j_a u . \tag{2.143}$$

The simplest way to deal with the action for this supermultiplet is to use harmonic superspace [32,33,13], but this approach isout of the scope of the present lectures.

Supermultiplet (2, 8, 6)

The $N = 8$ superfield formulation of this supermultiplet involves two scalar bosonic superfields $\mathcal{U}, \Phi$ obeying the constraints

$$\nabla^{(ai}\nabla^{b)j}\mathcal{U} = 0 , \qquad \nabla^{a(i}\nabla^{bj)}\Phi = 0 , \tag{2.144}$$
$$\nabla^{ai}\mathcal{U} = D^{ia}\Phi , \qquad \nabla^{ai}\Phi = -D^{ia}\mathcal{U} , \tag{2.145}$$

where we have identified the indices i and A, a and α of the covariant derivatives, thus retaining only two manifest $SU(2)$ automorphism groups. From (2.144), (2.145) some useful corollaries follow:

$$D^{ia}D^j_a\mathcal{U} + \nabla^{aj}\nabla^i_a\mathcal{U} = 0 , \qquad D^{i(a}D^{jb)}\mathcal{U} = 0 , \tag{2.146}$$
$$D^{ia}D^b_i\Phi + \nabla^{bi}\nabla^a_i\Phi = 0 , \qquad D^{(ia}D^{j)b}\Phi = 0 . \tag{2.147}$$

Comparing (2.35), (2.38) and (2.144) with (2.138), (2.139), we observe that the $N = 8$ supermultiplet with the field content $(\mathbf{2}, \mathbf{8}, \mathbf{6})$ can be obtained by combining two $(\mathbf{1}, \mathbf{8}, \mathbf{7})$ supermultiplets and imposing the additional relations (2.145) on the corresponding $N = 8$ superfields.

To construct the invariant actions and prove that the above $N = 8$ constraints indeed yield the multiplet $(\mathbf{2}, \mathbf{8}, \mathbf{6})$, we should reveal the structure of this supermultiplet in terms of $N = 4$ superfields, as we did in the previous cases. However, in the case at hand, we have two different choices for splitting the $(\mathbf{2}, \mathbf{8}, \mathbf{6})$ supermultiplet.

1. $(\mathbf{2}, \mathbf{8}, \mathbf{6}) = (\mathbf{1}, \mathbf{4}, \mathbf{3}) \oplus (\mathbf{1}, \mathbf{4}, \mathbf{3})$
2. $(\mathbf{2}, \mathbf{8}, \mathbf{6}) = (\mathbf{2}, \mathbf{4}, \mathbf{2}) \oplus (\mathbf{0}, \mathbf{4}, \mathbf{4})$

As it was already mentioned, the possibility to have a few different off-shell $N = 4$ decompositions of the same $N = 8$ multiplet is related to different choices of the manifest $N = 4$ supersymmetries, as subgroups of the $N = 8$ super-Poincaré group. We shall treat both options.

1. $(\mathbf{2}, \mathbf{8}, \mathbf{6}) = (\mathbf{1}, \mathbf{4}, \mathbf{3}) \oplus (\mathbf{1}, \mathbf{4}, \mathbf{3})$

To describe the $N = 8$ $(\mathbf{2}, \mathbf{8}, \mathbf{6})$ multiplet in terms of $N = 4$ superfields, we should choose the appropriate $N = 4$ superspace. The first (evident) possibility is to choose the $N = 4$ superspace with coordinates (t, θ_{ia}). In this superspace one $N = 4$ Poincaré supergroup is naturally realized, while the second one mixes two irreducible $N = 4$ superfields which comprise the $N = 8$ $(\mathbf{2}, \mathbf{8}, \mathbf{6})$ supermultiplet in question. Expanding the $N = 8$ superfields $\mathcal{U}, \Phi$ in ϑ^{ia}, one finds that the constraints (2.144), (2.145) leave in $\mathcal{U}$ and Φ as independent $N = 4$ projections only those of zeroth order in ϑ^{ia}

$$u = \mathcal{U}|_{\vartheta_{i\alpha}=0} \,, \qquad \phi = \Phi|_{\vartheta_{i\alpha}=0} \,. \tag{2.148}$$

Each $N = 4$ superfield proves to be subjected, in virtue of (2.144), (2.145), to the additional constraint

$$D^{i(a}D^{jb)}u = 0 \,, \qquad D^{(ia}D^{j)b}\phi = 0 \,. \tag{2.149}$$

Thus, we conclude that our $N = 8$ multiplet $\mathcal{U}, \Phi$, when rewritten in terms of $N = 4$ superfields, amounts to a direct sum of two $N = 4$ multiplets u and ϕ, both having the same off-shell field contents $(\mathbf{1}, \mathbf{4}, \mathbf{3})$.

The transformations of the implicit $N = 4$ Poincaré supersymmetry, completing the manifest one to the full $N = 8$ Poincaré supersymmetry, have the following form in terms of these $N = 4$ superfields:

$$\delta^* u = -\eta_{ia}D^{ia}\phi \,, \qquad \delta^* \phi = \eta_{ia}D^{ia}u \,. \tag{2.150}$$

It is rather easy to construct the action in terms of $N = 4$ superfields u and ϕ, such that it is invariant with respect to the implicit $N = 4$ supersymmetry (2.150). The generic action has the form

$$S = \int \mathrm{d}t\, \mathrm{d}^4\theta \mathcal{F}(u, \phi) \,, \tag{2.151}$$

where the function $\mathcal{F}$ obeys the Laplace equation

$$\mathcal{F}_{uu} + \mathcal{F}_{\phi\phi} = 0 . \tag{2.152}$$

2. $(2, 8, 6) = (2, 4, 2) \oplus (0, 4, 4)$

There is a more sophisticated choice of a $N = 4$ subspace in the $N = 8, d = 1$ superspace, which gives rise to the second possible $N = 4$ superfield splitting of the considered $N = 8$ supermultiplet, that is into the multiplets $(2, 4, 2)$ and $(0, 4, 4)$.

First of all, let us define a new set of covariant derivatives

$$\mathcal{D}^{ia} = \frac{1}{\sqrt{2}} \left(D^{ia} - i\nabla^{ai} \right) , \quad \overline{\mathcal{D}}^{ia} = \frac{1}{\sqrt{2}} \left(D^{ia} + i\nabla^{ai} \right) ,$$

$$\left\{ \mathcal{D}^{ia}, \overline{\mathcal{D}}^{jb} \right\} = 2i\epsilon^{ij}\epsilon^{ab}\partial_t , \tag{2.153}$$

and new $N = 8$ superfields $\mathcal{V}, \overline{\mathcal{V}}$ related to the original ones as

$$\mathcal{V} = \mathcal{U} + i\Phi, \quad \overline{\mathcal{V}} = \mathcal{U} - i\Phi . \tag{2.154}$$

In this basis the constraints (2.144), (2.145) read

$$\mathcal{D}^{ia}\mathcal{V} = 0, \quad \overline{\mathcal{D}}^{ia}\overline{\mathcal{V}} = 0 ,$$

$$\mathcal{D}^{i(a}\mathcal{D}^{jb)}\overline{\mathcal{V}} + \overline{\mathcal{D}}^{i(a}\overline{\mathcal{D}}^{jb)}\mathcal{V} = 0 , \quad \mathcal{D}^{(ia}\mathcal{D}^{j)b}\overline{\mathcal{V}} - \overline{\mathcal{D}}^{(ia}\overline{\mathcal{D}}^{j)b}\mathcal{V} = 0 . \tag{2.155}$$

Now, we split the complex quartet covariant derivatives (2.153) into two sets of the doublet $N = 4$ ones as

$$D^i = \mathcal{D}^{i1}, \quad \overline{D}^i = \overline{\mathcal{D}}^{i2}, \quad \nabla^i = \mathcal{D}^{i2}, \quad \overline{\nabla}^i = -\overline{\mathcal{D}}^{i1} \tag{2.156}$$

and cast the constraints (2.155) in the form

$$D^i\mathcal{V} = 0, \quad \nabla^i\mathcal{V} = 0, \quad \overline{D}_i\overline{\mathcal{V}} = 0, \quad \overline{\nabla}_i\overline{\mathcal{V}} = 0 ,$$

$$D^iD_i\overline{\mathcal{V}} - \nabla_i\overline{\nabla}^i\mathcal{V} = 0 , \quad D^i\nabla^j\overline{\mathcal{V}} - \overline{D}^i\overline{\nabla}^j\mathcal{V} = 0 . \tag{2.157}$$

Next, as an alternative $N = 4$ superspace, we choose the set of coordinates closed under the action of $D^i, \overline{D}^i$, i.e.,

$$(t , \theta_{i1} + i\vartheta_{i1} , \theta_{i2} - i\vartheta_{i2}) , \tag{2.158}$$

while the $N = 8$ superfields are expanded with respect to the orthogonal combinations $\theta_{i1} - i\vartheta_{i1}$, $\theta_{i2} + i\vartheta_{i2}$ which are annihilated by $D^i, \overline{D}^i$.

As a consequence of the constraints (2.157), the quadratic action of the derivatives ∇^i and $\overline{\nabla}^i$ on every $N = 8$ superfield $\mathcal{V}$, $\overline{\mathcal{V}}$ can be expressed as D^i, $\overline{D}^i$ of some other superfield. Therefore, only the zeroth and first-order components of each $N = 8$ superfield are independent $N = 4$ superfield projections. Thus, we are left with the following set of $N = 4$ superfields:

$$v = \mathcal{V}| , \quad \bar{v} = \overline{\mathcal{V}}| , \quad \psi^i = \overline{\nabla}^i\mathcal{V}| , \quad \bar{\psi}^i = -\nabla^i\overline{\mathcal{V}}| . \tag{2.159}$$

These $N = 4$ superfields prove to be subjected to the additional constraints which also follow from (2.157)

$$D^i v = 0 , \quad \overline{D}^i \bar{v} = 0 , \quad D^i \psi^j = 0 , \quad \overline{D}^i \bar{\psi}^j = 0 , \quad D^i \bar{\psi}^j = -\overline{D}^i \psi^j . \quad (2.160)$$

The $N = 4$ superfields $v, \bar{v}$ comprise the standard $N = 4, d = 1$ chiral multiplet **(2, 4, 2)**, while the $N = 4$ superfields $\psi^i, \bar{\psi}^j$ subjected to (2.160) and both having the off-shell contents **(0, 4, 4)** are recognized as the fermionic version of the $N = 4, d = 1$ hypermultiplet.

The implicit $N = 4$ supersymmetry is realized by the transformations

$$\delta v = -\bar{\eta}^i \psi_i , \qquad \delta \psi^i = -\frac{1}{2}\bar{\eta}^i D^2 \bar{v} - 2i\eta^i \dot{v} ,$$

$$\delta \bar{v} = \eta_i \bar{\psi}^i , \qquad \delta \bar{\psi}_i = \frac{1}{2}\eta_i \overline{D}^2 v + 2i\bar{\eta}_i \dot{\bar{v}} . \qquad (2.161)$$

The invariant free action has the following form:

$$S_f = \int \mathrm{dt}\, \mathrm{d}^4\theta v\bar{v} - \frac{1}{2}\int \mathrm{dt}\, \mathrm{d}^2\bar{\theta}\psi^i \psi_i - \frac{1}{2}\int \mathrm{dt}\, \mathrm{d}^2\theta \bar{\psi}_i \bar{\psi}^i . \qquad (2.162)$$

Let us note that this very simple form of the action for the $N = 4$ **(0, 4, 4)** supermultiplet $\psi_i, \bar{\psi}^j$ is related to our choice of the $N = 4$ superspace. It is worthwhile to emphasize that all differently looking superspace off-shell actions of the multiplet **(0, 4, 4)** yield the same component action for this multiplet.

Supermultiplet (3, 8, 5)

In the $N = 8$ superspace this supermultiplet is described by the triplet of bosonic superfields $\mathcal{V}^{ij}$ obeying the irreducibility constraints

$$D_a^{(i}\mathcal{V}^{jk)} = 0 , \qquad \nabla_\alpha^{(i}\mathcal{V}^{jk)} = 0 . \qquad (2.163)$$

So, three out of four original automorphism $SU(2)$ symmetries remain manifest in this description.

The $N = 8$ supermultiplet **(3, 8, 5)** can be decomposed into $N = 4$ supermultiplets in two ways.

1. **(3, 8, 5)** = **(3, 4, 1)** ⊕ **(0, 4, 4)**
2. **(3, 8, 5)** = **(1, 4, 3)** ⊕ **(2, 4, 2)**

As in the previous case, we discuss both options.

1. **(3, 8, 5)** = **(3, 4, 1)** ⊕ **(0, 4, 4)**

This splitting requires choosing the coordinate set (2.131) as the relevant $N = 4$ superspace. Expanding the $N = 8$ superfields $\mathcal{V}^{ij}$ in $\vartheta_{i\alpha}$, one finds that the constraints (2.163) leave in $\mathcal{V}^{ij}$ the following four bosonic and four fermionic $N = 4$ projections:

$$v^{ij} = \mathcal{V}^{ij}\big| , \quad \xi_\alpha^i \equiv \nabla_{j\alpha} \mathcal{V}^{ij}\big| , \quad A \equiv \nabla_i^\alpha \nabla_{j\alpha} \mathcal{V}^{ij}\big| , \tag{2.164}$$

where $|$ means the restriction to $\vartheta_{i\alpha} = 0$. As a consequence of (2.163), these $N = 4$ superfields obey the constraints

$$D_a^{(i} v^{jk)} = 0 , \quad D_a^{(i} \xi_\alpha^{j)} = 0 ,$$
$$A = 6m - D_i^a D_{aj} v^{ij} , \quad m = \text{const} . \tag{2.165}$$

Thus, for the considered splitting, the $N = 8$ tensor multiplet superfield $\mathcal{V}^{ij}$ amounts to a direct sum of the $N = 4$ "tensor" multiplet superfield v^{ij} with the off-shell content $(\mathbf{3, 4, 1})$ and a fermionic $N = 4$ hypermultiplet ξ_α^i with the off-shell content $(\mathbf{0, 4, 4})$, plus a constant m of the mass dimension.

2. $(\mathbf{3, 8, 5}) = (\mathbf{1, 4, 3}) \oplus (\mathbf{2, 4, 2})$

This option corresponds to another choice of $N = 4$ superspace, which amounts to dividing the $N = 8, d = 1$ Grassmann coordinates into doublets, with respect to some other $SU(2)$ indices. The relevant splitting of $N = 8$ superspace into the $N = 4$ subspace and the complement of the latter can be performed as follows. Firstly, we define the new covariant derivatives as

$$D^a \equiv \frac{1}{\sqrt{2}} \left(D^{1a} + i\nabla^{a1} \right) , \quad \bar{D}_a \equiv \frac{1}{\sqrt{2}} \left(D_a^2 - i\nabla_a^2 \right) ,$$
$$\nabla^a \equiv \frac{i}{\sqrt{2}} \left(D^{2a} + i\nabla^{a2} \right) , \quad \bar{\nabla}_a \equiv \frac{i}{\sqrt{2}} \left(D_a^1 - i\nabla_a^1 \right) . \tag{2.166}$$

Then we choose the set of coordinates closed under the action of $D^a, \bar{D}_a$, i.e.,

$$\left(t , \; \theta_{1a} - i\vartheta_{a1} , \; \theta^{1a} + i\vartheta^{a1} \right) , \tag{2.167}$$

while the $N = 8$ superfields are expanded with respect to the orthogonal combinations $\theta_2^a - i\vartheta_2^a$, $\theta_1^a + i\vartheta_1^a$ annihilated by $D^a, \bar{D}_a$.

The basic constraints (2.163), being rewritten in the basis (2.166), take the form

$$D^a \varphi = 0 , \quad D^a v - \nabla^a \varphi = 0 , \quad \nabla^a v + D^a \bar{\varphi} = 0 , \quad \nabla^a \bar{\varphi} = 0 ,$$
$$\bar{\nabla}_a \varphi = 0 , \quad \bar{\nabla}_a v + \bar{D}_a \varphi = 0 , \quad \bar{D}_a v - \bar{\nabla}_a \bar{\varphi} = 0 , \quad \bar{D}_a \bar{\varphi} = 0 , \tag{2.168}$$

where

$$v \equiv -2i\mathcal{V}^{12} , \quad \varphi \equiv \mathcal{V}^{11} , \quad \bar{\varphi} \equiv \mathcal{V}^{22} . \tag{2.169}$$

Due to the constraints (2.168), the derivatives ∇^a and $\bar{\nabla}_a$ of every $N = 8$ superfield in the triplet $\left(\mathcal{V}^{12}, \mathcal{V}^{11}, \mathcal{V}^{22} \right)$ can be expressed as $D^a, \bar{D}_a$ of some other superfield. Therefore, only the zeroth order (i.e., taken at $\theta_2^a - i\vartheta_2^a = \theta_1^a + i\vartheta_1^a = 0$) components of each $N = 8$ superfield are independent $N = 4$ superfield projections. These $N = 4$ superfields are subjected to the additional constraints which also follow from (2.168)

$$D^a D_a v = \overline{D}_a \overline{D}^a v = 0 \, , \qquad D^a \varphi = 0 \, , \quad \overline{D}_a \bar\varphi = 0 \, . \tag{2.170}$$

The $N = 4$ superfields $\varphi, \bar\varphi$ comprise the standard $N = 4, d = 1$ chiral multiplet $(\mathbf{2, 4, 2})$, while the $N = 4$ superfield v subjected to (2.170) has the needed off-shell content $(\mathbf{1, 4, 3})$.

The implicit $N = 4$ supersymmetry acts on the $N = 4$ superfields $v, \varphi, \bar\varphi$ as follows:

$$\delta^* v = \eta_a D^a \bar\varphi + \bar\eta^a \overline{D}_a \varphi \, , \quad \delta^* \varphi = -\eta_a D^a v \, , \quad \delta^* \bar\varphi = -\bar\eta^a \overline{D}_a v \, . \tag{2.171}$$

Invariant $N = 4$ superfield actions for both decompositions of the $N = 8$ multiplet $(\mathbf{3, 8, 5})$ were presented in [6].

2.4.3 Supermultiplet (4, 8, 4)

This supermultiplet can be described by a quartet of $N = 8$ superfields $\mathcal{Q}^{a\alpha}$ obeying the constraints

$$D_i^{(a} \mathcal{Q}^{b)\alpha} = 0, \qquad \nabla_i^{(\alpha} \mathcal{Q}_a^{\beta)} = 0 \, . \tag{2.172}$$

Let us note that the constraints (2.172) are manifestly covariant with respect to three $SU(2)$ subgroups realized on the indices i, a, and α.

From (2.172) some important relations follow:

$$D^{ia} D^{jb} \mathcal{Q}^{c\alpha} = 2i\epsilon^{ij} \epsilon^{cb} \dot{\mathcal{Q}}^{a\alpha}, \qquad \nabla^{i\alpha} \nabla^{j\beta} \mathcal{Q}^{a\gamma} = 2i\epsilon^{ij} \epsilon^{\gamma\beta} \dot{\mathcal{Q}}^{a\alpha} \, . \tag{2.173}$$

Using them, it is possible to show that the superfields $\mathcal{Q}^{a\alpha}$ contain the following independent components:

$$\mathcal{Q}^{a\alpha}|, \quad D_a^i \mathcal{Q}^{a\alpha}|, \quad \nabla_\alpha^i \mathcal{Q}^{a\alpha}|, \quad D_a^i \nabla_\alpha^j \mathcal{Q}^{a\alpha}| \, , \tag{2.174}$$

where $|$ means now restriction to $\theta_{ia} = \vartheta_{i\alpha} = 0$. This directly proves that we deal with the irreducible $(\mathbf{4, 8, 4})$ supermultiplet.

There are three different possibilities to split this $N = 8$ multiplet into the $N = 4$ ones.

1. $(\mathbf{4, 8, 4}) = (\mathbf{4, 4, 0}) \oplus (\mathbf{0, 4, 4})$
2. $(\mathbf{4, 8, 4}) = (\mathbf{3, 4, 1}) \oplus (\mathbf{1, 4, 3})$
3. $(\mathbf{4, 8, 4}) = (\mathbf{2, 4, 2}) \oplus (\mathbf{2, 4, 2})$

Once again, we shall consider all three cases separately.

1. $(\mathbf{4, 8, 4}) = (\mathbf{4, 4, 0}) \oplus (\mathbf{0, 4, 4})$

This case implies the choice of the $N = 4$ superspace (2.131). Expanding the $N = 8$ superfields $\mathcal{Q}^{a\alpha}$ in $\vartheta_{i\alpha}$, one may easily see that the constraints (2.172) leave in $\mathcal{Q}^{a\alpha}$ the following four bosonic and four fermionic $N = 4$ superfield projections:

$$q^{a\alpha} = \mathcal{Q}^{a\alpha}\big|\,, \qquad \psi^{ia} \equiv \nabla^i_\alpha \mathcal{Q}^{a\alpha}\big|\,. \tag{2.175}$$

Each $N = 4$ superfield is subjected, in virtue of (2.172), to an additional constraint

$$D^{i(a}q^{b)\alpha} = 0\,, \qquad D^{i(a}\psi^{b)i} = 0\,. \tag{2.176}$$

Consulting Sect. 2.2, we come to the conclusion that these are just the hypermultiplet $q^{i\alpha}$ with the off-shell field content $(\mathbf{4},\,\mathbf{4},\,\mathbf{0})$ and a fermionic analog of the $N = 4$ hypermultiplet ψ^{ia} with the field content $(\mathbf{0},\,\mathbf{4},\,\mathbf{4})$.

The transformations of the implicit $N = 4$ Poincaré supersymmetry have the following form in terms of these $N = 4$ superfields:

$$\delta^* q^{a\alpha} = \frac{1}{2}\eta^{i\alpha}\psi^a_i\,, \qquad \delta^* \psi^{ia} = -2i\eta^{i\alpha}\dot{q}^a_\alpha\,. \tag{2.177}$$

2. $(\mathbf{4},\,\mathbf{8},\,\mathbf{4}) = (\mathbf{3},\,\mathbf{4},\,\mathbf{1}) \oplus (\mathbf{1},\,\mathbf{4},\,\mathbf{3})$

To describe this $N = 4$ superfield realization of the $N = 8$ supermultiplet $(\mathbf{4},\,\mathbf{8},\,\mathbf{4})$, we introduce the $N = 8$ superfields $\mathcal{V}^{ab}, \mathcal{V}$ as

$$\mathcal{Q}^{a\alpha} \equiv \delta^\alpha_b \mathcal{V}^{ab} - \epsilon^{a\alpha}\mathcal{V}\,, \qquad \mathcal{V}^{ab} = \mathcal{V}^{ba}\,, \tag{2.178}$$

and use the covariant derivatives (2.153) to rewrite the basic constraints (2.172) as

$$\mathcal{D}^{i(a}\mathcal{V}^{bc)} = 0\,, \qquad \overline{\mathcal{D}}^{i(a}\mathcal{V}^{bc)} = 0\,, \tag{2.179}$$

$$\mathcal{D}^{ia}\mathcal{V} = \frac{1}{2}\overline{\mathcal{D}}^i_b\mathcal{V}^{ab}\,, \qquad \overline{\mathcal{D}}^{ia}\mathcal{V} = \frac{1}{2}\mathcal{D}^i_b\mathcal{V}^{ab}\,. \tag{2.180}$$

The constraints (2.179) define $\mathcal{V}^{ab}$ as the $N = 8$ superfield encompassing the off-shell multiplet $(\mathbf{3},\,\mathbf{8},\,\mathbf{5})$, while, as one can deduce from (2.179), (2.180), the $N = 8$ superfield $\mathcal{V}$ has the content $(\mathbf{1},\,\mathbf{8},\,\mathbf{7})$. Then the constraints (2.180) establish relations between the fermions in these two superfields and reduce the number of independent auxiliary fields to four, so that we end up, once again, with the irreducible $N = 8$ multiplet $(\mathbf{4},\,\mathbf{8},\,\mathbf{4})$.

Two sets of $N = 4$ covariant derivatives

$$\left(D^a, \overline{D}^a\right) \equiv \left(\mathcal{D}^{1a}, \overline{\mathcal{D}}^{2a}\right) \quad \text{and} \quad \left(\bar{\nabla}^a, \nabla^a\right) \equiv \left(\mathcal{D}^{2a}, \overline{\mathcal{D}}^{1a}\right)$$

are naturally realized in terms of the $N = 4$ superspaces $(t, \theta_{1a} + i\vartheta_{1a}, \theta_{2a} - i\vartheta_{2a})$ and $(t, \theta_{2a} + i\vartheta_{2a}, \theta_{1a} - i\vartheta_{1a})$. In terms of the new derivatives, the constraints (2.179), (2.180) become

$$D^{(a}\mathcal{V}^{bc)} = \overline{D}^{(a}\mathcal{V}^{bc)} = \nabla^{(a}\mathcal{V}^{bc)} = \bar{\nabla}^{(a}\mathcal{V}^{bc)} = 0\,,$$

$$D^a\mathcal{V} = \frac{1}{2}\nabla_b\mathcal{V}^{ab}\,, \qquad \overline{D}^a\mathcal{V} = \frac{1}{2}\bar{\nabla}_b\mathcal{V}^{ab}\,,$$

$$\nabla^a\mathcal{V} = \frac{1}{2}D_b\mathcal{V}^{ab}\,, \qquad \bar{\nabla}^a\mathcal{V} = \frac{1}{2}\overline{D}_b\mathcal{V}^{ab}\,. \tag{2.181}$$

Now, we see that the $\nabla^a, \bar{\nabla}_a$ derivatives of the superfields $\mathcal{V}, \mathcal{V}^{ab}$ are expressed as $D^a, \overline{D}^a$ of the superfields $\mathcal{V}^{ab}, \mathcal{V}$, respectively. Thus, in the $(\theta_{2a}+i\vartheta_{2a}, \theta_{1a}-i\vartheta_{1a})$ expansion of the superfields $\mathcal{V}, \mathcal{V}^{ab}$ only the first components (i.e., those of zero order in the coordinates $(\theta_{2a}+i\vartheta_{2a}, \theta_{1a}-i\vartheta_{1a})$) are independent $N = 4$ superfields. We denote them v, v^{ab}. The hidden $N = 4$ supersymmetry is realized on these $N = 4$ superfields as

$$\delta v = -\frac{1}{2}\eta_a D_b v^{ab} + \frac{1}{2}\bar{\eta}_a \overline{D}_b v^{ab}, \quad \delta v^{ab} = \frac{4}{3}\left(\eta^{(a}D^{b)}v - \bar{\eta}^{(a}\overline{D}^{b)}v\right), \quad (2.182)$$

while the superfields themselves obey the constraints

$$D^{(a}v^{bc)} = \overline{D}^{(a}v^{bc)} = 0, \quad D^{(a}\overline{D}^{b)}v = 0, \quad (2.183)$$

which are remnant of the $N = 8$ superfield constraints (2.181).

The invariant free action reads

$$S = \int dt\, d^4\theta \left(v^2 - \frac{3}{8}v^{ab}v_{ab}\right). \quad (2.184)$$

3. $(4, 8, 4) = (2, 4, 2) \oplus (2, 4, 2)$

This case is a little bit more tricky. First of all, we define the new set of $N = 8$ superfields $\mathcal{W}, \Phi$ in terms of $\mathcal{V}^{ij}, \mathcal{V}$ defined earlier in (2.178)

$$\mathcal{W} \equiv \mathcal{V}^{11}, \quad \overline{\mathcal{W}} \equiv \mathcal{V}^{22}, \quad \Phi \equiv \frac{2}{3}\left(\mathcal{V} + \frac{3}{2}\mathcal{V}^{12}\right), \quad \overline{\Phi} \equiv \frac{2}{3}\left(\mathcal{V} - \frac{3}{2}\mathcal{V}^{12}\right) \quad (2.185)$$

and construct two new sets of $N = 4$ derivatives D^i, ∇^i from those defined in (2.153)

$$D^i = \frac{1}{\sqrt{2}}\left(\mathcal{D}^{i1} + \overline{\mathcal{D}}^{i1}\right), \quad \overline{D}^i = \frac{1}{\sqrt{2}}\left(\mathcal{D}^{i2} + \overline{\mathcal{D}}^{i2}\right),$$

$$\nabla^i = \frac{1}{\sqrt{2}}\left(\mathcal{D}^{i1} - \overline{\mathcal{D}}^{i1}\right), \quad \overline{\nabla}^i = -\frac{1}{\sqrt{2}}\left(\mathcal{D}^{i2} - \overline{\mathcal{D}}^{i2}\right). \quad (2.186)$$

The basic constraints (2.179), (2.180) can be rewritten in terms of the superfields $\mathcal{W}, \Phi$ and the derivatives D^i, ∇^i as

$$D^i\mathcal{W} = \nabla^i\mathcal{W} = 0, \quad \overline{D}^i\overline{\mathcal{W}} = \overline{\nabla}^i\overline{\mathcal{W}} = 0,$$
$$D^i\Phi = \overline{\nabla}^i\Phi = 0, \quad \nabla^i\overline{\Phi} = \overline{D}^i\overline{\Phi} = 0,$$
$$\overline{D}^i\mathcal{W} - D^i\overline{\Phi} = 0, \quad D^i\overline{\mathcal{W}} + \overline{D}^i\Phi = 0,$$
$$\overline{\nabla}^i\mathcal{W} - \nabla^i\Phi = 0, \quad \nabla^i\overline{\mathcal{W}} + \overline{\nabla}^i\overline{\Phi} = 0. \quad (2.187)$$

The proper $N = 4$ superspace is defined as the one on which the covariant derivatives $D^1, \overline{D}^2, \nabla^1, \overline{\nabla}^2$ are naturally realized. The constraints (2.187) imply that the remaining set of covariant derivatives, i.e., $D^2, \overline{D}^1, \nabla^2, \overline{\nabla}^1$, when acting on every involved $N = 8$ superfield, can be expressed as spinor derivatives from the first set acting on some other $N = 8$ superfield. Thus, the first $N = 4$ superfield components of the $N = 8$ superfields $\mathcal{W}, \Phi$ are the only independent $N = 4$ superfield projections. The transformations of the implicit $N = 4$ Poincaré supersymmetry have the following form in terms of these $N = 4$ superfields:

$$\delta w = \bar{\epsilon} D^1 \bar{\phi} + \bar{\eta} \nabla^1 \phi \,, \qquad \delta \phi = -\eta \overline{\nabla}_1 w - \bar{\epsilon} D^1 \bar{w} \,,$$
$$\delta \bar{w} = \epsilon \overline{D}_1 \phi + \eta \overline{\nabla}_1 \bar{\phi} \,, \qquad \delta \bar{\phi} = -\epsilon \overline{D}_1 w - \bar{\eta} \nabla^1 \bar{w} \,. \qquad (2.188)$$

The free invariant action reads

$$S = \int dt \, d^4 \theta \left(w\bar{w} - \phi\bar{\phi} \right) \,. \qquad (2.189)$$

2.4.4 Supermultiplet (5, 8, 3)

This supermultiplet has been considered in detail in [6, 30]. It was termed there the "$N = 8$ vector multiplet". Here, we sketch its main properties.

To describe this supermultiplet, one should introduce five bosonic $N = 8$ superfields $\mathcal{V}_{\alpha a}, \mathcal{U}$ obeying the constraints

$$D^{ib} \mathcal{V}_{\alpha a} + \delta_a^b \nabla_\alpha^i \mathcal{U} = 0 \,, \qquad \nabla^{\beta i} \mathcal{V}_{\alpha a} + \delta_\alpha^\beta D_a^i \mathcal{U} = 0 \,. \qquad (2.190)$$

It is worth noting that the constraints (2.190) are covariant not only under three $SU(2)$ automorphism groups (realized on the doublet indices i, a, and α), but also under the $SO(5)$ automorphisms. These $SO(5)$ transformations mix the spinor derivatives D^{ia} and $\nabla^{\alpha i}$ in the indices α and a, while two $SU(2)$ groups realized on these indices constitute $SO(4) \subset SO(5)$. The superfields $\mathcal{U}, \mathcal{V}^{\alpha a}$ form an $SO(5)$ vector: under $SO(5)$ transformations belonging to the coset $SO(5)/SO(4)$ they transform as

$$\delta \mathcal{V}_{\alpha a} = a_{\alpha a} \mathcal{U} \,, \qquad \delta \mathcal{U} = -2a_{\alpha a} \mathcal{V}^{\alpha a} \,. \qquad (2.191)$$

As in the previous cases we may consider two different splittings of the $N = 8$ vector multiplet into irreducible $N = 4$ superfields.

1. $(\mathbf{5, 8, 3}) = (\mathbf{1, 4, 3}) \oplus (\mathbf{4, 4, 0})$
2. $(\mathbf{5, 8, 3}) = (\mathbf{3, 4, 1}) \oplus (\mathbf{2, 4, 2})$

Once again, they correspond to two different choices of the $N = 4, d = 1$ superspace as a subspace in the original $N = 8, d = 1$ superspace.

1. $(\mathbf{5, 8, 3}) = (\mathbf{1, 4, 3}) \oplus (\mathbf{4, 4, 0})$

The relevant $N = 4$ superspace is $\mathbb{R}^{(1|4)}$ parameterized by the coordinates (t, θ_{ia}) and defined in (2.131). As in the previous cases, it follows from the constraints (2.190) that the spinor derivatives of all involved superfields with respect to $\vartheta_{i\alpha}$ are expressed in terms of spinor derivatives with respect to θ_{ia}. Thus, the only essential $N = 4$ superfield components of $\mathcal{V}_{\alpha a}$ and $\mathcal{U}$ in their ϑ-expansion are the first ones

$$v_{\alpha a} \equiv \mathcal{V}_{\alpha a}|_{\vartheta=0} , \qquad u \equiv \mathcal{U}|_{\vartheta=0} . \tag{2.192}$$

They accommodate the whole off-shell component content of the $N = 8$ vector multiplet. These five bosonic $N = 4$ superfields are subjected, in virtue of (2.190), to the irreducibility constraints in $N = 4$ superspace

$$D^{i(a}v^{b)\alpha} = 0, \qquad D^{i(a}D_i^{b)}u = 0 . \tag{2.193}$$

Thus, from the $N = 4$ superspace standpoint, the vector $N = 8$ supermultiplet is the sum of the $N = 4, d = 1$ hypermultiplet $v_{\alpha a}$ with the off-shell component contents $(\mathbf{4, 4, 0})$ and the $N = 4$ "old" tensor multiplet u with the contents $(\mathbf{1, 4, 3})$.

The transformations of the implicit $N = 4$ Poincaré supersymmetry read

$$\delta v_{\alpha a} = \eta_{i\alpha}D_a^i u , \qquad \delta u = \frac{1}{2}\eta_{i\alpha}D^{ia}v_a^\alpha . \tag{2.194}$$

2. $(\mathbf{5, 8, 3}) = (\mathbf{3, 4, 1}) \oplus (\mathbf{2, 4, 1})$

Another interesting $N = 4$ superfield splitting of the $N = 8$ vector multiplet can be achieved by passing to the complex parameterization of the $N = 8$ superspace as

$$\left(t, \Theta_{i\alpha} = \theta_{i\alpha} + i\vartheta_{\alpha i}, \bar{\Theta}^{i\alpha} = \theta^{i\alpha} - i\vartheta^{\alpha i}\right)$$

where we have identified the indices a and α, thus reducing the number of manifest $SU(2)$ automorphism symmetries to just two. In this superspace the covariant derivatives $\mathcal{D}^{i\alpha}, \bar{\mathcal{D}}^{j\beta}$ defined in (2.153) (with the identification of indices just mentioned) are naturally realized. We are also led to define new superfields

$$\mathcal{V} \equiv -\epsilon_{\alpha a}\mathcal{V}^{\alpha a} , \quad \mathcal{W}^{\alpha\beta} \equiv \mathcal{V}^{(\alpha\beta)} = \frac{1}{2}\left(\mathcal{V}^{\alpha\beta} + \mathcal{V}^{\beta\alpha}\right) ,$$
$$\mathcal{W} \equiv \mathcal{V} + i\mathcal{U} , \quad \bar{\mathcal{W}} \equiv \mathcal{V} - i\mathcal{U} . \tag{2.195}$$

In this basis of $N = 8$ superspace the original constraints (2.190) amount to

$$\mathcal{D}^{i\alpha}\mathcal{W}^{\beta\gamma} = -\frac{1}{4}\left(\epsilon^{\beta\alpha}\bar{\mathcal{D}}^{i\gamma}\bar{\mathcal{W}} + \epsilon^{\gamma\alpha}\bar{\mathcal{D}}^{i\beta}\bar{\mathcal{W}}\right) ,$$

$$\bar{\mathcal{D}}^{i\alpha}\mathcal{W}^{\beta\gamma} = -\frac{1}{4}\left(\epsilon^{\beta\alpha}\mathcal{D}^{i\gamma}\mathcal{W} + \epsilon^{\gamma\alpha}\mathcal{D}^{i\beta}\mathcal{W}\right) ,$$

$$\mathcal{D}^{i\alpha}\bar{\mathcal{W}} = 0, \ \bar{\mathcal{D}}^{i\alpha}\mathcal{W} = 0 , \quad (\mathcal{D}^{k\alpha}\mathcal{D}_\alpha^i)\mathcal{W} = (\bar{\mathcal{D}}_\alpha^k\bar{\mathcal{D}}^{i\alpha})\bar{\mathcal{W}} . \tag{2.196}$$

Next, we single out the $N = 4, d = 1$ superspace as $(t, \theta_\alpha \equiv \Theta_{1\alpha}, \bar{\theta}^\alpha)$ and split our $N = 8$ superfields into the $N = 4$ ones in the standard way. As in all previous cases, the spinor derivatives of each $N = 8$ superfield with respect to $\overline{\Theta}^{2\alpha}$ and $\Theta_{2\alpha}$, as a consequence of the constraints (2.196), are expressed as derivatives of some other superfields with respect to $\bar{\theta}^\alpha$ and θ_α. Therefore, only the first (i.e., taken at $\overline{\Theta}^{2a} = 0$ and $\Theta_{2a} = 0$) $N = 4$ superfield components of the $N = 8$ superfields really matter. They accommodate the entire off-shell field content of the multiplet. These $N = 4$ superfields are defined as

$$\phi \equiv \mathcal{W}| \ , \quad \bar{\phi} \equiv \overline{\mathcal{W}}| \ , \quad w^{\alpha\beta} \equiv \mathcal{W}^{\alpha\beta}| \tag{2.197}$$

and satisfy the constraints following from (2.196)

$$\mathcal{D}^\alpha \bar{\phi} = 0 \ , \quad \overline{\mathcal{D}}_\alpha \phi = 0 \ , \quad \mathcal{D}^{(\alpha} w^{\beta\gamma)} = \overline{\mathcal{D}}^{(\alpha} w^{\beta\gamma)} = 0, \quad \mathcal{D}^\alpha \equiv \mathcal{D}^{1\alpha} \ , \quad \overline{\mathcal{D}}_\alpha \equiv \overline{\mathcal{D}}_{1\alpha} \ . \tag{2.198}$$

They tell us that the $N = 4$ superfields ϕ and $\bar{\phi}$ form the standard $N = 4$ chiral multiplet $(\mathbf{2}, \mathbf{4}, \mathbf{2})$, while the $N = 4$ superfield $w^{\alpha\beta}$ represents the $N = 4$ tensor multiplet $(\mathbf{3}, \mathbf{4}, \mathbf{1})$.

The implicit $N = 4$ supersymmetry is realized on $w^{\alpha\beta}$, ϕ and $\bar{\phi}$ as

$$\delta w^{\alpha\beta} = \frac{1}{2} \left(\eta^{(\alpha} \overline{\mathcal{D}}^{\beta)} \bar{\phi} - \bar{\eta}^{(\alpha} \mathcal{D}^{\beta)} \phi \right) \ , \quad \delta\phi = \frac{4}{3} \eta_\alpha \overline{\mathcal{D}}^\beta w_\beta^\alpha \ , \quad \delta\bar{\phi} = -\frac{4}{3} \bar{\eta}^\alpha \mathcal{D}_\beta w_\alpha^\beta \ . \tag{2.199}$$

An analysis of $N = 8$ supersymmetric actions for the $N = 8$ vector multiplet may be found in [6].

2.4.5 Supermultiplet (6, 8, 2)

This supermultiplet can be described by two $N = 8$ tensor multiplets $\mathcal{V}^{ij}$ and $\mathcal{W}^{ab}$,

$$D_a^{(i} \mathcal{V}^{jk)} = 0 \ , \quad \nabla_a^{(i} \mathcal{V}^{jk)} = 0 \ , \quad D_i^{(a} \mathcal{W}^{bc)} = 0 \ , \quad \nabla_i^{(a} \mathcal{W}^{bc)} = 0 \ , \tag{2.200}$$

with the additional constraints

$$D_j^a \mathcal{V}^{ij} = \nabla^{bi} \mathcal{W}_b^a \ , \quad \nabla_j^a \mathcal{V}^{ij} = -D_b^i \mathcal{W}^{ab} \ . \tag{2.201}$$

The role of the latter constraints is to identify the eight fermions, which are present in $\mathcal{V}^{ij}$, with the fermions from $\mathcal{W}^{ab}$, and to reduce the number of independent auxiliary fields in both superfields to two

$$F_1 = D_i^a D_{aj} \mathcal{V}^{ij}| \ , \quad F_2 = D_a^i D_{ib} \mathcal{W}^{ab}| \ , \tag{2.202}$$

where $|$ means here restriction to $\theta_{ia} = \vartheta_{ia} = 0$.

There are two different possibilities to split this $N = 8$ multiplet into $N = 4$ ones.

1. $(\mathbf{6}, \mathbf{8}, \mathbf{2}) = (\mathbf{3}, \mathbf{4}, \mathbf{1}) \oplus (\mathbf{3}, \mathbf{4}, \mathbf{1})$
2. $(\mathbf{6}, \mathbf{8}, \mathbf{2}) = (\mathbf{4}, \mathbf{4}, \mathbf{0}) \oplus (\mathbf{2}, \mathbf{4}, \mathbf{2})$

As before, we discuss the peculiarities of both decompositions.

1. $(\mathbf{6, 8, 2}) = (\mathbf{3, 4, 1}) \oplus (\mathbf{3, 4, 1})$

The corresponding $N = 4$ supersubspace is (2.131). The $N = 8$ constraints imply that the only essential $N = 4$ superfield components of $\mathcal{V}^{ij}$ and $\mathcal{W}^{ab}$ in their ϑ-expansion are the first ones

$$v^{ij} \equiv \mathcal{V}^{ij}| , \qquad w^{ab} \equiv \mathcal{W}^{ab}| . \tag{2.203}$$

These six bosonic $N = 4$ superfields are subjected, in virtue of (2.200), (2.201), to the irreducibility constraints in $N = 4$ superspace

$$D^{a(i}v^{jk)} = 0 , \qquad D^{i(a}w^{bc)} = 0 . \tag{2.204}$$

Thus, the $N = 8$ supermultiplet $(\mathbf{6, 8, 2})$ amounts to the sum of two $N = 4, d = 1$ tensor multiplets v^{ij}, w^{ab} with the off-shell field contents $(\mathbf{3, 4, 1}) \oplus (\mathbf{3, 4, 1})$.

The transformations of the implicit $N = 4$ Poincaré supersymmetry are

$$\delta v^{ij} = -\frac{2}{3}\eta_a^{(i}D_b^{j)}w^{ab}, \qquad \delta w^{ab} = \frac{2}{3}\eta_i^{(a}D_j^{b)}v^{ij} . \tag{2.205}$$

The free $N = 8$ supersymmetric action has the following form:

$$S = \int dt\, d^4\theta \left(v^2 - w^2\right) . \tag{2.206}$$

2. $(\mathbf{6, 8, 2}) = (\mathbf{4, 4, 0}) \oplus (\mathbf{2, 4, 2})$

In this case, to describe the $(\mathbf{6, 8, 2})$ multiplet, we combine two $N = 4$ superfields, i.e., the chiral superfield

$$D^i\phi = \overline{D}^i\bar{\phi} = 0 \tag{2.207}$$

and the hypermultiplet q^{ia}

$$D^{(i}q^{j)a} = \overline{D}^{(i}q^{j)a} = 0 . \tag{2.208}$$

The transformations of the implicit $N = 4$ supersymmetry read

$$\delta q^{ia} = \bar{\epsilon}^a D^i\bar{\phi} + \epsilon^a \overline{D}^i\phi , \quad \delta\phi = -\frac{1}{2}\bar{\epsilon}^a D^i q_{ia} , \quad \delta\bar{\phi} = -\frac{1}{2}\epsilon^a \overline{D}^i q_{ia} . \tag{2.209}$$

The invariant free action reads

$$S_{\text{free}} = \int dt\, d^4\theta \left(q^2 - 4\phi\bar{\phi}\right) . \tag{2.210}$$

2.4.6 Supermultiplet (7, 8, 1)

This supermultiplet has a natural description in terms of two $N = 8$ super-fields $\mathcal{V}^{ij}$ and $\mathcal{Q}^{a\alpha}$ satisfying the constraints

$$D^{(ia}\mathcal{V}^{jk)} = 0 , \quad \nabla^{\alpha(i}\mathcal{V}^{jk)} = 0 , \quad D^{i(a}\mathcal{Q}^{ab)} = 0, \quad \nabla_i^{(\alpha}\mathcal{Q}_a^{\beta)} = 0 , \tag{2.211}$$
$$D_j^a \mathcal{V}^{ij} = i\nabla_\alpha^i \mathcal{Q}^{a\alpha} , \quad \nabla_j^\alpha \mathcal{V}^{ij} = -iD_a^i \mathcal{Q}^{a\alpha} . \tag{2.212}$$

The constraints (2.211) leave in the superfields $\mathcal{V}^{ij}$ and $\mathcal{Q}^{a\alpha}$ the sets (**3, 8, 5**) and (**4, 8, 4**) of irreducible components, respectively. The role of the constraints (2.212) is to identify the fermions in the superfields $\mathcal{V}^{ij}$ and $\mathcal{Q}^{a\alpha}$ and reduce the total number of independent auxiliary components in both superfields to just one.

For this supermultiplet there is a unique splitting into $N = 4$ superfields as

$$(\mathbf{7, 8, 1}) = (\mathbf{3, 4, 1}) \oplus (\mathbf{4, 4, 0}).$$

The proper $N = 4$ superspace is parameterized by the coordinates (t, θ_{ia}). The constraints (2.211), (2.212) imply that the only essential $N = 4$ superfield components in the ϑ-expansion of $\mathcal{V}^{ij}$ and $\mathcal{Q}^{a\alpha}$ are the first ones

$$v^{ij} \equiv \mathcal{V}^{ij}|_{\vartheta=0} , \qquad q^{a\alpha} \equiv \mathcal{Q}^{a\alpha}|_{\vartheta=0} . \tag{2.213}$$

These seven bosonic $N = 4$ superfields are subjected, as a corollary of (2.211), (2.212), to the irreducibility constraints in $N = 4$ superspace

$$D^{a(i}v^{jk)} = 0 , \qquad D^{i(a}q^{b)\alpha} = 0 . \tag{2.214}$$

Thus, the $N = 8$ supermultiplet (**7, 8, 1**) amounts to the sum of the $N = 4, d = 1$ hypermultiplet $q^{a\alpha}$ with the (**4, 4, 0**) off-shell field content and the $N = 4$ tensor multiplet v^{ij} with the (**3, 4, 1**) content.

The implicit $N = 4$ Poincaré supersymmetry is realized by the transformations

$$\delta v^{ij} = -\frac{2i}{3} \eta_\alpha^{(i} D_a^{j)} q^{a\alpha} , \qquad \delta q^{a\alpha} = -\frac{i}{2} \eta^{i\alpha} D^{ja} v_{ij} . \tag{2.215}$$

The free action can be also easily written

$$S = \int dt\, d^4\theta \left[v^2 - \frac{4}{3}q^2 \right] . \tag{2.216}$$

2.4.7 Supermultiplet (8, 8, 0)

This supermultiplet is analogous to the supermultiplet (**0, 8, 8**): they differ in their overall Grassmann parity. It is described by the two real bosonic $N = 8$ superfields $\mathcal{Q}^{aA}, \Phi^{i\alpha}$ subjected to the constraints

$$D^{(ia}\Phi^{j)\alpha} = 0, \quad D^{i(a}\mathcal{Q}^{b)A} = 0, \quad \nabla^{(\alpha A}\Phi_i^{\beta)} = 0,$$

$$\nabla^{\alpha(A}\mathcal{Q}^{aB)} = 0, \quad \nabla^{\alpha A}\mathcal{Q}_A^a = -D^{ia}\Phi_i^\alpha, \tag{2.217}$$

$$\nabla^{\alpha A}\Phi_\alpha^i = D^{ia}\mathcal{Q}_a^A. \tag{2.218}$$

Analogously to the case of the supermultiplet (**8, 8, 0**), from the constraints (2.218) it follows that the spinor derivatives of all involved superfields with respect to $\vartheta_{\alpha A}$ are expressed in terms of spinor derivatives with respect to θ_{ia}. Thus, the only essential $N = 4$ superfield components in the ϑ-expansion of $\mathcal{Q}^{aA}$ and $\Phi^{i\alpha}$ are the first ones

$$q^{aA} \equiv \mathcal{Q}^{aA}|_{\vartheta=0}, \quad \phi^{i\alpha} \equiv \Phi^{i\alpha}|_{\vartheta=0}. \tag{2.219}$$

They accommodate the whole off-shell component content of the multiplet (**8, 8, 0**). These bosonic $N = 4$ superfields are subjected, as a consequence of (2.217), (2.218), to the irreducibility constraints in $N = 4$ superspace

$$D^{a(i}\phi^{j)\alpha} = 0, \quad D^{i(a}q^{b)A} = 0. \tag{2.220}$$

Thus, the $N = 8$ supermultiplet (**8, 8, 0**) can be represented as the sum of two $N = 4, d = 1$ hypermultiplets with the off-shell component contents (**4, 4, 0**) $\oplus$ (**4, 4, 0**).

The transformations of the implicit $N = 4$ Poincaré supersymmetry in this last case are as follows:

$$\delta q^{aA} = -\frac{1}{2}\eta^{Aa}D^{ia}\phi_{i\alpha}, \quad \delta\phi^{i\alpha} = \frac{1}{2}\eta_A^\alpha D_a^i q^{aA}. \tag{2.221}$$

The invariant free action reads

$$S = \int dt\, d^4\theta\, [q^2 - \phi^2]. \tag{2.222}$$

The most general action still respecting four $SU(2)$ automorphism symmetries has the following form:

$$S = \int dt\, d^4\theta F(q^2, \phi^2), \tag{2.223}$$

where, as a necessary condition of $N = 8$ supersymmetry, the function $F(q^2, \phi^2)$ should obey the equation

$$\frac{\partial^2}{\partial q^2 \partial q^2}\left(q^2 F(q^2, \phi^2)\right) + \frac{\partial^2}{\partial \phi^2 \partial \phi^2}\left(\phi^2 F(q^2, \phi^2)\right) = 0. \tag{2.224}$$

Thus, we presented superfield formulations of the full amount of off-shell $N = 8, d = 1$ supermultiplets with eight physical fermions, both in $N = 8$ and $N = 4$ superspaces. We listed all possible $N = 4$ superfield splittings of these multiplets.

2.5 Summary and Conclusions

In these lectures we reviewed the superfield approach to extended supersymmetric one-dimensional models. We presented superfield formulations of the full amount of off-shell $N = 4$ and $N = 8, d = 1$ supermultiplets with **4** and **8** physical fermions, respectively. We also demonstrated how to reproduce $N = 4$ supermultiplets from nonlinear realizations of the $N = 4, d = 1$ superconformal group.

It should be pointed out that here we addressed only those multiplets which satisfy linear constraints in superspace. As we know, there exist $N = 4, d = 1$ multiplets with nonlinear defining constraints (e.g., nonlinear versions of the chiral (**2, 4, 2**) multiplet, as well as of the hypermultiplet (**4, 4, 0**)). It would be interesting to construct analogous nonlinear versions of some $N = 8$ multiplets from the above set. Moreover, for all our linear supermultiplets the bosonic metrics of the general sigma-model type actions are proven to be conformally flat. This immediately raises the question – how to describe $N = 4$ and $N = 8, d = 1$ sigma models with hyper-Kähler metrics in the target space? For the $N = 8$ cases it seems the unique possibility is to use *infinite* dimensional supermultiplets, as in the case of $N = 2, d = 4$ supersymmetry [32, 33]. But for $N = 4, d = 1$ supersymmetric models infinite dimensional supermultiplets do not exist! Therefore, the unique possibility in this case is to use some nonlinear supermultiplets. In this respect, the harmonic superspace approach seems to yield the most relevant framework. So, all results we discussed should be regarded as preparatory for a more detailed study of $N = 4, 8$ $d = 1$ supersymmetric models.

Acknowledgments

We thank to E. Ivanov and O. Lechtenfeld in collaboration with whom some of the presented results were obtained. Useful conversations with A. Nersessian are acknowledged with pleasure. This research was partially supported by the INTAS-00-00254 grant, RFBR-DFG grant No 02-02-04002, grant DFG No 436 RUS 113/669, and RFBR grant No 03-02-17440. S.K. thanks INFN – Laboratori Nazionali di Frascati for the warm hospitality extended to him during the course of this work.

References

1. S.J. Gates, L. Rana: Phys. Lett. B **352**, 50 (1995)
2. A. Van Proeyen: Lectures given in the 1995 Trieste summer school in high energy physics and cosmology; hep-th/9512139
3. S. Hellerman, J. Polchinski: Supersymmetric quantum mechanics from light cone quantization In: Yuri Golfand memorial volume textit, ed by M.A. Shifman, "The Many Faces of the Superworld," pp. 142–155 [hep-th/9908202].

4. E. Ivanov, S. Krivonos, O. Lechtenfeld: Class. Quant. Grav. **21**, 1031 (2004) [hep-th/0310299]
5. S. Bellucci, E. Ivanov, S. Krivonos, O. Lechtenfeld: Nucl. Phys. B **699**, 226 (2004) [hep-th/0406015]
6. S. Bellucci, E. Ivanov, S. Krivonos, O. Lechtenfeld: Nucl. Phys. B **684**, 321 (2004) [hep-th/0312322]
7. S. Bellucci, S. Krivonos, A. Nersessian: Phys. Lett. B **605**, 181 (2005) [hep-th/0410029]; S. Bellucci, S. Krivonos, A. Nersessian, A. Shcherbakov: [hep-th/0410073]; S. Bellucci, S. Krivonos, A. Shcherbakov: Phys. Lett. B **612**, 283 (2005) [hep-th/0502245]
8. S. Bellucci, A. Beylin, S. Krivonos, A. Shcherbakov: Phys. Lett. B **663** (2006) 228 [hep-th/0511054]
9. V.A. Akulov, A.P. Pashnev: Teor. Mat. Fiz. **56**, 344 (1983)
10. R.A. Coles, G. Papadopoulos: Class. Quant. Grav. **7**, 427 (1990)
11. G.W. Gibbons, G. Papadopoulos, K.S. Stelle: Nucl. Phys. B **508**, 623 (1997); [hep-th/9706207]
12. C.M. Hull: [hep-th/9910028]
13. E. Ivanov, O. Lechtenfeld: JHEP **0309**, 073 (2003) [hep-th/0307111]
14. J. Wess: Acta Phys. Austriaca **41**, 409 (1975)
15. M. de Crombrugghe, V. Rittenberg: Ann. Phys. **151**, 99 (1983)
16. E.A. Ivanov, A.V. Smilga: Phys. Lett. B **257**, 79 (1991)
17. V.P. Berezovoj, A.I. Pashnev: Class. Quant. Grav. **8**, 2141 (1991)
18. A. Maloney, M. Spradlin, A. Strominger: JHEP **0204**, 003 (2002) [hep-th/9911001]
19. A. Pashnev, F. Toppan: J. Math. Phys. **42**, 5257 (2001) [hep-th/0010135]
20. V. Akulov, A. Pashnev: Teor. Mat. Fiz. **56**, 344 (1983)
21. S. Fubini, E. Rabinovici: Nucl. Phys. B **245**, 17 (1984)
22. E. Ivanov, S. Krivonos, V. Leviant: J. Phys. A: Math. Gen. **22**, 4201 (1989)
23. S. Bellucci, A. Beylin, S. Krivonos, A. Nersessian, E. Orazi: Phys. Lett. B **616**, 228 (2005) [hep-th/0503244]; S. Bellucci, S. Krivonos, A. Marrani, E. Orazi: Phys. Rev. D **73** (2006) 025011 [hep-th/0511249]
24. E. Ivanov, S. Krivonos, O. Lechtenfeld: JHEP **0303**, 014 (2003); [hep-th/0212303]; S. Bellucci, S. Krivonos and A. Sutulin: Phys. Lett. B **605**, 406 (2005) [hep-th/0410276]; S. Bellucci, E. Ivanov, A. Sutulin: Nucl. Phys. B **722**, 297 (2005) [hep-th/0504185]
25. S. Coleman, J. Wess, B. Zumino: Phys. Rev. B **177**, 2239 (1969); C. Callan, S. Coleman, J. Wess, B. Zumino: Phys. Rev. B **177**, 2279 (1969); D.V. Volkov: Sov. J. Part. Nucl. **4**, 3 (1973); V.I. Ogievetsky: In *Proceeding of X-th Winter School of Theoretical Physics in Karpach*, Wroclaw, Vol 1, p. 117 (1974)
26. E.A. Ivanov, V.I. Ogievetsky: Teor. Mat. Fiz. **25**, 164 (1975)
27. B. Zumino: Nucl. Phys. B **127**, 189 (1977)
28. V. De Alfaro, S. Fubini, G. Furlan: Nuovo Cim. A **34**, 569 (1974)
29. P. Claus, M. Derix, R. Kallosh, J. Kumar, P.K. Townsend, A. Van Proeyen: Phys. Rev. Lett. **81**, 4553 (1998) [hep-th/9804177]
30. D.-E. Diaconescu, R. Entin: Phys. Rev. D **56**, 8045 (1997) [hep-th/9706059]
31. B. Zupnik: Nucl. Phys. B **554**, 365 (1999); B **644**, 405E (2002) [hep-th/9902038]

32. A. Galperin, E. Ivanov, V. Ogievetsky, E. Sokatchev: Pis'ma ZhETF **40**, 155 (1984); JETP Lett. **40**, 912 (1984); A.S. Galperin, E.A. Ivanov, S. Kalitzin, V.I. Ogievetsky, E.S. Sokatchev: Class. Quant. Grav. **1**, 469 (1984)
33. A.S. Galperin, E.A. Ivanov, V.I. Ogievetsky, E.S. Sokatchev: *Harmonic Super-Space* (Cambridge University Press, Cambridge, 2001), 306 p

3

Noncommutative Mechanics, Landau Levels, Twistors, and Yang–Mills Amplitudes

V.P. Nair

City College of the CUNY, New York, NY 10031
vpn@sci.ccny.cuny.edu

These lectures fall into two distinct, although tenuously related, parts. The first part is about fuzzy and noncommutative spaces, and particle mechanics on such spaces, in other words, noncommutative mechanics. The second part is a discussion/review of the use of twistors in calculating Yang–Mills amplitudes. The point of connection between these two topics is in the realization of holomorphic maps as the lowest Landau level wave functions, or as wave functions of the Hilbert space used for the fuzzy version of the two-sphere.

3.1 Fuzzy Spaces

3.1.1 Definition and Construction of $\mathcal{H}_N$

Fuzzy spaces have been an area of research for a number of years by now [1]. They have proved to be useful in some physical problems. Part of the motivation for this has been the discovery that noncommutative spaces, and more specifically fuzzy spaces, can arise as solutions in string and M-theories [2]. For example, in the matrix model version of M-theory, noncommutative spaces can be obtained as $(N \times N)$-matrix configurations whose large N-limit will give smooth manifolds. Fluctuations of branes are described by gauge theories and, with this motivation, there has recently been a large number of papers dealing with gauge theories, and more generally field theories, on such spaces [3]. There is also an earlier line of development, motivated by quantum gravity, using the Dirac operator to characterize the manifold and using "spectral actions" [4].

Even apart from their string and M-theory connections, fuzzy spaces are interesting for other reasons. Because these spaces are described by finite-dimensional matrices, the number of possible modes for fields on such spaces is limited by the Cayley–Hamilton theorem, and so, one has a natural ultraviolet cutoff. We may think of such field theories as a finite-mode approximation to commutative continuum field theories, providing, in some sense, an alternative

V.P. Nair: *Noncommutative Mechanics, Landau Levels, Twistors, and Yang–Mills Amplitudes*, Lect. Notes Phys. **698**, 97–138 (2006)
DOI 10.1007/3-540-33314-2_3

to lattice gauge theories. Indeed, this point of view has been pursued in some recent work [5].

Analysis of fuzzy spaces and particle dynamics on such spaces are also closely related to the quantum Hall effect. The dynamics of charged particles in a magnetic field can be restricted to the lowest Landau level, if the field is sufficiently strong, and this is equivalent to dynamics on a fuzzy version of the underlying spatial manifold. (The fact that the restriction to the lowest Landau level gives noncommutativity of coordinates has been known for a long time; for a recent review focusing on the fuzzy aspects, see [6].)

The main idea behind fuzzy spaces is the standard correspondence principle of the quantum theory, which is as shown below.

Quantum Theory	$\hbar \to 0$	*Classical Theory*
Hilbert space $\mathcal{H}$		Phase space M
	$\longrightarrow$	
Operators on $\mathcal{H}$		Functions on M

This correspondence suggests a new paradigm. Rather than dealing with theories on a continuous manifold M, we take the Hilbert space $\mathcal{H}$ and the algebra of operators on it as the fundamental quantities and obtain the continuous manifold M as an approximation. Generally, instead of $\hbar$, we use an arbitrary deformation parameter θ, so that the continuous manifold emerges not as the classical limit in the physical sense, but as some other limit when $\theta \to 0$, which will mathematically mimic the transition from quantum mechanics to classical mechanics. The point of view where the space-time manifold is not fundamental can be particularly satisfying in the context of quantum gravity, and in fact, it was in this context that the first applications of noncommutative spaces to physics was initiated [4].

Now passing to more specific details, by a fuzzy space, we mean a sequence $(\mathcal{H}_N, Mat_N, \mathcal{D}_N)$, where $\mathcal{H}_N$ is an N-dimensional Hilbert space, Mat_N is the matrix algebra of $N \times N$-matrices which act on $\mathcal{H}_N$, and $\mathcal{D}_N$ is a matrix analog of the Dirac operator or, in many instances, just the matrix analog of the Laplacian. The inner product on the matrix algebra is given by $\langle A, B \rangle = \frac{1}{N}\mathrm{Tr}(A^\dagger B)$. The Hilbert space $\mathcal{H}_N$ leads to some smooth manifold M as $N \to \infty$. The matrix algebra Mat_N approximates to the algebra of functions on M. The operator $\mathcal{D}_N$ is needed to recover metrical and other geometrical properties of the manifold M. For example, information about the dimension of M is contained in the growth of the number of eigenvalues. More generally, noncommutative spaces are defined in a similar way, with a triple $(\mathcal{H}, \mathcal{A}, \mathcal{D})$, where $\mathcal{H}$ can be infinite-dimensional, $\mathcal{A}$ is the algebra of operators on $\mathcal{H}$ and $\mathcal{D}$ is a Dirac operator on $\mathcal{H}$ [4]. For fuzzy spaces the dimensionality of $\mathcal{H}$ is finite.

Rather than discuss generalities, we will consider the construction of some noncommutative and fuzzy spaces. Consider the flat $2k$-dimensional space

$\mathbf{R}^{2k}$. We build up the coherent state representation by considering the particle action [7]

$$\frac{\mathcal{S}}{\theta} = -\frac{i}{\theta} \int dt \, \bar{Z}_\alpha \dot{Z}^\alpha \,, \tag{3.1}$$

where $\alpha = 1, 2, \ldots, k$. Evidently, the time-evolution of the variables Z^α, $\bar{Z}_\alpha$ is trivial, and so, the theory is entirely characterized by the phase space, or upon quantization, by the specification of the Hilbert space. From the action, we can identify the canonical commutation rules as

$$[\bar{Z}_\beta, Z^\alpha] = \theta \, \delta_\beta^\alpha \,. \tag{3.2}$$

It is then possible to choose states, which are eigenstates of Z^α, defined by $\langle z | Z^\alpha = \langle z | z^\alpha$, so that wave functions $f(z) = \langle z | f \rangle$ can be taken to be holomorphic. The operators Z^α, $\bar{Z}_\beta$ are realized on these by

$$\begin{aligned} Z^\alpha \, f(z) &= z^\alpha \, f(z) \\ \bar{Z}_\beta \, f(z) &= \theta \frac{\partial}{\partial z^\beta} \, f(z) \,. \end{aligned} \tag{3.3}$$

The inner product for the wave functions should be of the form

$$\begin{aligned} \langle f | h \rangle &= \int d\mu \, C(z, \bar{z}) \, \bar{f} \, h \\ d\mu &= \prod_\alpha \frac{dz^\alpha d\bar{z}_\alpha}{(-2i)} \equiv \prod_\alpha d^2 z_\alpha \,. \end{aligned} \tag{3.4}$$

By imposing the adjointness condition $\langle f | Zh \rangle = \langle \bar{Z} f | h \rangle$, we get

$$\theta \frac{\partial C}{\partial z^\alpha} = -\bar{z}_\alpha C \tag{3.5}$$

which can be solved to yield

$$\langle f | h \rangle = \int \prod_\alpha \frac{d^2 z_\alpha}{\pi \theta} \exp\left[-\frac{\bar{z}_\alpha z^\alpha}{\theta} \right] \bar{f} \, h \,. \tag{3.6}$$

The overall normalization has been chosen so that the state $f = 1$ has norm equal to 1.

Since $f(z)$ is holomorphic, a basis of states can be given by $f = 1$, z^α, $z^{\alpha_1} z^{\alpha_2}, \ldots$. The Hilbert space is infinite-dimensional and can be used for the noncommutative version of $\mathbf{R}^{2k}$.

3.1.2 Star Products

The star product is very helpful in discussing the large N limit. (Star products have along history going back to Moyal and others. The books and reviews quoted, [1, 3, 7] and others, contain expositions of the star product.) We shall consider the two-dimensional case first, generalization to arbitrary even dimensions will be straightforward. A basis for the Hilbert space is given by 1, z, z^2, etc., and using this basis, we can represent an operator as a matrix A_{mn}. Associated to such a matrix, we define a function $A(z, \bar{z}) = (A)$, known as the symbol for A, by

$$(A) = A(z, \bar{z}) \equiv \sum_{mn} A_{mn} \frac{z^m \, \bar{z}^n}{\sqrt{m! \, n!}} \, e^{-z\bar{z}/\theta}$$

$$= \sum_{mn} A_{mn} \psi_m \psi_n^*, \tag{3.7}$$

where ψ_n are given by

$$\psi_n = e^{-z\bar{z}/2\theta} \frac{z^n}{\sqrt{n!}}. \tag{3.8}$$

These are normalized functions obeying the equation

$$\int \frac{d^2 z}{\theta \pi} \, \psi_n^* \, \psi_m = \delta_{mn}. \tag{3.9}$$

The symbol corresponding to the product of two operators (or matrices) A and B may be written as

$$(AB) = \sum \psi_m A_{mn} B_{nk} \psi_k^*$$

$$= \sum \psi_m(z) A_{mn} \left[\int \frac{d^2 w}{\theta \pi} \psi_n^*(z + w)\psi_r(z + w) \right] B_{rk} \psi_k^*(z)$$

$$= \int \frac{d^2 w}{\theta \pi} \, e^{-w\bar{w}/\theta} \, A(z, \bar{z} + \bar{w}) \, B(z + w, \bar{z}) \tag{3.10}$$

$$\equiv (A) * (B)$$

$$= (A)(B) + \theta \frac{\partial(A)}{\partial \bar{z}} \frac{\partial(B)}{\partial z} + \cdots. \tag{3.11}$$

Functions on $M = \mathbf{C}$, under the star product, form an associative but noncommutative algebra. As θ becomes small, we may approximate the star product by the first two terms, giving

$$([A, B]) = (A) * (B) - (B) * (A) = \theta \left(\frac{\partial(A)}{\partial \bar{z}} \frac{\partial(B)}{\partial z} - \frac{\partial(B)}{\partial \bar{z}} \frac{\partial(A)}{\partial z} \right). \tag{3.12}$$

The r.h.s. is the Poisson bracket of A and B, and this relation is essentially the standard result that the commutators of operators tend to (i times) the Poisson bracket of the corresponding functions (symbols) for small values of the deformation parameter. In particular, we find

$$Z * \bar{Z} = z\bar{z} + \cdots$$

$$\bar{Z} * Z = \bar{z}z + \theta + \cdots \tag{3.13}$$

$$Z * \bar{Z} - \bar{Z} * Z = \theta + \cdots.$$

We can interpret $Z, \bar{Z}$ as the coordinates of the space; they are noncommuting. The noncommutativity is characterized by the parameter θ, as we can use the equations given above in terms of symbols to obtain the small θ-limit.

These considerations can be generalized in an obvious way to $M = \mathbf{C}^k$.

3.1.3 Complex Projective Space CP^k

We shall now discuss the fuzzy version of $\mathbf{CP}^k$. Unlike the case of flat space, we will get a finite number of states for $\mathbf{CP}^k$, say, N, so this will be a truly fuzzy space, rather than just noncommutative. The continuous manifold $\mathbf{CP}^k$ can be obtained as $N \to \infty$. In the previous discussion, we started with continuous $\mathbf{R}^{2k}$, set up the quantum theory for the action (3.1), and the resulting Hilbert space could be interpreted as giving the noncommutative version of $\mathbf{R}^{2k}$. We can follow the same strategy for $\mathbf{CP}^k$. In fact, we can adapt the coherent state construction to obtain the fuzzy version of $\mathbf{CP}^k$. For a more detailed and group theoretic approach, see [6–8].

Continuous $\mathbf{CP}^k$ is defined as the set of $k+1$ complex variables Z^α, with the identification of Z^α and λZ^α where λ is any nonzero complex number, i.e., we start with $\mathbf{C}^{k+1}$ and make the identification $Z^\alpha \sim \lambda Z^\alpha$, $\lambda \in \mathbf{C} - \{0\}$. Based on the fact that there is natural action of $SU(k+1)$ on Z^α given by

$$Z^\alpha \longrightarrow Z'^\alpha = g^\alpha{}_\beta \, Z^\beta, \quad g \in SU(k+1), \tag{3.14}$$

we can show that $\mathbf{CP}^k$ can be obtained as the coset

$$\mathbf{CP}^k = \frac{SU(k+1)}{U(k)}. \tag{3.15}$$

In fact, we may take (3.15) as the definition of $\mathbf{CP}^k$. The division by $U(k)$ suggests that we can obtain $\mathbf{CP}^k$ by considering a "gauged" version of the action (3.1), where the gauge group is taken to be $U(k)$. Replacing the time-derivative by the covariant derivative, the action becomes

$$S = -i \int dt \, \bar{Z}_\alpha (\partial_0 Z^\alpha - iA_0 Z^\alpha) - n \int dt \, A_0, \tag{3.16}$$

where we have also included a term for the gauge field. (From now on, we will not display θ explicitly.) This action is easily checked to be invariant under the $U(1)$ gauge transformation

$$Z^\alpha \to e^{i\varphi} Z^\alpha, \qquad A_0 \to A_0 + \partial_0 \varphi. \tag{3.17}$$

The pure gauge field part of the action is the one-dimensional Chern–Simons term. The coefficient n has to be quantized, following the usual arguments. For example, we can consider the transformation where $\varphi(t)$ obeys $\varphi(\infty) - \varphi(-\infty) = 2\pi$. The action then changes by $-2\pi n$, and since $\exp(iS)$ has to be single-valued to have a well-defined quantum theory, n has to be an integer.

The variation of the action with respect to A_0 leads to the Gauss law for the theory,

$$\bar{Z}_\alpha \, Z^\alpha - n \approx 0\,, \tag{3.18}$$

where the weak equality (denoted by $\approx$) indicates, as usual, that this condition is to be imposed as a constraint. This is a first-class constraint in the Dirac sense, and hence it removes two degrees of freedom. Thus, from $\mathbf{C}^{k+1}$, we go to a space with k complex dimensions. Given the $U(k)$ invariance, this can be identified as $\mathbf{CP}^k$.

The time-evolution of Z^α is again trivial and we are led to the complete characterization of the theory by the Hilbert space, which must be obtained taking account of the constraint (3.18). In the quantum theory, the allowed physical states must be annihilated by the Gauss law. Using the realization of the Z^α, $\bar{Z}_\alpha$ given in (3.2), this becomes

$$\left(z^\alpha \frac{\partial}{\partial z^\alpha} - n \right) f(z) = 0\,. \tag{3.19}$$

Thus, the allowed functions $f(z)$ must have n powers of z's. They are of the form

$$f(z) = \frac{1}{\sqrt{n!}} \, z^{\alpha_1} z^{\alpha_2} \cdots z^{\alpha_n}\,. \tag{3.20}$$

There are $N = (n+k)!/n!k!$ independent functions. The Hilbert space of such functions form the carrier space of a completely symmetric rank n irreducible representation of $SU(k+1)$. A simple parameterization in terms of local coordinates on $\mathbf{CP}^k$ can be obtained by writing $z^{k+1} = \lambda$, $z^i = \lambda \xi^i$, where $\xi^i = z^i/z^{k+1} = z^i/\lambda$, for $i = 1, 2, \ldots, k$. Correspondingly, the wave functions have the form $f(z) = \lambda^n \, f(\xi)$. The inner product for two such wave functions can be obtained from the inner product (3.6). We get

$$
\begin{aligned}
\langle f | h \rangle &= \frac{1}{n!} \int \frac{\mathrm{d}^2 \lambda}{\pi} \prod_i \frac{\mathrm{d}^2 \xi_i}{\pi} \; e^{-\lambda\bar{\lambda}(1+\bar{\xi}\cdot\xi)} \, (\lambda\bar{\lambda})^{k+n} \; \bar{f} \, h \\
&= \frac{(n+k)!}{n!\, k!} \int \left[\frac{k! \prod \mathrm{d}^2 \xi_i}{\pi^k (1+\bar{\xi}\cdot\xi)^{k+1}} \right] \frac{\overline{f(\xi)}}{(1+\bar{\xi}\cdot\xi)^{n/2}} \, \frac{h(\xi)}{(1+\bar{\xi}\cdot\xi)^{n/2}} \\
&= N \int \mathrm{d}\mu(\mathbf{CP}^k) \, \frac{\overline{f(\xi)}}{(1+\bar{\xi}\cdot\xi)^{n/2}} \, \frac{h(\xi)}{(1+\bar{\xi}\cdot\xi)^{n/2}}\,. \tag{3.21}
\end{aligned}
$$

Here, N is the dimension of the Hilbert space and $\mathrm{d}\mu(\mathbf{CP}^k)$ is the standard volume element for $\mathbf{CP}^k$ in the local coordinates ξ^i, $\bar{\xi}_i$. Now, an $SU(k+1)$ matrix g can be parameterized in such a way that the last column g^α_{k+1} is given in terms of ξ^i, and the factor $\sqrt{1 + \bar{\xi}\cdot\xi}$, as

$$
g = \begin{bmatrix} \cdot & \cdot & \cdot & \cdot & \xi^1 \\ \cdot & \cdot & \cdot & \cdot & \xi^2 \\ \cdot & \cdot & \cdot & \cdot & \cdot \\ \cdot & \cdot & \cdot & \cdot & \xi^k \\ \cdot & \cdot & \cdot & \cdot & 1 \end{bmatrix} \frac{1}{\sqrt{1+\bar{\xi}\cdot\xi}}\,. \tag{3.22}
$$

The states f are thus of the form

$$\frac{f(\xi)}{(1 + \bar{\xi} \cdot \xi)^{n/2}} = g^{\alpha_1}_{k+1} g^{\alpha_2}_{k+1} \cdots g^{\alpha_n}_{k+1}. \tag{3.23}$$

Let $|n, r\rangle$, $r = 1, 2, \ldots, N$, denote the states of the rank n symmetric representation of $SU(k + 1)$. Then the Wigner $\mathcal{D}$-function corresponding to g in this representation is defined by $\mathcal{D}^{(n)}_{rs}(g) = \langle n, r | \hat{g} | n, s \rangle$; it is the matrix representative of the group element g in this representation. One can then check easily that

$$g^{\alpha_1}_{k+1} g^{\alpha_2}_{k+1} \cdots g^{\alpha_n}_{k+1} = \mathcal{D}^{(n)}_{r,w} = \langle n, r | \hat{g} | n, w \rangle, \tag{3.24}$$

where the state $|n, w\rangle$ is the lowest weight state obeying

$$T_{k^2+2k} |n, w\rangle = -n \frac{k}{\sqrt{2k(k+1)}} |n, w\rangle$$
$$T_a |n, w\rangle = 0. \tag{3.25}$$

Here, T_a are the generators of the $SU(k)$ subalgebra and T_{k^2+2k} is the generator of the $U(1)$ algebra, both for the subgroup $U(k)$ of $SU(k + 1)$. The normalized wave functions for the basis states are thus $\Psi_r = \sqrt{N} \, \mathcal{D}^{(n)}_{r,w}(g)$. Notice that $\mathcal{D}^{(n)}_{n,w}$ are invariant under right translations of g by $SU(k)$ transformations, and under the $U(1)$ defined by T_{k^2+2k} they have a definite charge n, up to the k-dependent normalization factor. Since they are not $U(1)$ invariant, they are really not functions on $\mathbf{CP}^k$, but sections of a line bundle on $SU(k+1)/U(k)$, the rank of the bundle being n. This is exactly what we should expect for quantization of $\mathbf{CP}^k$ since this space is given as $SU(k + 1)/U(k)$.

3.1.4 Star Products for Fuzzy $\mathbf{CP}^k$

As for the flat case, we can construct a star product for functions on $\mathbf{CP}^k$ which captures the noncommutative algebra of functions [8, 9]. First, we need to establish some notation. Let t_A denote the generators of the Lie algebra of $SU(k + 1)$, realized as $(k + 1 \times k + 1)$-matrices. (The T's given in (3.25) correspond to the generators of $U(k) \subset SU(k + 1)$ in the rank n symmetric representation; they are the rank n representatives of t_{k^2+2k} and t_a. The remaining generators are of two types, t_{-i}, $i = 1, 2, \ldots, k$, which are lowering operators and t_{+i} which are raising operators.) Left and right translation operators on g are defined by the equations

$$\hat{L}_A \, g = t_A \, g, \qquad \hat{R}_A \, g = g \, t_A \tag{3.26}$$

If g is parameterized by φ^A, some of which are the ξ's, then we write

$$g^{-1} dg = -i t_A E^A_{\ B} \, d\varphi^B, \qquad dg g^{-1} = -i t_A \tilde{E}^A_{\ B} \, d\varphi^B \tag{3.27}$$

The operators $\hat{L}_A$ and $\hat{R}_A$ are then realized as differential operators

$$\hat{L}_A = i(\tilde{E}^{-1})^B{}_A \frac{\partial}{\partial \varphi^B}\,, \qquad \hat{R}_A = i(E^{-1})^B{}_A \frac{\partial}{\partial \varphi^B}\,. \qquad (3.28)$$

The state $|n, w\rangle$, used in $\mathcal{D}^{(n)}_{rw}(g) = \langle n, r|\hat{g}|n, w\rangle$, is the lowest weight state, which means that we have the condition

$$\hat{R}_{-i}\mathcal{D}^{(n)}_{r,w} = 0\,. \qquad (3.29)$$

This is essentially a holomorphicity condition. Notice that $f(\xi)$ are holomorphic in the ξ's; the $\mathcal{D}^{(n)}_{rw}$ have an additional factor $(1 + \bar{\xi} \cdot \xi)^{-n/2}$, which can be interpreted as due to the nonzero connection in $\hat{R}_{-i}$, ultimately due to the nonzero curvature of the bundle. Equation (3.29) tells us that $\mathcal{D}^{(n)}_{r,w}$ are sections of a rank n *holomorphic* line bundle.

We define the symbol corresponding to a matrix A_{ms} as the function

$$A(g) = A(\xi, \bar{\xi}) = \sum_{ms} \mathcal{D}^{(n)}_{m,w}(g) A_{ms} \mathcal{D}^{*(n)}_{s,w}(g)$$

$$= \langle w|\hat{g}^T \hat{A}\hat{g}^*|w\rangle\,. \qquad (3.30)$$

The symbol corresponding to the product of two matrices A and B can be simplified as follows.

$$(AB) = \sum_r A_{mr} B_{rs} \mathcal{D}^{(n)}_{m,w}(g) \mathcal{D}^{*(n)}_{s,w}(g)$$

$$= \sum_{rr'p} \mathcal{D}^{(n)}_{m,w}(g)\, A_{mr}\, \mathcal{D}^{*(n)}_{r,p}(g) \mathcal{D}^{(n)}_{r',p}(g)\, B_{r's}\, \mathcal{D}^{*(n)}_{s,w}(g)\,, \qquad (3.31)$$

where we use the fact that $g^* g^T = 1$, which reads in the rank n symmetric representation as $\delta_{rr'} = \sum_p \mathcal{D}^{*(n)}_{r,p}(g)\mathcal{D}^{(n)}_{r',p}(g)$. In the sum over p on the r.h.s. of (3.31), the term with $p = -n$ (corresponding to the lowest weight state $|n, w\rangle$) gives the product of the symbols for A and B. The terms with $p > -n$ may be written in terms of powers of the raising operators $R_{+1}, R_{+2}, \ldots, R_{+k}$, as

$$\mathcal{D}^{(n)}_{r,p}(g) = \left[\frac{(n-s)!}{n! i_1! i_2! \cdots i_k!}\right]^{\frac{1}{2}} \hat{R}^{i_1}_{+1} \hat{R}^{i_2}_{+2} \cdots \hat{R}^{i_k}_{+k}\, \mathcal{D}^{(n)}_{r,w}(g)\,. \qquad (3.32)$$

Here, $s = i_1 + i_2 + \cdots + i_k$ and the eigenvalue for the $U(1)$ generator T_{k^2+2k} for the state $|n, p\rangle$ is $(-nk + sk + s)/\sqrt{2k(k+1)}$.

We also get

$$\left[\hat{R}_{+i}\mathcal{D}^{(n)}_{r',w}(g)\right] B_{r's}\mathcal{D}^{*(n)}_{s,w}(g) = \left[\hat{R}_{+i}\mathcal{D}^{(n)}_{r',w} B_{r's}\mathcal{D}^{*(n)}_{s,w}(g)\right] = \hat{R}_{+i}B(g)\,, \qquad (3.33)$$

where we used the fact that $\hat{R}_{+i}\mathcal{D}^{*(n)}_{s,w} = 0$. Keeping in mind that $\hat{R}^*_+ = -\hat{R}_-$, (3.31)–(3.33) combine to give

$$(AB)(g) = \sum_s (-1)^s \left[\frac{(n-s)!}{n!s!} \right] \sum_{i_1 + \cdots + i_k = s}^{n} \frac{s!}{i_1! i_2! \cdots i_k!}$$

$$\times \hat{R}_{-1}^{i_1} \hat{R}_{-2}^{i_2} \cdots \hat{R}_{-k}^{i_k} A(g) \; \hat{R}_{+1}^{i_1} \hat{R}_{+2}^{i_2} \cdots \hat{R}_{+k}^{i_k} B(g)$$

$$\equiv A(g) * B(g). \tag{3.34}$$

This expression gives the star product for functions on $\mathbf{CP}^k$. As expected, the first term of the sum on the r.h.s. gives the ordinary product $A(g)B(g)$, successive terms involve derivatives and are down by powers of n, as $n \to \infty$. Since the dimension of the matrices is given by $N = (n+k)!/n!k!$, large n is what we need for the limit of the continuous manifold, and the star product, as written here, is suitable for extracting this limit for various quantities. For example, for the symbol corresponding to the commutator of A, B, we have

$$([A, B])(g) = -\frac{1}{n} \sum_{i=1}^{k} (\hat{R}_{-i} A \, \hat{R}_{+i} B - \hat{R}_{-i} B \, \hat{R}_{+i} A) + \mathcal{O}(1/n^2)$$

$$= \frac{i}{n} \{A, B\} + \mathcal{O}(1/n^2). \tag{3.35}$$

The term involving the action of $\hat{R}$'s on the functions can indeed be verified to be the Poisson bracket on $\mathbf{CP}^k$. Equation (3.35) is again the general correspondence of commutators and Poisson brackets, here realized for the specific case of $\mathbf{CP}^k$.

We also note that traces of matrices can be converted to phase space integrals. For a single matrix A, and for the product of two matrices A, B, we find

$$\mathrm{Tr} A = \sum_m A_{mm} = N \int d\mu(g) \mathcal{D}_{m,w}^{(n)} A_{mm'} \, \mathcal{D}_{m',w}^{*(n)}$$

$$= N \int d\mu(g) \, A(g)$$

$$\mathrm{Tr} AB = N \int d\mu(g) \, A(g) * B(g). \tag{3.36}$$

3.1.5 The Large n-Limit of Matrices

Consider the symbol for the product $\hat{T}_B \hat{A}$, where $\hat{T}_B$ are the generators of $SU(k+1)$, viewed as linear operators on the states. We can simplify it along the following lines.

$$
\begin{aligned}
(\hat{T}_B \hat{A})_{rs} &= \langle r | \, \hat{g}^T \, \hat{T}_B \, \hat{A} \, \hat{g}^* \, | s \rangle \\
&= S_{BC} \, \langle r | \, \hat{T}_C \, \hat{g}^T \, \hat{A} \, \hat{g}^* \, | s \rangle \\
&= S_{Ba}(T_a)_{rp} \, \langle p | \, \hat{g}^T \, \hat{A} \, \hat{g}^* \, | s \rangle + S_{B+i} \, \langle r | \, \hat{T}_{-i} \, \hat{g}^T \, \hat{A} \, \hat{g}^* \, | s \rangle \\
&\quad + S_{B \; k^2+2k} \, \langle r | \, \hat{T}_{k^2+2k} \, \hat{g}^T \, \hat{A} \, \hat{g}^* \, | s \rangle \\
&= \mathcal{L}_B \, \langle r | \, \hat{g}^T \, \hat{A} \, \hat{g}^* \, | s \rangle \\
&= \mathcal{L}_B \, A(g)_{rs} \,,
\end{aligned}
\tag{3.37}
$$

where we have used $\hat{g}^T \hat{T}_B \hat{g}^* = S_{BC}\hat{T}_C$, $S_{BC} = 2\mathrm{Tr}(g^T t_B g^* t_C)$. (Here, t_B, t_C and the trace are in the fundamental representation of $SU(k+1)$.) We have also used the fact that the states $|r\rangle$, $|s\rangle$ are $SU(k)$-invariant. (They are both equal to $|n, w\rangle$, but we will make this identification only after one more step of simplification.) $\mathcal{L}_B$ is defined as

$$
\mathcal{L}_B = -\frac{nk}{\sqrt{2k(k+1)}} S_{B \; k^2+2k} + S_{B+i}\hat{\tilde{R}}_{-i}
\tag{3.38}
$$

and $\hat{\tilde{R}}_{-i}$ is a differential operator defined by $\hat{\tilde{R}}_{-i}g^T = T_{-i}g^T$; it can be written in terms of $\hat{R}_{-i}$, but the precise formula is not needed here.

By taking $\hat{A}$ itself as a product of $\hat{T}$'s, we can iterate this calculation and obtain the symbol for any product of $\hat{T}$'s as

$$
(\hat{T}_A \hat{T}_B \cdots \hat{T}_M) = \mathcal{L}_A \mathcal{L}_B \cdots \mathcal{L}_M \cdot 1 \,,
\tag{3.39}
$$

where we have now set $|n, r\rangle = |n, s\rangle = |n, w\rangle$.

A function on fuzzy $\mathbf{CP}^k$ is an $N \times N$-matrix. It can be written as a linear combination of products of $\hat{T}$'s, and by using the above formula, we can obtain its large n limit. When n becomes very large, the term that dominates in $\mathcal{L}_A$ is $S_{A \; k^2+2k}$. We then see that for any matrix function we have the relation, $F(\hat{T}_A) \approx F(S_{A \; k^2+2k})$.

We are now in position to define a set of "coordinates" X_A by

$$
X_A = -\frac{1}{\sqrt{C_2(k+1, n)}} T_A \,,
\tag{3.40}
$$

where T_A is the matrix corresponding to $\hat{T}_A$ and

$$
C_2(k+1, n) = \frac{n^2 k^2}{2k(k+1)} + \frac{nk}{2}
\tag{3.41}
$$

is the value of the quadratic Casimir for the symmetric rank n representation. The coordinates X_A are $N \times N$-matrices and can be taken as the coordinates of fuzzy $\mathbf{CP}^k$, embedded in $\mathbf{R}^{k^2+2k}$. In the large n limit, we evidently have $X_A \approx S_{A \; k^2+2k} = 2\mathrm{Tr}(g^T t_A g^* t_{k^2+2k})$. From the definition, we can see that $S_{A \; k^2+2k}$ obey algebraic constraints which can be verified to be the correct ones for describing $\mathbf{CP}^k$ as embedded in $\mathbf{R}^{k^2+2k}$ [10].

3.2 Noncommutative Plane, Fuzzy CP^1, CP^2, etc.

The noncommutative plane has already been described. The basic commutation rules are given by (3.2), with the indices taking only one value, 1. The star product is given by (3.10). While the coherent state basis is very ideal for considering the commutative limit $\theta \to 0$, for many purposes, it is easy enough to deal with the representation of $Z, \bar{Z}$ as infinite-dimensional matrices. In fact, one can also use real coordinates and characterize them by the commutation rules

$$[X_i, X_j] = i\theta\, \epsilon_{ij}\,. \tag{3.42}$$

More generally, one may consider $\mathbf{R}^{2k}$, with the commutation rules

$$[X_i, X_j] = i\, \theta_{ij}\,, \tag{3.43}$$

where the constant matrix θ_{ij} characterizes the noncommutativity.

Fuzzy $\mathbf{CP}^1$ is the same as the fuzzy two-sphere and has been studied for a long time [11]. It can be treated as the special case $k = 1$ of our analysis. The Hilbert space corresponds to representations of $SU(2)$, and they are given by the standard angular momentum theory. Representations are labeled by the maximal angular momentum $j = \frac{n}{2}$, with $N = 2j+1 = n+1$. The generators are the angular momentum matrices, and the coordinates of fuzzy S^2 are given by $X_i = J_i/\sqrt{j(j+1)}$, as in (3.40). These coordinate matrices obey the commutation rule

$$[X_i, X_j] = \frac{i}{\sqrt{j(j+1)}}\, \epsilon_{ijk} X_k\,. \tag{3.44}$$

We get commuting coordinates only at large n.

If g is an element of $SU(2)$ considered as a 2×2-matrix, we can parameterize it, apart from an overall $U(1)$ factor and along the lines of (3.22), as

$$g = \frac{1}{\sqrt{(1 + \bar{\xi}\xi)}} \begin{bmatrix} -1 & \xi \\ \bar{\xi} & 1 \end{bmatrix}\,. \tag{3.45}$$

The large n limit of the coordinates is given by $X_i \approx S_{i3}(g)$, which can be worked out as

$$S_{13} = -\frac{\xi + \bar{\xi}}{(1 + \xi\bar{\xi})}\,, \quad S_{23} = -i\,\frac{\xi - \bar{\xi}}{(1 + \xi\bar{\xi})}\,, \quad S_{33} = \frac{\xi\bar{\xi} - 1}{\xi\bar{\xi} + 1}\,. \tag{3.46}$$

The quantities S_{i3} obey the condition $S_{i3}S_{i3} = 1$ corresponding to a unit two-sphere embedded in $\mathbf{R}^3$; ξ, $\bar{\xi}$ are the local complex coordinates for the sphere. The matrix coordinates obey the condition $X_i X_i = 1$. Thus, we may regard them as giving the fuzzy two-sphere, which approximates to the continuous two-sphere as $n \to \infty$.

We can also study functions on fuzzy S^2, which are given as $N \times N$-matrices. At the matrix level, there are $N^2 = (n+1)^2$ independent "functions."

A basis for them is given by $\mathbf{1}$, X_i, $X_{(i}X_{j)}$, etc., where $X_{(i}X_{j)}$ denotes the symmetric part of the product X_iX_j with all contractions of indices i,j removed; i.e., $X_{(i}X_{j)} = \frac{1}{2}(X_iX_j + X_jX_i) - \frac{1}{3}\delta_{ij}\mathbf{1}$. Since we have finite-dimensional matrices, the last independent function corresponds to the symmetric n-fold product of X_i's with all contractions removed.

On the smooth S^2, a basis for functions is given by the spherical harmonics, labeled by the integer $l = 0, 1, 2, \ldots$. They are given by the products of S_{i3} with all contractions of indices removed. There are $(2l + 1)$ such functions for each value of l. If we consider a truncated set of functions with a maximal value of l equal to n, the number of functions is $\sum_0^n(2l+1) = (n+1)^2$. Notice that this number coincides with the number of "functions" at the matrix level. There is one-to-one correspondence with the spherical harmonics, for $l = 0, 1, 2$, etc., up to $l = n$. Further, by using the relation $X_i \approx S_{i3}$, we can see that the matrix functions, $\mathbf{1}$, X_i, $X_{(i}X_{j)}$, etc., in the large n limit, approximate to the spherical harmonics. The set of functions at the matrix level go over to the set of functions on the smooth S^2 as $n \to \infty$. Fuzzy S^2 may thus be viewed as a regularized version of the smooth S^2 where we impose a cut-off on the number of modes of a function; n is the regulator or cut-off parameter.

Fuzzy $\mathbf{CP}^2$ is the case $k = 2$ of our general analysis. The coordinates are given by

$$X_A = -\frac{3}{\sqrt{n(n+3)}}T_A. \tag{3.47}$$

The large n limit of the coordinates X_A are $S_{A8} = 2\text{Tr}(g^T t_A g^* t_8)$. In this limit, the coordinates obey the condition

$$X_A X_A = 1$$
$$d_{ABC}X_B X_C = -\frac{1}{\sqrt{3}}X_C, \tag{3.48}$$

where $d_{ABC} = 2\text{Tr}\, t_A(t_B t_C + t_C t_B)$. These conditions are known to be the equations for $\mathbf{CP}^2$ as embedded in $\mathbf{R}^8$. Thus, our definition of fuzzy $\mathbf{CP}^2$ does approximate to the smooth $\mathbf{CP}^2$ in the large n limit. Equation (3.48) can also be imposed at the level of matrices to get a purely matrix-level definition of fuzzy $\mathbf{CP}^2$ [9, 10].

The dimension of the Hilbert space is given by $N = \frac{1}{2}(n+1)(n+2)$. Matrix functions are $N \times N$-matrices; a basis for them is given by products of the T's with up to $N - 1$ factors. There are N^2 independent functions possible. On the smooth $\mathbf{CP}^2$, a basis of functions is given by products of the form $\bar{u}_{\beta_1}\bar{u}_{\beta_2}\cdots\bar{u}_{\beta_l}u^{\alpha_1}u^{\alpha_2}\cdots u^{\alpha_l}$, where $u^\alpha = g^\alpha_3$. The number of such functions, for a given value of l, is

$$\left[\frac{1}{2}(l+1)(l+2)\right]^2 - \left[\frac{1}{2}l(l+1)\right]^2 = (l+1)^3 \tag{3.49}$$

(All traces for these functions correspond to lower ones and can be removed from the counting at the level l.) If we consider a truncated set of functions, with values of l going up to n, the number of independent functions will be

$$\sum_0^n (l+1)^3 = \frac{1}{4}(n+1)^2(n+2)^2 = N^2. \qquad (3.50)$$

It is thus possible to consider the fuzzy $\mathbf{CP}^2$ as a regularization of the smooth $\mathbf{CP}^2$ with a cut-off on the number of modes of a function. Since any matrix function can be written as a sum of products of $\hat{T}$'s, the corresponding large n limit has a sum of products of S_{A8}'s. The independent basis functions are thus given by representations of $SU(3)$ obtained from reducing symmetric products of the adjoint representation with itself. These are exactly what we expect based on the fact the smooth $\mathbf{CP}^2$ is given by the embedding conditions (3.48). The algebra of matrix functions for the fuzzy $\mathbf{CP}^2$, as we have constructed it, does go over to the algebra of functions on the smooth $\mathbf{CP}^2$.

Since the fuzzy spaces, the fuzzy $\mathbf{CP}^k$ in particular, can be regarded as a regularization of the smooth $\mathbf{CP}^k$ with a cut-off on the number of modes of a function, they can be used for regularization of field theories, in much the same way that lattice regularization of field theories is carried out. There are some interesting features or fuzzy regularization; for example, it may be possible to evade fermion doubling problem on the lattice [5].

3.3 Fields on Fuzzy Spaces, Schrödinger Equation

A scalar field on a fuzzy space can be written as $\Phi(X)$, indicating that it is a function of the coordinate matrices X_A. Thus, Φ is an $N \times N$-matrix. Further, (3.35) tells us that

$$[T_A, \Phi] \approx -\frac{i}{n}\frac{nk}{\sqrt{2k(k+1)}}\{S_{A\ k^2+2k}, \Phi\}$$

$$\equiv -iD_A\Phi. \qquad (3.51)$$

D_A, as defined by this equation, are the derivative operators on the space of interest. For example, for the fuzzy S^2, they are given by

$$D_1 = \frac{1}{2}\left(\bar{\xi}^2\partial_{\bar{\xi}} + \partial_\xi - \xi^2\partial_\xi - \partial_{\bar{\xi}}\right)$$

$$D_2 = -\frac{i}{2}\left(\bar{\xi}^2\partial_{\bar{\xi}} + \partial_\xi + \xi^2\partial_\xi + \partial_{\bar{\xi}}\right) \qquad (3.52)$$

$$D_3 = \bar{\xi}\partial_{\bar{\xi}} - \xi\partial_\xi.$$

These obey the $SU(2)$ algebra, $[D_A, D_B] = i\epsilon_{ABC}D_C$. They are the translation operators on the two-sphere and correspond to the three isometry transformations.

These equations show that we can define the derivative, at the matrix level, as the commutator $i[T_A, \Phi]$, which is the adjoint action of T_A on Φ. The Laplacian on Φ is then given by $-\Delta \cdot \Phi = [T_A, [T_A, \Phi]]$. The Euclidean action for a scalar field can be taken as

$$S = \frac{1}{N} \text{Tr}[\Phi^\dagger [T_A, [T_A, \Phi]] + V(\Phi)], \qquad (3.53)$$

where $V(\Phi)$ is a potential term; it does not involve derivatives.

The identification of derivatives also leads naturally to gauge fields. We introduce a gauge field $\mathcal{A}_A$ by defining the covariant derivative as

$$-i\mathcal{D}_A\Phi = [T_A, \Phi] + \mathcal{A}_A\Phi, \qquad (3.54)$$

where $\mathcal{A}_A$ is a set of hermitian matrices. In the absence of the gauge field, we have the commutation rules $[T_A, T_B] = if_{ABC}T_C$. The field strength tensor $\mathcal{F}_{AB}$, which is the deviation from this algebra, is thus given by

$$-i\mathcal{F}_{AB} = [T_A + \mathcal{A}_A, T_B + \mathcal{A}_B] - if_{ABC}(T_C + \mathcal{A}_C). \qquad (3.55)$$

The action for Yang–Mills theory on a fuzzy space is then given by

$$S = \frac{1}{N} \text{Tr}\left[\frac{1}{4} \mathcal{F}_{AB}\mathcal{F}_{AB}\right]. \qquad (3.56)$$

The quantum theory of these fields can be defined by the functional integral over actions such as (3.53) and (3.56). Perturbation theory, Feynman diagrams, etc., can be worked out. Our main focus will be on particle dynamics, so we will not do this here. However, some of the relevant literature can be traced from [3, 5, 12].

One can also write down the Schrödinger equation for particle quantum mechanics on a fuzzy space [13]. The wave function $\Psi(X)$ is matrix and its derivative is given by $-i\mathcal{D}_A\Psi = [T_A, \Psi]$. Coupling to an external potential may be taken to be of the form $V(X)\Psi$. The Schrödinger equation is then given by

$$i\frac{\partial \Psi}{\partial t} + \mathcal{D}_A(\mathcal{D}_A\Psi) - V\Psi = 0. \qquad (3.57)$$

When it comes to gauge fields, there is a slight subtlety. The covariant derivative is of the form (3.54). To distinguish the action of the gauge field from the potential V, for the covariant derivative we may use the definition $-i\mathcal{D}_A\Psi = [T_A, \Psi] + \Psi\mathcal{A}_A$. (One could also change the action of the potential.) The Schrödinger equation retains the usual form,

$$i\mathcal{D}_0\Psi + \frac{1}{2m} \mathcal{D}_A(\mathcal{D}_A\Psi) - V\Psi = 0. \qquad (3.58)$$

3.4 The Landau Problem on R^2_{NC} and S^2_F

As a simple example of the application of the ideas given above, we shall now work out the quantum mechanics of a charged particle in a magnetic field on the fuzzy two-plane [13, 14]. This is the fuzzy version of the classic Landau problem. We shall also include an oscillator potential to include the case of an ordinary potential as well. At the operator level, the inclusion of a background magnetic field is easily achieved by changing the commutation rules for the momenta. The modified algebra of observables is given by

$$[X_1, X_2] = i\,\theta$$
$$[X_i, P_j] = i\,\delta_{ij} \tag{3.59}$$
$$[P_1, P_2] = i\,B\,,$$

where $i, j = 1, 2$, and B is the magnetic field. The Hamiltonian may be taken as

$$H = \frac{1}{2}\left[P_1^2 + P_2^2 + \omega^2(X_1^2 + X_2^2)\right]\,. \tag{3.60}$$

We have chosen the isotropic oscillator (with frequency ω), and H is invariant under rotations. The form of various operators can be slightly different from the usual ones because of the noncommutativity of the coordinates. The angular momentum is given by

$$L = \frac{1}{1-\theta B}\left[X_1 P_2 - X_2 P_1 + \frac{B}{2}(X_1^2 + X_2^2) + \frac{\theta}{2}(P_1^2 + P_2^2)\right] \tag{3.61}$$

L commutes with H, as can be checked easily.

The strategy for solving this problem involves expressing X_i, P_i in terms of a usual canonical set, so that thereafter, it can be treated as an ordinary quantum mechanical system. This change of variables will be different for $B < 1/\theta$ and for $B > 1/\theta$. For $B < 1/\theta$, we define a change of variables

$$X_1 = l\alpha_1\,, \qquad P_1 = \frac{1}{l}\beta_1 + q\alpha_2$$
$$\tag{3.62}$$
$$X_2 = l\beta_1\,, \qquad P_2 = \frac{1}{l}\alpha_1 - q\beta_2\,,$$

where $l^2 = \theta$ and $q^2 = (1 - B\theta)/\theta$. α_i, β_i form a standard set of canonical variables, with

$$[\alpha_i, \alpha_j] = 0$$
$$[\alpha_i, \beta_j] = i\,\delta_{ij} \tag{3.63}$$
$$[\beta_i, \beta_j] = 0\,.$$

The Hamiltonian is now given by

$$H = \frac{1}{2}\left[\left(\omega^2 l^2 + \frac{1}{l^2}\right)(\alpha_1^2 + \beta_1^2) + q^2(\alpha_2^2 + \beta_2^2) + \frac{2q}{l}(\alpha_1\beta_2 + \alpha_2\beta_1)\right]\,. \tag{3.64}$$

We can eliminate the mixing of the two sets of variables in the last term, and diagonalize H, by making a Bogoliubov transformation which will express α_i, β_i in terms of a canonical set q_i, p_i as

$$\begin{pmatrix} \alpha_1 \\ \alpha_2 \\ \beta_1 \\ \beta_2 \end{pmatrix} = \cosh\lambda \begin{pmatrix} q_1 \\ q_2 \\ p_1 \\ p_2 \end{pmatrix} + \sinh\lambda \begin{pmatrix} p_2 \\ p_1 \\ q_2 \\ q_1 \end{pmatrix}. \tag{3.65}$$

The Hamiltonian can be diagonalized by the choice

$$\tanh 2\lambda = -\frac{2ql}{1 + \omega^2 l^4 + q^2 l^2} \tag{3.66}$$

and is given by

$$H = \frac{1}{2}\left[\Omega_+ \, (p_1^2 + q_1^2) + \Omega_-(p_2^2 + q_2^2)\right] \tag{3.67}$$

with

$$\Omega_\pm = \frac{1}{2}\sqrt{(\omega^2\theta - B)^2 + 4\omega^2} \pm \frac{1}{2}(\omega^2\theta + B). \tag{3.68}$$

From (3.67), we see that the problem is equivalent to that of two harmonic oscillators with frequencies Ω_+ and Ω_-.

The case of $B > 1/\theta$ can be treated in a similar way. We again have two oscillators, with the Hamiltonian (3.67), but with frequencies given as

$$\Omega_\pm = \pm\frac{1}{2}\sqrt{(\omega^2\theta - B)^2 + 4\omega^2} + \frac{1}{2}(\omega^2\theta + B). \tag{3.69}$$

Notice that $B\theta = 1$ is a special value, for both regions of B, with one of the frequencies becoming zero. The symplectic two-form which leads to the commutation rules (3.59) is given by

$$\Omega = \frac{1}{1 - B\theta}\,(dP_1\,dX_1 + dP_2\,dX_2 + \theta dP_1\,dP_2 + BdX_1\,dX_2) \tag{3.70}$$

The phase space volume is given by

$$d\mu = \frac{1}{|1 - B\theta|}\,d^2X d^2P. \tag{3.71}$$

A semiclassical estimate of the number of states is given by the volume divided by $(2\pi)^2$. The formula (3.71) shows that the density of states diverges at $B\theta = 1$, again indicating that it is a special value.

The Landau problem on the fuzzy sphere can be formulated in a similar way. On the sphere, the translation operators are the angular momenta J_i and the algebra of observables is given by

$$\begin{aligned} [X_i, X_j] &= 0 \\ [J_i, X_j] &= i\epsilon_{ijk}X_k \\ [J_i, \, J_j] &= i\epsilon_{ijk}J_k\,. \end{aligned} \tag{3.72}$$

X^2 commutes with all operators and its value can be fixed to be a^2, where a is the radius of the sphere. The other Casimir operator is $X \cdot J$; its value is written as $-a(n/2)$, where n must be an integer and gives the strength of the magnetic field; it is the charge of the monopole at the center of the sphere (if we think of it as being embedded in $\mathbf{R}^3$).

For the fuzzy case, the coordinates themselves are noncommuting and are given, up to normalization, by $SU(2)$ operators R_i as $X_i = aR_i/\sqrt{C_2}$. The algebra of observables becomes

$$
\begin{aligned}
[R_i, R_j] &= i\epsilon_{ijk}R_k \\
[J_i, R_j] &= i\epsilon_{ijk}R_k \\
[J_i, J_j] &= i\epsilon_{ijk}J_k \,.
\end{aligned}
\tag{3.73}
$$

The Casimir operators are now R^2 and $R \cdot J - \frac{1}{2}J^2$, the latter being related to the strength of the magnetic field. The algebra (3.73) can be realized by two independent $SU(2)$ algebras $\{R_i\}$ and $\{K_i\}$, with $J_i = R_i + K_i$. The two Casimirs are now R^2 and K^2, which we fix to the values $r(r+1)$ and $k(k+1)$, r, k being positive half-integers. The difference $k - r = n/2$. The limit of the smooth sphere is thus obtained by taking $k, r \to \infty$, with $k - r$ fixed. As the generalization of $P^2/2m$, we take the Hamiltonian as

$$
H = \frac{\gamma}{2a^2} \, J^2 \,,
\tag{3.74}
$$

where γ is some constant. The spectrum of the Hamiltonian is now easily calculated as

$$
E = \frac{\gamma}{2a^2} \, j(j+1) \,, \qquad j = \frac{|n|}{2}, \frac{|n|}{2} + 1, \ldots, j + k \,.
\tag{3.75}
$$

From the commutation rule for the coordinates $X_i = aR_i/\sqrt{r(r+1)}$, we may identify the noncommutativity parameter as $\theta \approx a^2/r$, for large r. The limit of this problem to the noncommutative plane can be obtained by taking r large, but keeping θ fixed. Naturally, this will require a large radius for the sphere. The strength of the magnetic field in the plane is related to n by $(1 - B\theta)n = 2Ba^2$. For more details, see [13]; also the Landau problem on general noncommutative Riemann surfaces has been analyzed, see [15].

3.5 Lowest Landau Level and Fuzzy Spaces

There is an interesting connection between the Landau problem on a smooth manifold M and the construction of the fuzzy version of M; we shall explain this now.

The splitting of Landau levels is controlled by the magnetic field and, if the field is sufficiently strong, transitions between levels are suppressed and the dynamics is restricted to one level, say, the lowest. The observables are given as hermitian operators on this subspace of the Hilbert space corresponding to the lowest Landau level; they can be obtained by projecting the full operators to this subspace. The commutation rules can change due to this projection. The position coordinates, for example, when projected to the lowest Landau level (or any other level), are no longer mutually commuting. The dynamics restricted to the lowest Landau level is thus dynamics on a noncommutative space. In fact, the Hilbert (sub)space of the lowest Landau level can be taken as the Hilbert space $\mathcal{H}_N$ used to define the fuzzy version of M. Thus the solution of the Landau problem on smooth M gives a construction of the fuzzy version of M.

We can see how this is realized explicitly by analyzing the two-sphere [16]. Since $S^2 = SU(2)/U(1)$, the wave functions can be obtained in terms of functions on the group $SU(2)$, i.e., in terms of the Wigner functions $\mathcal{D}_{rs}^{(j)}(g)$. We need two derivative operators which can be taken as two of the right translations of g, say, $R_\pm = R_1 \pm iR_2$. With the correct dimensions, the covariant derivatives can be written as

$$D_\pm = i \, \frac{R_\pm}{a} . \tag{3.76}$$

The $SU(2)$ commutation rule $[R_+, \ R_-] = 2R_3$, shows that the covariant derivatives do not commute and we may identify the value of R_3 as the field strength. In fact, comparing this commutation rule to $[D_+, D_-] = 2B$, we see that R_3 should be taken to be $-(n/2)$, where n is the monopole number, $n = 2Ba^2$. Thus, the wave functions on S^2 with the magnetic field background are of the form $\Psi_m \sim \mathcal{D}_{m,-\frac{n}{2}}^{(j)}(g)$.

The one-particle Hamiltonian is given by

$$H = -\frac{1}{4\mu}(D_+D_- + D_-D_+) = \frac{1}{2\mu a^2} \left(\sum_{A=1}^{3} R_A^2 - R_3^2 \right) , \tag{3.77}$$

where μ is the particle mass. The eigenvalue $-\frac{n}{2}$ must occur as one of the possible values for R_3, so that we can form $\mathcal{D}_{m,-\frac{n}{2}}^{(j)}(g)$. This means that j should be of the form $j = \frac{|n|}{2} + q$, $q = 0, 1, \ldots$. Since $R^2 = j(j+1)$, the energy eigenvalues are easily obtained as

$$E_q = \frac{1}{2\mu a^2} \left[\left(\frac{n}{2} + q \right) \left(\frac{n}{2} + q + 1 \right) - \frac{n^2}{4} \right] = \frac{B}{2\mu}(2q + 1) + \frac{q(q+1)}{2\mu a^2} . \tag{3.78}$$

The integer q is the Landau level index, $q = 0$ being the lowest energy state or the ground state. The gap between levels increases as B increases, and, in the limit of large magnetic fields, it is meaningful to restrict dynamics to one level, say the lowest, if the available excitation energies are small compared to $B/2\mu$. In this case, $j = \frac{|n|}{2}$, $R_3 = -\frac{n}{2}$, so that we have the lowest weight state for the right action of $SU(2)$, taking n to be positive. The condition for the lowest Landau level is $R_-\Psi = 0$.

The Hilbert space of the lowest Landau level is spanned by $\Psi_m \sim \mathcal{D}^{(\frac{n}{2})}_{m,-\frac{n}{2}}$. Notice that this is exactly the Hilbert space for fuzzy S^2. Hence, all observables for the lowest Landau level correspond to the observables of the fuzzy S^2.

This correspondence can be extended to the Landau problem on other spaces, say, $\mathbf{CP}^k$ with a $U(1)$ background field, for example. The background field specifies the choice of the representation of T_a and T_{k+2k}, the $U(k)$ sub-algebra of $SU(k+1)$, in the Wigner $\mathcal{D}$-functions. For zero $SU(k)$ background field, $T_a\Psi = 0$ and the eigenvalue of T_{k^2+2k} gives the magnetic field, which must obey appropriate quantization conditions. In fact, we get the (3.25) and the Hilbert subspace of the lowest Landau level is the same as the Hilbert space $\mathcal{H}_N$ used for the construction of fuzzy $\mathbf{CP}^k$ [6, 8].

The lowest Landau level wave functions are holomorphic, except possibly for a common prefactor, which has to do with the inner product. This is also seen from (3.20). In fact, the condition (3.29), namely, $R_{-i}\Psi = 0$, which selects the lowest level, are the holomorphicity conditions. The higher levels are not necessarily holomorphic. This will be useful later in writing the Yang–Mills amplitudes in terms of a Landau problem on $\mathbf{CP}^1 = S^2$.

3.6 Twistors, Supertwistors

3.6.1 The Basic Idea of Twistors

The idea of twistors is due to Roger Penrose, many years ago, in 1967 [17]. There are many related ways of thinking about twistors, but a simple approach is in terms of constructing solutions to massless wave equations or their Euclidean counterparts.

We start by considering the two-dimensional Laplace equation, which may be written in complex coordinates as

$$\partial \bar{\partial} f = 0, \tag{3.79}$$

where $z = x_1 + ix_2$. The solution is then obvious, $f(x) = h(z) + g(\bar{z})$, where $h(z)$ is a holomorphic function of z and $g(\bar{z})$ is antiholomorphic. For a given physical problem such as electrostatics or two-dimensional hydrodynamics, we then have to simply guess the holomorphic function with the required singularity structure. Further, the problem has conformal invariance and one can use the techniques of conformal mapping to simplify the problem.

We now ask the question: Can we do an analogous trick to find solutions of the four-dimensional problem, say, the Dirac or Laplace equations on S^4? Clearly, this is not so simple as in two dimensions, there are some complications. First of all, S^4 does not admit a complex structure. Even if we consider $\mathbf{R}^4$, which is topologically equivalent to S^4 with a point removed, there is no natural choice of complex coordinates. We can, for example, combine the four coordinates into two complex ones as in

$$\begin{pmatrix} z_1 \\ z_2 \end{pmatrix} = \begin{pmatrix} x_1 + ix_2 \\ x_3 + ix_4 \end{pmatrix}. \tag{3.80}$$

Equally well we could have considered

$$\begin{pmatrix} z_1' \\ z_2' \end{pmatrix} = \begin{pmatrix} x_1 + ix_3 \\ x_2 + ix_4 \end{pmatrix} \tag{3.81}$$

or, in fact, an infinity of other choices. Notice that any particular choice will destroy the overall $O(4)$-symmetry of the problem. We may now ask: How many inequivalent choices can be made, subject to, say, preserving $x^2 = \bar{z}_1 z_1 + \bar{z}_2 z_2$? Given one choice, as in (3.80), we can do an $O(4)$ rotation of x_μ which will generate other possible complex combinations with the same value of x^2. However, if we do a $U(2)$-transformation of (z_1, z_2), this gives us a new combination of the z's preserving holomorphicity. In particular, a holomorphic function of the z_i will remain a holomorphic function after a $U(2)$ rotation. Thus, the number of inequivalent choices of local complex structure is given by $O(4)/U(2) = S^2 = \mathbf{CP}^1$. The idea now is to consider S^4 with the set of all possible local complex structures at each point, in other words, a $\mathbf{CP}^1$ bundle over S^4. This bundle is $\mathbf{CP}^3$. The case of $\mathbf{R}^4$ is similar to considering a neighborhood of S^4.

An explicit realization of this is as follows. We represent $\mathbf{CP}^1$ by a two-spinor U^A, $A = 1, 2$, with the identification $U^A \sim \lambda U^A$, where $\lambda \in \mathbf{C} - \{0\}$. We now take a four-spinor with complex element Z^α, $\alpha = 1, 2, 3, 4$, and write it as $Z^\alpha = (W_{\dot{A}}, U^A)$, where U^A describes $\mathbf{CP}^1$ as above. The relation between $W_{\dot{A}}$ and U^A is taken as

$$W_{\dot{A}} = x_{\dot{A}A} \, U^A \tag{3.82}$$

$x_{\dot{A}A}$ defined by this equation may be taken as the local coordinates on S^4. We can even write this out as

$$x_{\dot{A}A} = \begin{pmatrix} x_4 + ix_3 & x_2 + ix_1 \\ -x_2 + ix_1 & x_4 - ix_3 \end{pmatrix} = x_\mu e^\mu \tag{3.83}$$

where $e^i = \sigma^i$ are the Pauli matrices and $e^4 = \mathbf{1}$, so that x_μ are the usual coordinates. One can read (3.82) in another way, namely, as combining the x's into complex combinations W_1 and W_2, in a manner specified by the choice of U^A. Thus a point on $\mathbf{CP}^1$, namely, a choice of U^A, gives a specific combination of complex coordinates.

We have the identification $Z^\alpha \sim \lambda Z^\alpha$, which follows from $U^A \sim \lambda U^A$ and the definition of $x_{\dot{A}A}$ as in (3.82). This means that Z^α define $\mathbf{CP}^3$. Further the indices $\dot{A}$, A correspond to $SU(2)$ spinor indices, right and left, in the splitting $O(4) \sim SU_L(2) \times SU_R(2)$. Z^α are called twistors.

Given the above-described structure, there is a way of constructing solutions to massless wave equations (or their Euclidean versions), in terms of holomorphic functions defined on a neighborhood of $\mathbf{CP}^3$. Evidently, preserving the $O(4)$ symmetry requires some sort of integration over all U's, consistent with holomorphicity. There is a unique holomorphic differential we can make out of the U's which is $O(4)$ invariant, namely, $U \cdot dU = \epsilon_{AB} U^A dU^B$. We will now do a contour integration of holomorphic functions using this. Let $f(Z)$ be a holomorphic function of Z^α defined on some region in twistor space. We can then construct the contour integral

$$\tilde{f}^{A_1 A_2 \cdots A_n}(x) = \oint_C U \cdot dU \; U^{A_1} U^{A_2} \cdots U^{A_n} \; f(Z). \qquad (3.84)$$

For this to make sense on a neighborhood of $\mathbf{CP}^3$, $f(Z)$ should have degree of homogeneity $-n - 2$, so that the integrand is invariant under the scaling $Z^\alpha \to \lambda Z^\alpha$, $U^A \to \lambda U^A$, and thus projects down to a proper differential on $\mathbf{CP}^3$. The contour C will be taken to enclose some of the poles of the function $f(Z)$. Since we write $W_{\dot{A}} = x_{\dot{A}A} U^A$, after integration, we are left with a function of the x's; $\tilde{f}$ is a function of the S^4 or $\mathbf{R}^4$ coordinates; it is also a multispinor of $SU_L(2)$.

Consider now the action of the chiral Dirac operator on this, namely, $\epsilon_{CA_1} \nabla^{\dot{B}C} \tilde{f}^{A_1 A_2 \cdots A_n}$. Since x_μ appear in $f(Z)$ only via the combination $x_{\dot{A}A} U^A$, we can write

$$\epsilon_{CA_1} \nabla^{\dot{B}C} \tilde{f}^{A_1 A_2 \cdots A_n} = \epsilon_{CA_1} \oint_C U \cdot dU \; U^{A_1} U^{A_2} \cdots U^{A_n} \; \nabla^{\dot{B}C} f(Z)$$

$$= \epsilon_{CA_1} \oint_C U \cdot dU \; U^{A_1} U^{A_2} \cdots U^{A_n} \; U^C \frac{\partial f(Z)}{\partial W_{\dot{B}}}$$

$$= 0 \qquad (3.85)$$

since $\epsilon_{CA_1} U^C U^{A_1} = 0$ by antisymmetry. Thus, $\tilde{f}^{A_1 A_2 \cdots A_n}(x)$ is a solution to the chiral Dirac equation in four dimensions.

In a similar way, one can define

$$\tilde{g}^{\dot{A}_1 \dot{A}_2 \cdots \dot{A}_n}(x) = \oint_C U \cdot dU \; \frac{\partial}{\partial W_{\dot{A}_1}} \frac{\partial}{\partial W_{\dot{A}_1}} \cdots \frac{\partial}{\partial W_{\dot{A}_1}} \; g(Z), \qquad (3.86)$$

where $g(Z)$ has degree of homogeneity equal to $n - 2$. It is then easy to check that

$$\epsilon_{\dot{B}\dot{A}_1} \nabla^{B\dot{B}} \tilde{g}^{\dot{A}_1 \dot{A}_2 \cdots \dot{A}_n} = 0. \qquad (3.87)$$

The two sets of functions, $\tilde{f}^{A_1 A_2 \cdots A_n}(x)$ and $\tilde{g}^{\dot{A}_1 \dot{A}_2 \cdots \dot{A}_n}(x)$, give a complete set of solutions to the chiral Dirac equation in four dimensions. This is essentially Penrose's theorem, for this case. (The theorem is more general, applicable to other manifolds which admit twistor constructions.) The mapping between holomorphic functions in twistor space and massless fields in space-time is known as the Penrose transform. (Strictly speaking, we are not concerned with holomorphic functions. They are holomorphic in some neighborhood in twistor space, and further, they are not really defined on $\mathbf{CP}^3$, since they have nontrivial degree of homogeneity. The proper mathematical characterization would be as sections of holomorphic sheaves of appropriate degree of homogeneity.)

3.6.2 An Explicit Example

As an explicit example of the Penrose transform, consider the holomorphic function

$$f(Z) = \frac{1}{a \cdot W \; b \cdot W \; c \cdot U}, \tag{3.88}$$

where $a \cdot W = a^{\dot{A}} x_{\dot{A}A} U^A \equiv U^1 w^2 - U^2 w^1$, $b \cdot W = b^{\dot{A}} x_{\dot{A}A} U^A \equiv U^1 v^2 - U^2 v^1$. Defining $z = U^2/U^1$, we find, for the Penrose integral,

$$
\begin{aligned}
\psi^A &= \oint U \cdot dU \frac{u^A}{a \cdot W \; b \cdot W \; c \cdot U} \\
&= \oint dz \frac{U^A}{U^1} \frac{1}{(w^2 - zw^1)(v^2 - zv^1)c^2 - zc^1)}.
\end{aligned} \tag{3.89}
$$

Taking the contour to enclose the pole at w^2/w^1, we find

$$
\begin{aligned}
\psi^A &= \epsilon^{AB} \frac{a^{\dot{A}} x_{\dot{A}B}}{x^2 w \cdot c \; a \cdot b} \frac{1}{} \\
&= \epsilon^{AB} \frac{a^{\dot{A}} x_{\dot{A}B}}{x^2 (axc) \; a \cdot b},
\end{aligned} \tag{3.90}
$$

where $axc = a^{\dot{A}} x_{\dot{A}A} c^A$. (We take $a \cdot b \neq 0$.) One can check directly that this obeys the equation

$$\nabla_{\dot{A}A} \psi^A = 0. \tag{3.91}$$

3.6.3 Conformal Transformations

There is a natural action of conformal transformations on twistors. We can consider Z^α as a four-spinor of $SU(4)$, the latter acting as linear transformations on Z^α, explicitly given by

$$Z^\alpha \longrightarrow Z'^\alpha = (gZ)^\alpha = g^\alpha{}_\beta \, Z^\beta, \tag{3.92}$$

where $g \in SU(4)$. The generators of infinitesimal $SU(4)$ transformations are thus given by

$$L^\alpha{}_\beta = Z^\alpha \frac{\partial}{\partial Z^\beta} - \frac{1}{4} \delta^\alpha{}_\beta \left(Z^\gamma \frac{\partial}{\partial Z^\gamma} \right). \tag{3.93}$$

This may be split into different types of transformations as follows.

$$J_{AB} = U_A \frac{\partial}{\partial U^B} + U_B \frac{\partial}{\partial U^A} \quad SU_L(2)$$

$$J_{\dot{A}\dot{B}} = U_{\dot{A}} \frac{\partial}{\partial U^{\dot{B}}} + U_{\dot{B}} \frac{\partial}{\partial U^{\dot{A}}} \quad SU_R(2)$$

$$P^{A\dot{A}} = U^A \frac{\partial}{\partial W_{\dot{A}}} \quad \text{Translation}$$

$$K_{\dot{A}A} = W_{\dot{A}} \frac{\partial}{\partial U^A} \quad \text{Special conformal transformation} \qquad (3.94)$$

$$D = W_{\dot{A}} \frac{\partial}{\partial W_{\dot{A}}} - U^A \frac{\partial}{\partial U^A} \quad \text{Dilatation} ,$$

where we have also indicated the interpretation of each type of generators. We see that the $SU(4)$ group is indeed the Euclidean conformal group; it is realized in a linear and homogeneous fashion on the twistor variables Z^α. At the level of the purely holomorphic transformations, one can also choose the Minkowski signature, where upon the transformations given above become conformal transformations in Minkowski space, forming the group $SU(2,2)$.

3.6.4 Supertwistors

One can generalize the twistor space to an $\mathcal{N}$-extended supertwistor space by adding fermionic or Grassman coordinates ξ_i, $i = 1, 2, \ldots, \mathcal{N}$; thus, supertwistor space is parameterized by (Z^α, ξ_i), with the identification $Z^\alpha \sim \lambda Z^\alpha$, $\xi_i \sim \lambda \xi_i$, where λ is any nonzero complex number [18]. λ is bosonic, so only one of the bosonic dimensions is removed by this identification. Thus, the supertwistor space is $\mathbf{CP}^{3|\mathcal{N}}$.

The case of $\mathcal{N} = 4$ is special. In this case, one can form a top-rank holomorphic form on the supertwistor space; it is given by

$$\Omega = \frac{1}{4!} \epsilon_{\alpha\beta\gamma\delta} Z^\alpha \, dZ^\beta \, dZ^\gamma \, dZ^\delta \, d\xi_1 \, d\xi_2 \, d\xi_3 \, d\xi_4 . \qquad (3.95)$$

Notice that the bosonic part gets a factor of λ^4 under the transformation $Z^\alpha \to \lambda Z^\alpha$, $\xi_i \sim \lambda \xi_i$, while the fermionic part has a factor of λ^{-4}. Ω is thus invariant under such scalings and becomes a differential form on the supermanifold $\mathbf{CP}^{3|4}$.

At this point, it is worth recalling the Calabi–Yau theorem [19].

Theorem. *For a given complex structure and Kähler class on a Kähler manifold, there exists a unique Ricci flat metric if and only if the first Chern class of the manifold vanishes or if and only if there is a globally defined top-rank holomorphic form on the manifold.*

This is for an ordinary manifold. For the supersymmetric case, we will define a Calabi–Yau supermanifold as one which admits a globally defined top-rank holomorphic differential form [20]. Whether such spaces admit a generalization of the Calabi–Yau theorem is not known. (For $\mathcal{N} = 1$ spaces, a counterexample is known. However, super-Ricci flatness may follow from the vanishing of the first Chern class for $\mathcal{N} \geq 2$ [21].)

3.6.5 Lines in Twistor Space

Holomorphic lines in twistor space will turn out to be important for the construction of Yang–Mills amplitudes. First, we will consider a holomorphic straight line, or a curve of degree one, in twistor space, giving the generalization to supertwistor space later. Since $\mathbf{CP}^3$ has three complex dimensions, we need two complex conditions to reduce to a line in twistor space. Thus, we can specify a line in twistor space as the solution set of the equations

$$A_\alpha Z^\alpha = 0 , \qquad B_\alpha Z^\alpha = 0 , \qquad (3.96)$$

where A_α, B_α are constant twistors which specify the placement of the line in twistor space. These equations can be combined as

$$a_A^i U^A + b_{\dot{A}}^i W^{\dot{A}} = 0 , \qquad (3.97)$$

where $A_\alpha = (a_A^1, b_{\dot{A}}^1)$, $B_\alpha = (a_A^2, b_{\dot{A}}^2)$. a, b can be considered as (2×2)-matrices; $\det a$ and $\det b$ may both be nonzero, but both cannot be zero simultaneously, since (3.96) are then not sufficient to reduce to a line. We will take $\det b \neq 0$ in the following. (The arguments presented will go through with appropriate relabelings if $\det b = 0$, but $\det a \neq 0$.) In this case, b is invertible and we can solve the (3.97) by

$$W_{\dot{A}} = -(b^{-1}a)_{\dot{A}A} U^A$$
$$\equiv x_{\dot{A}A} U^A . \qquad (3.98)$$

This shows that the condition (3.82) identifying the space-time coordinates may be taken as defining a line in twistor space. In fact, here, $x_{\dot{A}A}$ specify the placement and orientation of the line in twistor space, in other words, they are the moduli of the line. We see that the moduli space of straight lines (degree-one curves) in twistor space is space-time.

There is another way to write (3.98). Recall that a line in a real space M can be defined as a mapping of the interval $[0, 1]$ into the space M, $L : [0, 1] \to M$. We can do a similar construction for the complex case. We will define an abstract $\mathbf{CP}^1$ space by a two-spinor u^a with the identification $u^a \sim \rho u^a$ for any nonzero complex number ρ. Then we can regard a holomorphic line in twistor space as a map $\mathbf{CP}^1 \to \mathbf{CP}^3$, realized explicitly as

$$U^A = (a^{-1})_a^A u^a, \qquad W_{\dot{A}} = (b^{-1})_{\dot{A}a} u^a . \qquad (3.99)$$

One can do $SL(2, \mathbf{C})$ transformations on the coordinates u^a of $\mathbf{CP}^1$; using this freedom, we can set $a = 1$, or equivalently,

$$U^A = u^A, \qquad W_{\dot{A}} = x_{\dot{A}A} u^A .\tag{3.100}$$

This is identical to (3.98).

The generalization of this to supertwistor space is now obvious. We will consider a map of $\mathbf{CP}^1$ to the supertwistor space $\mathbf{CP}^{3|4}$, given explicitly by

$$U^A = u^A, \qquad W_{\dot{A}} = x_{\dot{A}A} u^A$$
$$\xi^\alpha = \theta_A^\alpha u^A .\tag{3.101}$$

We have the fermionic moduli θ_A^α in addition to the bosonic ones $x_{\dot{A}A}$.

The construction of curves of higher degree can be done along similar lines. A curve of degree d is given by

$$Z^\alpha = \sum_{\{a\}} a_{a_1 a_2 \cdots a_d}^\alpha \, u^{a_1} u^{a_2} \cdots u^{a_d}$$
$$\xi^\alpha = \sum_{\{a\}} \gamma_{a_1 a_2 \cdots a_d}^\alpha \, u^{a_1} u^{a_2} \cdots u^{a_d} .\tag{3.102}$$

The coefficients $a_{a_1 a_2 \cdots a_d}^\alpha$, $\gamma_{a_1 a_2 \cdots a_d}^\alpha$ give the moduli of the curve. One can use $SL(2, \mathbf{C})$ to set three of the coefficients to fixed values.

Given that each index a takes values $1, 2$, and the fact that the coefficients are symmetric in $a_1, a_2, \ldots, a_n$, we see that there are $4(d+1)$ bosonic and fermionic coefficients. The identification of u^a and ρu^a and $Z \sim \lambda Z$, $\xi \sim \lambda \xi$ tells us that we can remove an overall scale degree of freedom. In other words, for the moduli, we have the identification,

$$a_{a_1 a_2 \cdots a_d}^\alpha \sim \lambda \, a_{a_1 a_2 \cdots a_d}^\alpha$$
$$\gamma_{a_1 a_2 \cdots a_d}^\alpha \sim \lambda \, \gamma_{a_1 a_2 \cdots a_d}^\alpha \tag{3.103}$$

for $\lambda \in \mathbf{C} - \{0\}$. Thus, the moduli space of the curves may be taken as $\mathbf{CP}^{4d+3|4d+4}$. For the expressions of interest, as we shall see later, there is an overall $SL(2, \mathbf{C})$ invariance, and hence three of the bosonic parameters can be fixed to arbitrarily chosen values.

3.7 Yang–Mills Amplitudes and Twistors

3.7.1 Why Twistors Are Useful

In this section, we will start our discussion of the twistor approach to amplitudes in Yang–Mills theory [22, 23], or multigluon scattering amplitudes, as they are often referred to.

We begin with the question of why the calculation of multigluon amplitudes is interesting. One of the motivations in seeking a twistor string theory was to obtain a weak coupling version of the standard duality between string theory on anti-de Sitter space and $\mathcal{N} = 4$ supersymmetric Yang–Mills theory [23]. However, developments in the subject over the last year or so have focused on, and yielded, many interesting results on the calculation of the scattering amplitudes themselves, so we shall concentrate on this aspect of twistors [24]. Scattering amplitudes in any gauge theory are not only interesting in a general sense of helping to clarify a complicated interacting theory, but also there is a genuine need for them from a very practical point of view. This point can be illustrated by taking the experimental determination of the strong coupling constant as an example. We quote three values from the Particle Data Group based on three different processes.

$$\begin{aligned}
\alpha_s = 0.116 &\ + 0.003 \ (\text{expt.}) \pm 0.003 \ (\text{theory}) \\
&\ - 0.005 \\
= 0.120 &\ \pm 0.002 \ (\text{expt.}) \pm 0.004 \ (\text{theory}) \\
= 0.1224 &\ \pm 0.002 \ (\text{expt.}) \pm 0.005 \ (\text{theory}) \, .
\end{aligned} \tag{3.104}$$

These values are for the momentum scale corresponding to the mass of the Z-boson, namely, $\alpha_s(M_Z)$; they are based on the Bjorken spin sum rule, jet rates in e-p collisions and the photoproduction of two or more jets, respectively. Notice that the theoretical uncertainty is comparable to, or exceeds, the experimental errors. The major part of this comes from lack of theoretical calculations (to the order required) for the processes from which this value is extracted. Small as it may seem, this uncertainty can affect the hadronic background analysis at the Large Hadron Collider (currently being built at CERN), for instance. The relative signal strength for processes of interest, such as the search for the Higgs particle, can be improved if this uncertainty is reduced. This can also affect the estimate of the grand unification scale and theoretical issues related to it.

One could then ask the question: Since we know the basic vertices involved, and these are ultimately perturbative calculations, why not just do the calculations, to whatever order is required? Unfortunately, the direct calculation of the amplitudes is very, very difficult since there are large numbers, of the order of millions, of Feynman diagrams involved. (It is easy to see that the number of diagrams involved increases worse than factorially as the number of external lines increases.) Twistors provide a way to improve the situation.

A natural next question is then: What can twistors do, what has been accomplished so far? The progress so far may be summarized as follows.

1. It has been possible to write down a formula for all the tree-level amplitudes in $\mathcal{N} = 4$ supersymmetric Yang–Mills theory [25, 26]. Being at the tree level, this formula applies to the tree amplitudes of the nonsupersymmetric Yang–Mills theory as well. The formula reduces the calculation to the evaluation of the zeros of a number of polynomial equations and the evaluation of an integral. A certain analogy with instantons might be helpful in explaining the nature of this formula. To find the instanton field configurations, one must solve the self-duality conditions which are a set of coupled first-order

differential equations. However, the ADHM procedure reduces this to an algebraic problem, namely, of solving a set of matrix equations, which lead to the construction of the appropriate holomorphic vector bundles. This algebraic problem is still difficult for large instanton numbers, nevertheless, an algebraicization has been achieved. In a similar way, the formula for the tree amplitudes replaces the evaluation of large numbers of Feynman diagrams, or an equivalent functional integral, by an ordinary integral whose evaluation requires the solution of some polynomial equations. This may still be difficult for cases with large numbers of negative helicity gluons; nevertheless, it is a dramatic simplification.

2. At the one-loop level, a similar formula has been obtained for all the so-called maximally helicity violating (MHV) amplitudes [27]. A number of results for next-to-MHV amplitudes have been obtained [28].

3. A set of new diagrammatic rules based on the MHV vertices has been developed [29]. Also, new types of recursion rules have been developed [30]. These are promising new directions for perturbative analysis of a field theory.

4. Twistor-inspired techniques have been used for some processes involving massive particles, particularly for the electroweak calculations [16].

3.7.2 The MHV Amplitudes

We will begin with a discussion of the maximally helicity violating (MHV) amplitudes. These refer to n gluon scattering amplitudes with $n-2$ gluons of positive helicity and 2 gluons of negative helicity. With n gluons, $n-2$ is the maximum number of positive helicity possible by conservation laws and such amplitudes are often referred to as the MHV amplitudes. The analysis of these amplitudes will lead the way to the generalization for all amplitudes.

The gluons are massless and have momenta p_μ which obey the condition $p^2 = 0$; i.e., p_μ is a null vector. Using the identity and the Pauli matrices, we can write the four-vector p_μ as a 2×2-matrix

$$p^A_{\dot{A}} = (\sigma^\mu)^A_{\dot{A}} \, p_\mu = \begin{pmatrix} p_0 + p_3 & p_1 - ip_2 \\ p_1 + ip_2 & p_0 - p_3 \end{pmatrix} \tag{3.105}$$

This matrix has a zero eigenvalue and, using this fact, we can see that it can be written as

$$p^A_{\dot{A}} = \pi^A \bar{\pi}_{\dot{A}} \tag{3.106}$$

There is a phase ambiguity in the definition of π, $\bar{\pi}$; $\pi' = e^{i\theta}\pi$, $\bar{\pi}' = e^{-i\theta}\bar{\pi}$ give the same momentum vector p_μ. Thus, physical results should be independent of this phase transformation. For a particular choice of this phase, an explicit realization of π, $\bar{\pi}$ is given by

$$\pi = \frac{1}{\sqrt{p_0 - p_3}} \begin{pmatrix} p_1 - ip_2 \\ p_0 - p_3 \end{pmatrix}, \qquad \bar{\pi} = \frac{1}{\sqrt{p_0 - p_3}} \begin{pmatrix} p_1 + ip_2 \\ p_0 - p_3 \end{pmatrix} \tag{3.107}$$

The fact that p_μ is real gives a condition between π and $\bar{\pi}$, which may be taken as $\bar{\pi}_{\dot{A}} = (\pi^A)^*$. We see that, for the momentum for each massless particle, we can associate a spinor momentum π.

There is a natural action of the Lorentz group $SL(2, \mathbf{C})$ on the dotted and undotted indices, given by

$$\pi^A \to \pi'^A = (g\pi)^A, \qquad \bar{\pi}_{\dot{A}} \to \bar{\pi}'_{\dot{A}} = (g^*\bar{\pi})_{\dot{A}}, \qquad (3.108)$$

where $g \in SL(2, \mathbf{C})$. The scalar product which preserves this symmetry is given by $\langle 12 \rangle = \pi_1 \cdot \pi_2 = \epsilon_{AB}\pi_1^A\pi_2^B$, $[12] = \epsilon^{\dot{A}\dot{B}}\bar{\pi}_{1\dot{A}}\bar{\pi}_{2\dot{B}}$. At the level of vectors, this corresponds to the Minkowski product; i.e., $\eta^{\mu\nu}p_{1\mu}p_{2\nu} = p_1 \cdot p_2 = \langle 12 \rangle [12]$. Because of the ϵ-tensor, $\langle 11 \rangle = 0$ and the factorization (3.106) is consistent with $p^2 = 0$. (More generally, $\langle 12 \rangle = 0$ and $[12] = 0$ if π_1 and π_2 are proportional to each other.) The scattering amplitudes can be simplified considerably when expressed in terms of these invariant spinor products. We will also define raising and lowering of the spinorial indices using the ϵ-tensor.

It is also useful to specify the polarization states of the gluons by helicity. The polarization vector ϵ_μ may then be written as

$$\epsilon_\mu \to \epsilon^A_{\dot{A}} = (\sigma^\mu)^A_{\dot{A}} \, \epsilon_\mu = \begin{cases} \lambda^A \bar{\pi}_{\dot{A}}/\pi \cdot \lambda & +1 \text{ helicity} \\ \pi^A \bar{\lambda}_{\dot{A}}/\bar{\pi} \cdot \bar{\lambda} & -1 \text{ helicity} \end{cases} \qquad (3.109)$$

The spinor $(\lambda^A, \bar{\lambda}_{\dot{A}})$ characterizes the choice of helicity.

We can now state the MHV amplitude for scattering of n gluons, originally obtained by Parke and Taylor [32]. They carried out the explicit calculation of Feynman diagrams, for small values of n, using some supersymmetry tricks for simplifications. Based on this, they guessed the general form of the amplitude; this guess was proved by Berends and Giele by using recursion rules for scattering amplitudes [32]. The results are the following:

$$\mathcal{A}(1_+^{a_1}, 2_+^{a_2}, 3_+^{a_3}, \ldots, n_+^{a_n}) = 0$$
$$\mathcal{A}(1_-^{a_1}, 2_+^{a_2}, 3_+^{a_3}, \ldots, n_+^{a_n}) = 0$$
$$\mathcal{A}(1_-^{a_1}, 2_-^{a_2}, 3_+^{a_3}, \ldots, n_+^{a_n}) = ig^{n-2}(2\pi)^4\delta(p_1 + p_2 + \ldots + p_n) \, \mathcal{M} \quad (3.110)$$
$$+\text{noncyclic permutations}$$
$$\mathcal{M}(1_-^{a_1}, 2_-^{a_2}, 3_+^{a_3}, \ldots, n_+^{a_n}) = \langle 12 \rangle^4 \frac{\text{Tr}(t^{a_1}t^{a_2}\ldots t^{a_n})}{\langle 12 \rangle \langle 23 \rangle \ldots \langle n-1 \, n \rangle \langle n1 \rangle}.$$

The first nonvanishing amplitude with the maximum difference of helicities has two negative helicity gluons and $n - 2$ positive helicity gluons. This is what is usually called the MHV amplitude. In (3.110), g is the gauge coupling constant. Notice that the amplitude $\mathcal{M}$ is cyclically symmetric in all the particle labels except for the prefactor $\langle 12 \rangle^4$. The latter refers to the momenta of the two negative helicity gluons. The summation over the noncyclic permutations makes the full amplitude symmetric in the gluon labels. We have taken all gluons as incoming. One can use the standard crossing symmetry to write down the corresponding amplitudes, with appropriate change of helicities, if some of the gluons are outgoing.

We will now carry out three steps of simplification of this result to bring out the twistor connection.

The First Step: The Chiral Dirac Determinant on $\mathbf{CP}^1$

Consider the functional determinant of the Dirac operator of a chiral fermion coupled to a gauge field $A_{\bar{z}}$ in two dimensions. By writing $\log \det D_{\bar{z}} = \mathrm{Tr} \log D_{\bar{z}}$ and expanding the logarithm, we find

$$
\begin{aligned}
\mathrm{Tr} \log D_{\bar{z}} &= \mathrm{Tr} \log(\partial_{\bar{z}} + A_{\bar{z}}) \\
&= \mathrm{Tr} \log\left(1 + \frac{1}{\partial_{\bar{z}}} A_{\bar{z}}\right) + \text{constant} \\
&= \sum_n \int \frac{d^2 x_1}{\pi} \frac{d^2 x_2}{\pi} \cdots \frac{(-1)^{n+1}}{n} \frac{\mathrm{Tr}[A_{\bar{z}}(1) A_{\bar{z}}(2) \cdots A_{\bar{z}}(n)]}{z_{12} z_{23} \cdots z_{n-1\,n} z_{n1}},
\end{aligned}
$$
(3.111)

where $z_{12} = z_1 - z_2$, etc. In writing this formula we have used the result

$$
\left(\frac{1}{\partial_{\bar{z}}}\right)_{12} = \frac{1}{\pi(z_1 - z_2)}.
$$
(3.112)

We can regard the z's as local coordinates on $\mathbf{CP}^1$. Recall that $\mathbf{CP}^1$ is defined by two complex variables α and β, which may be regarded as a two-spinor u^a, $u^1 = \alpha$, $u^2 = \beta$, with the identification $u^a \sim \rho u^a$, $\rho \in \mathbf{C} - \{0\}$. On the coordinate patch with $\alpha \neq 0$, we take $z = \beta/\alpha$ as the local coordinate. We can then write

$$
\begin{aligned}
z_1 - z_2 &= \frac{\beta_1}{\alpha_1} - \frac{\beta_2}{\alpha_2} = \frac{\beta_1 \alpha_2 - \beta_2 \alpha_1}{\alpha_1 \alpha_2} \\
&= \frac{\epsilon_{ab} u_1^a u_2^b}{\alpha_1 \alpha_2} = \frac{u_1 \cdot u_2}{\alpha_1 \alpha_2}
\end{aligned}
$$
(3.113)

Further, if we define $\alpha^2 A_{\bar{z}} = \bar{A}$, (3.111) becomes

$$
\mathrm{Tr} \log D_{\bar{z}} = -\sum_n \frac{1}{n} \int \frac{\mathrm{Tr}[\bar{A}(1)\bar{A}(2) \cdots \bar{A}(n)]}{(u_1 \cdot u_2)(u_2 \cdot u_3) \cdots (u_n \cdot u_1)}
$$
(3.114)

Notice that if u^a is replaced by the spinor momentum π^A, the denominator is exactly what appears in (3.110). The factor of $1/n$ gets cancelled out because (3.114) generates all permutations which gives n times the sum over all noncyclic permutations.

The Second Step: The Helicity Factors

The denominator for the MHV amplitude can be related to the chiral Dirac determinant as above. The factor $\langle 12 \rangle^4$ can also be obtained if we introduce supersymmetry. We take up this second step of simplification now.

The transformation (3.108) shows that the Lorentz generator for the π's is given by

$$J_{AB} = \frac{1}{2} \left(\pi_A \frac{\partial}{\partial \pi^B} + \pi_B \frac{\partial}{\partial \pi^A} \right), \tag{3.115}$$

where $\pi_A = \epsilon_{AB} \pi^B$. The spin operator is given by $S_\mu \sim \epsilon_{\mu\nu\alpha\beta} J^{\nu\alpha} p^\beta$, where $J^{\mu\nu}$ is the full Lorentz generator. This works out to $S_{A\dot{A}} = J^A_B \pi^B \bar{\pi}_{\dot{A}} = -p^A_{\dot{A}} s$, identifying the helicity as

$$s = -\frac{1}{2} \pi^A \frac{\partial}{\partial \pi^A}. \tag{3.116}$$

Thus s is, up to a minus sign, half the degree of homogeneity in the π's. If we start with a positive helicity gluon, which would correspond to two negative powers of the corresponding spinor momentum, then we should expect an additional four factors of π for a negative helicity gluon. Notice that there are two factors of spinor momenta in the denominator of the scattering amplitude (3.110) for each positive helicity gluon; for the two negative helicity gluons, because of the extra factor of $\langle 12 \rangle^4$, the net result is two positive powers of π.

We now notice that if we have an anticommuting spinor θ_A, $\int d^2\theta\, \theta_A \theta_B = \epsilon_{AB}$, so that

$$\int d^2\theta\, (\pi\theta)(\pi'\theta) = \int d^2\theta\, (\pi^A \theta_A)(\pi'^B \theta_B) = \pi \cdot \pi'. \tag{3.117}$$

We see that an $\mathcal{N} = 4$ theory is what we need to get four such factors, so as to get a term like $\langle 12 \rangle^4$. Therefore, we define an $\mathcal{N} = 4$ superfield

$$\bar{A}^a(\pi, \bar{\pi}) = a^a_+ + \xi^\alpha a^a_\alpha + \frac{1}{2} \xi^\alpha \xi^\beta a^a_{\alpha\beta} + \frac{1}{3!} \xi^\alpha \xi^\beta \xi^\gamma \epsilon_{\alpha\beta\gamma\delta} \bar{a}^{a\delta} + \xi^1 \xi^2 \xi^3 \xi^4 a^a_-, \tag{3.118}$$

where $\xi^\alpha = (\pi\theta)^\alpha = \pi^A \theta^\alpha_A$, $\alpha = 1, 2, 3, 4$. We can interpret a^a_+ as the classical version of the annihilation operator for a positive helicity gluon (whose gauge charge is specified by the Lie algebra index a), a^a_- as the annihilation operator for a negative helicity gluon; $a^a_\alpha, \bar{a}^{a\alpha}$ correspond to four spin-$\frac{1}{2}$ particles and $a^a_{\alpha\beta}$ correspond to six spin-zero particles. This is exactly the particle content of $\mathcal{N} = 4$ Yang–Mills theory.

We now choose the gauge potential in (3.114) to be given by

$$\bar{A} = g t^a \bar{A}^a \exp(ip \cdot x) \tag{3.119}$$

Using this in the chiral Dirac determinant (3.114), we construct the expression

$$\Gamma[a] = \frac{1}{g^2} \int d^8\theta\, d^4x\, \mathrm{Tr} \log D_{\bar{z}} \Bigg]_{u^a \to \pi^A}. \tag{3.120}$$

It is then clear that the MHV amplitude can be written as

$$A(1_-^{a_1}, 2_-^{a_2}, 3_+^{a_3}, \ldots, n_+^{a_n}) = i \left[\frac{\delta}{\delta a_-^{a_1}(p_1)} \frac{\delta}{\delta a_-^{a_2}(p_2)} \frac{\delta}{\delta a_+^{a_3}(p_3)} \cdots \frac{\delta}{\delta a_+^{a_n}(p_n)} \Gamma[a] \right]_{a=0}.$$

(3.121)

An alternate representation involves introducing a supersymmetric version of the factor $\exp(ip \cdot x)$ for the $\mathcal{N} = 4$ supermultiplet. Consider the function

$$\exp(i\eta \cdot \xi) = 1 + i\eta \cdot \xi + \frac{1}{2!} i\eta \cdot \xi \, i\eta \cdot \xi + \frac{1}{3!} i\eta \cdot \xi \, i\eta \cdot \xi \, i\eta \cdot \xi$$

$$+ \frac{1}{4!} i\eta \cdot \xi \, i\eta \cdot \xi \, i\eta \cdot \xi \, i\eta \cdot \xi , \quad (3.122)$$

where $\eta \cdot \xi = \eta_\alpha \xi^\alpha$. Here, η_α are four Grassman variables. We may think of them as characterizing the state of the external particles, specifically, their helicity. (The state is thus specified by the spinor momentum π and η.) Since each ξ carries one power of π, we see that we can associate the first term with a positive helicity gluon, the last with a negative helicity gluon, and the others with the superpartners of gluons, accordingly. We can then take

$$\bar{A} = g t^a \phi^a \exp(ip \cdot x + i\eta \cdot \xi). \quad (3.123)$$

The scattering amplitudes are given by $\Gamma[a]$ again, where, to get negative helicity for particles labeled 1 and 2 we take the coefficient of the factor $\eta_{11}\eta_{21}\eta_{31}\eta_{41}\eta_{12}\eta_{22}\eta_{32}\eta_{42}$, where the first subscript gives the component of η and the second refers to the particle. We should also look at the term with n factors of ϕ^a for the n-gluon amplitude.

The Third Step: Reduction to a Line in Twistor Space

The results (3.121)–(3.123) were known for a long time. The importance of supertwistor space was also recognized [22]. (Some of the earlier developments, with connections to the self-dual Yang–Mills theory, etc., can be traced from [33].) Notice that with $U^A = u^A$, $W = xu$ and ξ, alongwith the condition $\pi^A = U^A$, we are close to the usual variables of supertwistor space. More recently, Witten achieved enormous advances in this field by relating this formula to twistor string theory and curves in twistor space [23]. To arrive at this generalization, first of all, we notice that the amplitude is holomorphic in the spinor momenta except for the exponential factor $\exp(ip \cdot x)$. We can rewrite this factor as follows.

$$\exp(ip \cdot x) = \exp\left(\frac{i}{2} \bar{\pi}^{\dot{A}} x_{\dot{A}A} \pi^A \right)$$

$$= \exp\left(\frac{i}{2} \bar{\pi}^{\dot{A}} W_{\dot{A}} \right) \Big|_{u^A = \pi^A}, \quad (3.124)$$

where $W_{\dot{A}} = x_{\dot{A}A}\pi^A$. The strategy is now to regard $W_{\dot{A}}$ as a free variable, interpreting the condition $W_{\dot{A}} = x_{\dot{A}A}u^A$ as the restriction to a line in twistor space. We shall also use $U^A = u^A$, which is the other condition defining a line in twistor space; see (3.100). We can then write

$$\int d\sigma \, \delta\left(\frac{\pi^2}{\pi^1} - \frac{U^2}{U^1}\right) \exp\left(\frac{i}{2}\bar{\pi}^{\dot{A}}\pi^1 \frac{W_{\dot{A}}}{U^1}\right) = \exp\left(\frac{i}{2}\bar{\pi}^{\dot{A}}x_{\dot{A}A}\pi^A\right) = \exp(ip \cdot x) \,,$$
(3.125)

where $\sigma = u^2/u^1$, and we have used the restriction to the line $W_{\dot{A}} = x_{\dot{A}A}u^A$, $U^A = u^A$. The integration is along a line which contains the support of the δ-function.

We can also treat ξ as an independent variable, interpreting the condition $\xi^\alpha = \theta^\alpha_A u^A$ as part of the line in supertwistor space, as in (3.101). The amplitude for n particle scattering, with particle momenta labeled by π_i^A, $\bar{\pi}_i^{\dot{A}}$ and helicity factors $\eta_{\alpha i}$, can then be written as

$$\mathcal{A} = ig^{n-2}\int d^4x d^8\theta \int d\sigma_1 \cdots d\sigma_n \frac{\mathrm{Tr}(t^{a_1}\cdots t^{a_n})}{(\sigma_1 - \sigma_2)(\sigma_2 - \sigma_3)\cdots(\sigma_n - \sigma_1)}$$
$$\times \prod_i \delta\left(\frac{\pi_i^2}{\pi_i^1} - \frac{U^2(\sigma_i)}{U^1(\sigma_i)}\right) \exp\left(\frac{i}{2}\bar{\pi}_i^{\dot{A}}\pi_i^1 \frac{W_{\dot{A}}(\sigma_i)}{U^1(\sigma_i)} + i\pi_i^1 \eta_{\alpha i} \frac{\xi^\alpha(\sigma_i)}{U^1(\sigma_i)}\right)$$

+noncyclic permutations ,
(3.126)

where the functions $W_{\dot{A}}$, U^A, ξ^α are given by

$$U^A = u^A, \qquad W_{\dot{A}} = x_{\dot{A}A}u^A$$
$$\xi^\alpha = \theta^\alpha_A u^A$$
(3.127)

exactly as in (3.101). The variable σ is given by $\sigma = u^2/u^1$. Notice that the overall factor of u^1 in $W_{\dot{A}}$, U^A, ξ^α, cancels out in the formula (3.126).

3.7.3 Generalization to Other Helicities

The amplitude, in the form given in (3.126), shows a number of interesting properties. First of all, the amplitude is entirely holomorphic in the twistor variables $Z^\alpha = (W_{\dot{A}}, U^A)$, ξ^α. It is also holomorphic in the variable σ or u^A. (It is not holomorphic in π since there is $\bar{\pi}$ in the exponentials, but this is immaterial for our arguments given below.) Secondly, the amplitude is invariant under the scalings $Z^\alpha \to \lambda Z^\alpha$, $\xi^\alpha \to \lambda \xi^\alpha$, so that it is a properly defined function on some neighborhood in the supertwistor space. Further, the amplitude has support only on a curve of degree one in supertwistor space given by (3.127). The moduli of this curve are given by $x_{\dot{A}A}$ and θ^α_A; there is integration over all these in the amplitude.

We may interpret this as follows. We consider a holomorphic map $\mathbf{CP}^1 \to \mathbf{CP}^{3|4}$ which is of degree one. We pick n points $\sigma_1, \sigma_2, \ldots, \sigma_n$ and then evaluate the integral in (3.126) over all σ's and the moduli of the chosen curve.

The generalization of the formula suggested by Witten is to use curves of higher degree [23]. In fact, Witten argued, based on twistor string theory, that one should consider curves of degree d and genus g, with

$$d = q - 1 + l, \qquad g \le l \tag{3.128}$$

for l-loop Yang–Mills amplitudes with q gluons of negative helicity. This generalization has been checked for various cases as mentioned before. For the tree amplitudes, the generalized formula reads [23, 25]

$$A = ig^{n-2} \int d\mu \int d\sigma_1 \cdots d\sigma_n \frac{\text{Tr}(t^{a_1} \cdots t^{a_n})}{(\sigma_1 - \sigma_2)(\sigma_2 - \sigma_3) \cdots (\sigma_n - \sigma_1)}$$

$$\times \prod_i \delta \left(\frac{\pi_i^2}{\pi_i^1} - \frac{U^2(\sigma_i)}{U^1(\sigma_i)} \right) \exp \left(\frac{i}{2} \bar{\pi}_i^{\dot{A}} \pi_i^1 \frac{W_{\dot{A}}(\sigma_i)}{U^1(\sigma_i)} + i\pi_i^1 \eta_{\alpha i} \frac{\xi^\alpha(\sigma_i)}{U^1(\sigma_i)} \right)$$

+noncyclic permutations , \qquad\qquad\qquad (3.129)

where the curves of degree d are

$$W_{\dot{A}}(\sigma) = (u^1)^d \sum_0^d b_{\dot{A}k} \sigma^k, \qquad U^A(\sigma) = (u^1)^d \sum_0^d a_k^A \sigma^k$$

$$\xi^\alpha(\sigma) = (u^1)^d \sum_0^d \gamma_k^\alpha \sigma^k . \tag{3.130}$$

This is exactly as in (3.102). The measure of integration for the moduli in (3.129) is given by

$$d\mu = \frac{d^{2d+2}a \ d^{2d+2}b \ d^{4d+4}\gamma}{vol[GL(2, \mathbf{C})]} . \tag{3.131}$$

The division by the volume of $GL(2, \mathbf{C})$ arises as follows. There is an overall scale invariance for the integrand in (3.129), which means that we can remove one complex scale factor, corresponding to the moduli space being $\mathbf{CP}^{4d+3|4d+4}$. The integrand is also holomorphic in σ and so has invariance under the $SL(2, \mathbf{C})$ transformations $u^a \to u'^a = (gu)^a$ where g is a (2×2)-matrix with unit determinant, or an element of $SL(2, \mathbf{C})$. We must remove this factor to get an integral which does not diverge.

The actual evaluation of the integral can still be quite involved. One has to identify the zeros of the functions $U^2(\sigma)/U^1(\sigma)$ to integrate over the δ-functions. This can be difficult to do explicitly for arbitrary values of the moduli. This is then followed by the integration over the moduli. Nevertheless, the formulae (3.129), (3.130) constitute a significant achievement. They reduce the problem of amplitude calculations in the gauge theory to an ordinary, multidimensional integral. As mentioned at the beginning of this section, this reduction has a status somewhat similar to what the ADHM construction has achieved for instantons.

Another important qualification about the formula (3.129) is that the integrals have to be defined by a continuation to real variables. The space-time signature has to be chosen to be $(--++)$ to be compatible with this. One has to carry out the analytic continuation after the integrals are done.

The justification for the generalization embodied in (3.129) comes from twistor string theory. We shall now briefly review this connection following Witten's construction of twistor string theory [23]; there is an alternative string theory proposed by Berkovits which can also be used [34]. Some of the structure of the latter will be used in Sect. 3.9.

3.8 Twistor String Theory

As mentioned in Sect. 3.6, supertwistor space $\mathbf{CP}^{3|4}$ is a Calabi–Yau space. This allows the construction of a topological B-model with $\mathbf{CP}^{3|4}$ as the target space. In this theory, one considers open strings which end on $D5$-branes, with the condition $\bar{\xi} = 0$. The gauge fields which characterize the dynamics of the ends of the open strings is then a potential $\bar{A}(Z, \bar{Z}, \xi)$ which can be checked to have the same content as the $\mathcal{N} = 4$ gauge theory. An effective action for the topological sector can be written down; it is given by

$$\mathcal{I} = \frac{1}{2} \int_Y \Omega \wedge \mathrm{Tr}\left(\bar{A}\,\bar{\partial}\,\bar{A} + \frac{2}{3}\bar{A}^3\right). \tag{3.132}$$

Here Y is a submanifold of $\mathbf{CP}^{3|4}$ with $\bar{\xi} = 0$. To linear order in the fields, the equations of motion for (3.132) correspond to $\bar{\partial}\bar{A} = 0$. Thus, the fields are holomorphic in the Z's and, via the Penrose transform, they correspond to massless fields in ordinary space-time. Including the nonlinear terms, one can ask for an action in terms of the fields in space-time which is equivalent to (3.132). This is given by

$$\mathcal{I} = \int \mathrm{Tr}\left[G^{AB}F_{AB} + \bar{\chi}^{A\alpha}D_{A\dot{A}}\chi_\alpha^{\dot{A}} + \cdots\right], \tag{3.133}$$

where G^{AB} is a self-dual field, F_{AB} is the self-dual part of the usual field strength $F_{\mu\nu}$ of an ordinary gauge potential, $\chi, \bar{\chi}$ are fermionic fields, etc. In terms of helicities G^{AB} corresponds to -1, the nonvanishing field strength $F_{\dot{A}\dot{B}}$ corresponds to $+1$ and so on. The action (3.133) cannot generate amplitudes with arbitrary number of negative helicity gluons; it is not the (super)-Yang–Mills action either. The usual Yang–Mills term can be generated by the effect of a $D1$-instanton, which can lead to a term of the form $\frac{1}{2}\int \epsilon G^2$, where ϵ is related to the action for the instanton. Integrating out the G-field, we then get a Yang–Mills term of the form $\int F^2/4g^2$ with a Yang–Mills coupling constant $g^2 \sim \epsilon$. A term with q factors of G, corresponding to q particles of helicity -1, will require $q - 1$ powers of ϵ. This corresponds to instanton number $q - 1 = d$. Such $D1$-instantons are described by holomorphic curves of degree d. This is in agreement with the formula (3.128). This is the basic argument, schematically, why we should expect curves of degree d to lead to tree-level amplitudes in the $\mathcal{N} = 4$ Yang–Mills theory.

We have not discussed amplitudes at the loop level yet. A similar approach with curves of genus one has been verified at the one-loop level [27]. However, the general formula in terms of higher genus curves has not been very useful for actual computations. This is partially due to the complexity of the formula. A more relevant reason is the emergence of new rules to calculate both tree level and loop level amplitudes using a sewing procedure with the basic MHV amplitudes as the vertices [29]. The MHV amplitudes have to be continued off-shell for this reason. An off-shell extension has been proposed and used in [29].

The one-loop amplitudes which emerge naturally are the amplitudes for the $\mathcal{N} = 4$ Yang–Mills theory. If the external (incoming and outgoing) particles are gluons, the superpartners can only occur in loops. As a result, at the tree-level, one can get the amplitudes in the pure Yang–Mills theory with no supersymmetry by restricting the external lines to be gluons. But at the one-loop level all superpartners can contribute. While the one-loop amplitudes for the $\mathcal{N} = 4$ Yang–Mills theory are interesting in their own right, the corresponding amplitudes in the nonsupersymmetric theory are of even greater interest, since they pertain to processes which are experimentally accessible. It is not trivial to extract the nonsupersymmetric amplitudes from the $\mathcal{N} = 4$ theory. One approach is to subtract out the contributions of the superpartners. An alternative is to build up the one-loop amplitude from the unitarity relation, using tree amplitudes. In principle, this can only yield the imaginary part of the one-loop amplitude. (One could attempt to construct the real part via dispersion relations. While this is usually ambiguous due to subtractions needed for the dispersion integrals, the $\mathcal{N} = 4$ theory, which is finite, is special. There are relations among amplitudes which can be used for this theory.) If one makes an ansatz for some off-shell extensions of the tree amplitudes, one can obtain, via unitarity relations, some of the one-loop amplitudes in the nonsupersymmetric theory as well. The off-shell extensions can be checked for consistency in soft-gluon limits, etc., so they are fairly unambiguous. The results quoted in the beginning of Sect. 3.7 emerged from such analyses. Notice that the state of the art here is a combination of rules emerging from the twistor approach and unitarity relations and a bit of guess work. The new set of recursion rules also has been very useful [30].

3.9 Landau Levels and Yang–Mills Amplitudes

3.9.1 The General Formula for Amplitudes

There is an interesting relationship between the amplitudes of the Yang–Mills theory and the Landau problem or the problem of quantum Hall effect. (This is also related to Berkovits' twistor string theory [34].) To see how this connection arises, we start by rewriting the formula (3.126) in more compact form as follows. Define a one-particle wave function for the $\mathcal{N} = 4$ supermultiplet by

$$\Phi(\pi, \bar{\pi}, \eta) = \delta\left(\frac{\pi^2}{\pi^1} - \frac{U^2(\sigma)}{U^1(\sigma)}\right) \exp\left(\frac{i}{2}\bar{\pi}^{\dot{A}}\pi^1 \frac{W_{\dot{A}}(\sigma)}{U^1(\sigma)} + i\pi^1\eta_\alpha \frac{\xi^\alpha(\sigma)}{U^1(\sigma)}\right)$$

$$= \delta(\Pi \cdot Z(\sigma))\frac{Z(\sigma) \cdot A}{\Pi \cdot A} \exp\left(\frac{i}{2}\frac{\bar{\Pi} \cdot Z(\sigma)}{Z(\sigma) \cdot A}\frac{\Pi \cdot A}{\Pi \cdot A} + i\frac{\Pi \cdot A}{Z(\sigma) \cdot A}\eta \cdot \xi(\sigma)\right),$$

$$(3.134)$$

where we introduced the twistors,

$$\Pi^\alpha = (0, \pi^A) = (0, 0, \pi^1, \pi^2), \qquad A_\alpha = (0, 0, 1, 0) \qquad (3.135)$$

which gives $Z \cdot A = U^1$, $\Pi \cdot A = \pi^1$. Notice again that Φ is holomorphic in the twistor variables (Z, ξ) and is invariant under the scaling $Z^\alpha \to \lambda Z^\alpha$, $\xi^\alpha \to \lambda \xi^\alpha$. It is also invariant under the scaling of the twistor A_α. There is also an obvious $SU(4)$ or $SU(2,2)$ invariance, if we transform A_α as well. Thus, the expression for Φ can be used with a more general choice of A_α than the one given in (3.135). The twistor A_α plays the role of the reference momentum which has been used in many discussions of scattering amplitudes; ultimately, it drops out of the physical results due to conservation of momentum.

On the $\mathbf{CP}^1$ with the homogeneous coordinates u^a, we define the holomorphic differential

$$(u\,du) \equiv u \cdot du = \epsilon_{ab}\,u^a du^b = (u^1)^2\,d\sigma\,. \qquad (3.136)$$

As a result we can write

$$\int d\sigma_1\,d\sigma_2 \cdots d\sigma_n \frac{1}{(\sigma_1 - \sigma_2)(\sigma_2 - \sigma_3)\cdots(\sigma_n - \sigma_1)}$$

$$= \int (udu)_1(udu)_2 \cdots (udu)_n \frac{1}{(u_1 \cdot u_2)(u_2 \cdot u_3)\cdots(u_n \cdot u_1)} \qquad (3.137)$$

Using the formulae (3.134), (3.137), we can write the amplitude (3.129) as

$$\mathcal{A} = \int d\mu \int \prod_i (u\,du)_i \Phi(\pi_i, \bar{\pi}_i, \eta_i) \frac{\mathrm{Tr}(t^{a_1}\cdots t^{a_n})}{(u_1 \cdot u_2)(u_2 \cdot u_3)\cdots(u_n \cdot u_1)}$$

$$+\text{noncyclic permutations.} \qquad (3.138)$$

Finally, the denominators arise from the chiral Dirac determinant, so we can also write this as

$$\mathcal{A} = \int d\mu \int \prod_i (u\,du)_i \Phi(\pi_i, \bar{\pi}_i, \eta_i) \left(\frac{\delta}{\delta\bar{A}^{a_1}(u_1)} \cdots \frac{\delta}{\delta\bar{A}^{a_n}(u_n)}\right) \mathrm{Tr}\,\log D_{\bar{z}}\Bigg]_{\bar{A}=0}.$$

$$(3.139)$$

This formula takes care of the permutations as well. The holomorphic curves of degree d are as given in (3.102).

3.9.2 A Field Theory on $\mathbf{CP}^1$

We now consider a field theory on $\mathbf{CP}^1$ or the two-sphere. The action is given by

$$\mathcal{S} = \int d\mu(\mathbf{CP}^1) \left[\bar{q}(\bar{\partial} + \bar{A})q + \bar{Y}(\bar{D}Q) \right]. \tag{3.140}$$

Here, q and $\bar{q}$ are standard fermionic fields, so that the first term is the chiral Dirac action on $\mathbf{CP}^1$. These fields are analogous to what generates the extra current algebra in Berkovits' paper [34]. In the second term, Q stands for the supertwistor variables (Z^α, ξ^α). $\bar{Y}$ is another field with values in twistor variables again. Thus, the second term corresponds to a two-dimensional action with target space $\mathbf{C}^{4|4}$. Further, $\bar{D} = \bar{\partial} + \bar{A}$, where $\bar{A}$ is a $GL(1, \mathbf{C})$ gauge field. This term has invariance under the scaling $Q \to \lambda Q$, $\bar{Y} \to \lambda^{-1}\bar{Y}$, with the gauge transformation $\bar{A} \to \bar{A} - \bar{\partial} \log \lambda$. Since there are equal number of fermions and bosons, the $GL(1, \mathbf{C})$ transformation has no anomaly and the use of the chiral action for the second term is consistent with this symmetry at the quantum level.

We will consider the functional integral of $e^{-\mathcal{S}}$. The integration over the fermions $q, \bar{q}$ leads to an effective action $\mathrm{Tr} \log \bar{D}$, and hence, the connected correlator with n factors of $\bar{A}$ is given by

$$M = \int [d\bar{Y} \, dZ \, d\bar{A}] \exp\left(-\int \bar{Y} \bar{D} Z \right)$$
$$\times \int \frac{d^2\sigma_1}{\pi} \cdots \frac{d^2\sigma_n}{\pi} \frac{\mathrm{Tr}[\bar{A}(1)\bar{A}(2)\cdots\bar{A}(n)]}{(\sigma_1 - \sigma_2)(\sigma_2 - \sigma_3)\cdots(\sigma_n - \sigma_1)}. \tag{3.141}$$

We will now take $\bar{A}$ to be of the form $\bar{A} = \bar{\partial}\Phi$. We can use this in (3.141) and carry out the integration over $d\bar{\sigma}$'s. Partial integration can produce δ-functions like $\delta^{(2)}(z_1 - z_2)$. If we exclude coincidence of points, which will correspond to coincidence of external momenta in the context of multigluon scattering, then these δ-functions have no support and the only contribution is from the boundary. The correlator (3.141) then becomes

$$M = \int [d\bar{Y} \, dZ \, d\bar{A}] \exp\left(-\int \bar{Y} \bar{D} Z \right)$$
$$\times \oint \frac{d\sigma_1}{2\pi i} \cdots \frac{d\sigma_n}{2\pi i} \frac{\mathrm{Tr}[\Phi(1)\Phi(2)\cdots\Phi(n)]}{(\sigma_1 - \sigma_2)(\sigma_2 - \sigma_3)\cdots(\sigma_n - \sigma_1)}. \tag{3.142}$$

Using the result (3.137), we can rewrite this as

$$M = \int [d\bar{Y} \, dZ \, d\bar{A}] \exp\left(-\int \bar{Y} \bar{D} Z \right)$$
$$\times \oint \frac{(u \cdot du)_1}{2\pi i} \cdots \frac{(u \cdot du)_n}{2\pi i} \frac{\mathrm{Tr}[\Phi(1)\Phi(2)\cdots\Phi(n)]}{(u_1 \cdot u_2)(u_2 \cdot u_3)\cdots(u_n \cdot u_1)}. \tag{3.143}$$

We now turn to the integration over the gauge field $\bar{A}$. There are nontrivial $U(1)$ bundles over $S^2 \sim \mathbf{CP}^1$, corresponding to Dirac monopoles. The same result holds for $GL(1, \mathbf{C})$. The space of gauge potentials then splits up into a set of disconnected pieces, one for each monopole number. In each sector, we can write $\bar{A} = \bar{A}_d + \delta\bar{A}$, where $\bar{A}_d$ is a fixed configuration of monopole number d and $\delta\bar{A}$ is a fluctuation (of zero monopole number). In two dimensions, we can always write $\delta\bar{A} = \bar{\partial}\Theta$ for some complex function Θ on $\mathbf{CP}^1$. The measure of integration splits up as $[d\bar{A}] = [d\Theta] \det \bar{\partial}$. The determinant can contribute to the conformal anomaly, if we interpret this as the world-sheet formulation of a string theory calculation. For us, thinking of this as a two-dimensional field theory, this anomaly is not relevant. We can now eliminate Θ by absorbing it into the definition of the fields. Φ will be chosen to be $GL(1, \mathbf{C})$ gauge-invariant, so this will not affect it. The integration over the fields $\bar{A}$ is thus reduced to a summation over the different monopole numbers, with a fixed representative background field $\bar{A}_d$ for each value of d.

For the integration over the twistor fields, we need a mode expansion. For the sector with monopole number d, we can expand the fields as

$$Z^\alpha = \sum_{\{a\}} a^\alpha_{a_1 a_2 \cdots a_d} \, u^{a_1} u^{a_2} \cdots u^{a_d} + \text{higher Landau levels}$$

$$\xi^\alpha = \sum_{\{a\}} \gamma^\alpha_{a_1 a_2 \cdots a_d} \, u^{a_1} u^{a_2} \cdots u^{a_d} + \text{higher Landau levels} \qquad (3.144)$$

with similar expansions for $\bar{Y}$. The first set of terms correspond to the lowest Landau level, the higher terms, which we have not displayed explicitly, correspond to higher Landau levels. The wave functions (or mode functions) for the lowest Landau level are holomorphic in the u's, the higher levels involve $\bar{u}$'s as well. The functional integration is now over the coefficients $a^\alpha_{a_1 a_2 \cdots a_d}$, $\gamma^\alpha_{a_1 a_2 \cdots a_d}$, etc. Notice that the zero modes define a holomorphic curve of degree d in supertwistor space. Our choice of Φ will have no $\bar{Y}$, so, in integrating over the nonzero modes, one cannot have any propagators or loops generated by $\bar{Y} - Z$ Wick contractions. As a result, the nonzero modes give only an overall normalization factor. In any correlator involving only Z's and ξ's, we can saturate the fields by the zero modes. Then, apart from constant factors, the correlator (3.143) becomes

$$M = \sum_d C_d \mathcal{M}_d$$

$$\mathcal{M}_d = \int d\mu(a, \gamma) \oint \frac{(u \cdot du)_1}{2\pi i} \cdots \frac{(u \cdot du)_n}{2\pi i} \frac{\text{Tr}[\Phi(1)\Phi(2) \cdots \Phi(n)]}{(u_1 \cdot u_2)(u_2 \cdot u_3) \cdots (u_n \cdot u_1)}.$$
$$(3.145)$$

We now choose the Φ's as given in (3.134), and the correlator (3.145) becomes the multigluon amplitude given in (3.139), once we can show that the integration over the moduli space is given by the invariant measure (3.131).

To see how this arises, consider an $SL(2, \mathbf{C})$ transformation of the coefficients of the mode expansion given by

$$a'^{\alpha}_{a_1 a_2 \cdots a_d} = a^{\alpha}_{b_1 b_2 \cdots b_d} g^{b_1}_{a_1} g^{b_2}_{a_2} \cdots g^{b_n}_{a_n}$$
$$\gamma'^{\alpha}_{a_1 a_2 \cdots a_d} = \gamma^{\alpha}_{a_1 a_2 \cdots a_d} g^{b_1}_{a_1} g^{b_2}_{a_2} \cdots g^{b_n}_{a_n} , \tag{3.146}$$

where $g \in SL(2, \mathbf{C})$. If we use (a', γ'), this is equivalent to using (a, γ) and redefined u's, $u'^a = g^a_b u^b$ in Φ. Since $u \cdot du$ and scalar products like $(u_1 \cdot u_2)$ are invariant under such transformations, we can change the variables $u' \to u$; the integrand for the integration over the moduli (a, γ) is thus $SL(2, \mathbf{C})$-invariant. We must therefore consider the measure (3.131) to obtain well-defined correlators.

What we have shown is that there is a set of correlators in the two-dimensional problem defined by (3.140) which can be calculated exactly, being saturated by the lowest Landau levels, and which give the multigluon amplitudes. The YM amplitudes are thus obtained as a set of "holomorphic" correlators of the two-dimensional problem. The fermionic integration obviously leads to an expression which is similar to the Laughlin wave function for the $\nu = 1$ quantum Hall state. If we consider just the value $\alpha = 1$, the fermionic integration has the form

$$\int [d\gamma] \prod_i \gamma^1_{a_1 a_2 \cdots a_n} u_i^{a_1} \cdots u_i^{a_n} \sim \prod_{i<j} (u_i \cdot u_j) . \tag{3.147}$$

(With four sets of such terms for $\alpha = 1$ to $\alpha = 4$, we get the fourth power of the r.h.s.) In some sense, we can interpret the integration over the bosonic moduli as defining the bosonic version of the Laughlin wave function, since it is related to the fermionic one by the natural supersymmetry of the expression Φ in (3.134). Whether this Landau level point of view for the amplitudes can lead to new insights into the YM problem is yet to be seen.

Acknowledgment

This workwas supported by the National Science Foundation grant number PHY-0244873 and by a PSC-CUNY grant.

References

1. A. Connes: *Nocommutative Geometry* (Academic Press, New York, 1994); J. Madore: *An Introduction to Noncommutative Geometry and Its Physical Applications*, LMS Lecture Notes 206 (1995); G. Landi: *An Introduction to Noncommutative Spaces and Their Geometry*, Lecture Notes in Physics, Monographs m51 (Springer, Berlin Heidelberg New York, 1997); J. Fröhlich, K. Gawedzki:

Conformal field theory and the geometry of strings. In: *Mathematical Quantum Theory*, Proceedings of the conference on Mathematical Quantum Theory, Vancouver, 1993 [hep-th/9310187]

2. E. Witten: Nucl. Phys. B **443**, 85 (1995); T. Banks, W. Fischler, S. Shenker, L. Susskind: Phys. Rev. D **55**, 5112 (1997); N. Ishibashi, H. Kawai, Y. Kitazawa, A. Tsuchiya: Nucl. Phys. B **498**, 467 (1997); W. Taylor IV: Rev. Mod. Phys. **73**, 419 (2001)

3. M.R. Douglas, N.A. Nekrasov: Rev. Mod. Phys. **73**, 977 (2001); J.A. Harvey: Lectures at Komaba 2000: Nonperturbative dynamics in string theory [hep-th/0102076]; A.P. Balachandran: Pramana **59**, 359 (2002) [hep-th/0203259], and references therein

4. A. Connes: Publ. I.H.E.S. **62**, 257 (1986); A. Connes: Commun. Math. Phys. **117**, 673 (1988); A.H. Chamseddine, A. Connes: Phys. Rev. Lett. **77**, 4868 (1996); A.H. Chamseddine, A. Connes: Commun. Math. Phys. **186**, 731 (1997); A.H. Chamseddine, G. Felder, J. Fröhlich: Commun. Math. Phys. **155**, 205 (1993); A. Connes, J. Lott: Nucl. Phys. B (Proc. Suppl.) **18**, 29 (1990; S. Doplicher, K. Fredenhagen, J.E. Roberts: Phys. Lett. B **331**, 39 (1994); D. Kastler: Rev. Math. Phys. **5**, 477 (1993); D. Kastler: Rev. Math. Phys. **8**, 103 (1996); G. Landi, C. Rovelli: Phys. Rev. Lett. **78**, 3051 (1997)

5. U. Carow-Watamura, S. Watamura: Commun. Math. Phys. **212**, 395 (2000); S. Baez, A.P. Balachandran, S. Vaidya, B. Ydri: Commun. Math. Phys. **208**, 787 (2000); A.P. Balachandran, T.R. Govindarajan, B. Ydri: Mod. Phys. Lett. A **15**, 1279 (2000); A.P. Balachandran, T.R. Govindarajan, B. Ydri: [hep-th/0006216]; G. Landi: Rev. Math. Phys. **12**, 1367 (2000); A.P. Balachandran, S. Vaidya: Int. J. Mod. Phys. A **16**, 17 (2001); H. Grosse, C.W. Rupp, A. Strohmaier: J. Geom. Phys. **42**, 54 (2002); B.P. Dolan, D. O'Connor, P. Prešnajder: JHEP **0203**, 013 (2002); S. Iso et al.: Nucl. Phys. B **604**, 121 (2001); P. Valtancoli: Mod. Phys. Lett. A **16**, 639 (2001); A.P. Balachandran, G. Immirzi: Phys. Rev. D **68**, 065023 (2003); Int. J. Mod. Phys. A **18**, 5931 (2003); S. Vaidya, B. Ydri: Nucl. Phys. B **671**, 401 (2003); B. Ydri, Nucl. Phys. B **690**, 230 (2004) [hep-th/0403233]; H. Aoki, S. Iso, K. Nagao: Phys. Rev. D **67**, 065018 (2003); H. Aoki, S. Iso, K. Nagao: Phys. Rev. D **67**, 085005 (2003); H. Aoki, S. Iso, K. Nagao: Nucl. Phys. B **684**, 162 (2004)

6. D. Karabali, V.P. Nair, R. Randjbar-Daemi: Fuzzy spaces, the M(atrix) model and the quantum hall effect. In: *From Fields to Strings: Circumnavigating Theoretical Physics*, Ian Kogan Memorial Collection, ed by M. Shifman, A. Vainshtein, J. Wheater (World Scientific, Singapore, 2004), pp. 831–876

7. J. Sniatycki: *Geometric Quantization and Quantum Mechanics* (Springer, Berlin Heidelberg New York, 1980); N.M.J. Woodhouse: *Geometric Quantization* (Clarendon Press, Oxford, 1992); A.M. Perelomov: *Generalized Coherent States and Their Applications* (Springer, Berlin Heidelberg New York, 1996).

8. D. Karabali, V.P. Nair: Nucl. Phys. B **679**, 427 (2004); D. Karabali, V.P. Nair: Nucl. Phys. B **697**, 513 (2004); S. Randjbar-Daemi: A. Salam, J. Strathdee, Phys. Rev. B **48**, 3190 (1993)

9. G. Alexanian, A. Pinzul, A. Stern: Nucl. Phys. B **600**, 531 (2001); A.P. Balachandran et al.: J. Geom. Phys. **43**, 184 (2002); G. Alexanian et al.: J. Geom. Phys. **42**, 28 (2002); V.P. Nair: Nucl. Phys. B **651**, 313 (2003)

10. V.P. Nair, S. Randjbar-Daemi: Nucl. Phys. B **533**, 333 (1998); H. Grosse, A. Strohmaier: Lett. Math. Phys. **48**, 163 (1999); G. Alexanian et al.: J. Geom. Phys. **42**, 28 (2002)

11. J. Madore: Class. Quant. Grav. **9**, 69 (1992); D. Kabat, W. Taylor IV: Adv. Theor. Math. Phys. **2**, 181 (1998); D. Kabat, W. Taylor IV: Phys. Lett. B **426**, 297 (1998); S.-J. Rey: hep-th/9711081; T. Banks, N. Seiberg, S. Shenker: Nucl. Phys. B **490**, 91 (1997); O.J. Ganor, S. Ramgoolam, W. Taylor IV: Nucl. Phys. B **492**, 191 (1997)

12. H. Grosse, C. Klimčik, P. Prešnajder: Int. J. Theor. Phys. **35**, 231 (1996); H. Grosse, C. Klimčik, P. Prešnajder: Commun. Math. Phys. **178**, 507 (1996); H. Grosse, P. Prešnajder: Lett. Math. Phys. **46**, 61 (1998); P. Prešnajder, J. Math. Phys. **41**, 2789 (2000)

13. V.P. Nair: Phys. Lett. B **505**, 249 (2001); V.P. Nair, A.P. Polychronakos: Phys. Lett. **505**, 267 (2001); D. Karabali, V.P. Nair, A.P. Polychronakos: Nucl. Phys. B **627**, 565 (2002)

14. C. Duval, P.A. Horvàthy: Phys. Lett. B **479**, 284 (2000)

15. B. Morariu, A.P. Polychronakos: Nucl. Phys. B **610**, 531 (2001); B. Morariu, A.P. Polychronakos: Nucl. Phys. B **634**, 326 (2002)

16. F.D. Haldane: Phys. Rev. Lett. **51**, 605 (1983); D. Karabali, V.P. Nair: Nucl. Phys. B **641**, 533 (2002)

17. R. Penrose: J. Math. Phys. **8**, 345 (1967); R. Penrose, M.A.H. MacCallum: Phys. Rep. **6**, 241 (1972); R. Penrose, W. Rindler: *Spinors and Space-Time*, 2 vol (Cambridge University Press, Cambridge, 1984, 1987); S.A. Hugget, K.P. Tod: *An Introduction to Twistor Theory* (Cambridge, University Press, Cambridge 1993)

18. A. Ferber: Nucl. Phys. B **132**, 55 (1978); J. Isenberg, P.B. Yasskin, P.S. Green: Phys. Lett. B **78**, 462 (1978)

19. M.B. Green, J.H. Schwarz, E. Witten: *Superstring Theory*, Vol 2 (Cambridge University Press, Cambridge, 1987)

20. S. Sethi: Nucl. Phys. B **430**, 31 (1994); A. Schwarz: hep-th/9506070

21. M. Roček, N. Wadhwa: hep-th/0408188; C. Zhou, hep-th/0410047

22. V.P. Nair: Phys. Lett. B **214**, 215 (1988)

23. E. Witten: Commun. Math. Phys. **252**, 189 (2004)

24. F. Cachazo, P. Svrček: hep-th/0504194; V.V. Khoze: Gauge theory amplitudes, scalar graphs and twistor space. In: *From Fields to Strings: Circumnavigating Theoretical Physics*, Ian Kogan Memorial Collection, ed by M. Shifman, A. Vainshtein, J. Wheater (World Scientific, Singapore, 2004)

25. R. Roiban, M. Spradlin, A. Volovich: JHEP **0404**, 012 (2004); R. Roiban, M. Spradlin, A. Volovich: Phys. Rev. D **70**, 026009 (2004); R. Roiban, A. Volovich: hep-th/0402121; G. Georgiou, V.V. Khoze: JHEP **0405**, 070 (2004); C.-J. Zhu: JHEP **0404**, 032 (2004); R. Britto et al.: hep-th/0503198

26. I. Bena, Z. Bern, D.A. Kosower: Phys. Rev. D **71**, 045008 (2005); D.A. Kosower: Phys. Rev. D **71**, 045007 (2005); G. Georgiou, E.W.N. Glover, V.V. Khoze: JHEP **0407**, 048 (2004)

27. A. Brandhuber, B. Spence, G. Travaglini: Nucl. Phys. B **706**, 150 (2005); J. Bedford, A. Brandhuber, B. Spence, G. Travaglini: Nucl. Phys. B **706**, 100 (2005); M.X. Luo, C.K. Wen: JHEP **0411**, 004 (2004); M.X. Luo, C.K. Wen: Phys. Lett. B **609**, 86 (2005); F. Cachazo, P. Svrček, E. Witten: JHEP **0410**, 074 (2004); F. Cachazo, P. Svrček, E. Witten: JHEP **0410**, 077 (2004); I. Bena, Z. Bern, D.A. Kosower, R. Roiban: hep-th/0410054; R. Roiban, M. Spradlin, A. Volovich: Phys. Rev. Lett. **94**, 102002 (2005); R. Britto, F. Cachazo, B. Feng: hep-th/0412103; R. Britto, F. Cachazo, B. Feng: Phys. Rev. D **71**, 025012 (2005)

28. Z. Bern, V. Del Duca, L. Dixon, D. Kosower: Phys. Rev. D **71**, 045006 (2005); F. Cachazo: hep-th/0410077; R. Britto, E. Buchbinder, F. Cachazo, B. Feng: hep-th/0503132; Z. Bern, L. Dixon, D. Kosower: hep-th/0412210

29. F. Cachazo, P. Svrcek, E. Witten: JHEP **0409**, 006 (2004); J.-B. Wu, C.-J. Zhu: JHEP **0407**, 032 (2004); J.-B. Wu, C.-J. Zhu: JHEP **0409**, 063 (2004); A. Brandhuber, B. Spence, G. Travaglini: Nucl. Phys. B **706**, 150 (2005); S. Gukov, L. Motl, A. Neitzke: hep-th/0404085; C. Quigley, M. Rozali: JHEP **0501**, 053 (2005)

30. R. Britto, F. Cachazo, B. Feng: hep-th/0412308; R. Britto, F. Cachazo, B. Feng, E. Witten: hep-th/0501052; Z. Bern, L.J. Dixon, D.A. Kosower: hep-th/0501240; M.X. Luo, C.K. Wen: hep-th/0501121

31. L.J. Dixon, E.W.N. Glover, V.V. Khoze: JHEP **0412**, 015 (2004); Z. Bern, D. Forde, D.A. Kosower, P. Mastrolia: hep-ph/0412167; S.D. Badger, E.W.N. Glover, V.V. Khoze: JHEP **0503**, 023 (2005); S.D. Badger, E.W.N. Glover, V.V. Khoze, P. Svrček: hep-th/0504159

32. S. Parke, T. Taylor: Phys. Rev. Lett. **56**, 2459 (1986); Z. Xu, D.-H. Zhong, L. Chang: Nucl. Phys. B **291**, 392 (1987); F.A. Berends, W. Giele: Nucl. Phys. B **294**, 700 (1987); M.L. Mangano, S.J. Parke: Phys. Rep. **200**, 301 (1991); D. Kosower, B.-H. Lee, V.P. Nair: Phys. Lett. B **201**, 85 (1988); M. Mangano, S. Parke, Z. Xu: Nucl. Phys. B **298**, 653 (1988). Z. Bern, L. Dixon, D. Kosower: Ann. Rev. Nucl. Part. Sci. **36**, 109 (1996)

33. W.A. Bardeen: Prog. Theor. Phys. Supp. **123**, 1 (1996); K.G. Selivanov: hep-ph/9604206; G. Chalmers, W. Siegel: Phys. Rev. D **54**, 7628 (1996); A.A. Rosly, K.G. Selivanov: Phys. Lett. B **339**, 135 (1997); D. Cangemi: Nucl. Phys. B **484**, 521 (1996); D. Cangemi: Int. J. Mod. Phys. A **12**, 1215 (1997); C. Kim, V.P. Nair: Phys. Rev. D **55**, 3851 (1997)

34. N. Berkovits: Phys. Rev. Lett. **93**, 011601 (2004); N. Berkovits: hep-th/0403280; N. Berkovits, L. Motl: JHEP **0404**, 056 (2004)

4

Elements of (Super-)Hamiltonian Formalism

A. Nersessian

Artsakh State University, Stepanakert and Yerevan State University, Yerevan, Armenia

Abstract. In these lectures we discuss some basic aspects of Hamiltonian formalism, which usually do not appear in standard textbooks on classical mechanics for physicists. We pay special attention to the procedure of Hamiltonian reduction illustrating it by the examples related to Hopf maps. Then we briefly discuss the supergeneralization(s) of the Hamiltonian formalism and present some simple models of supersymmetric mechanics on Kähler manifolds.

4.1 Introduction

The goal of these lectures is to convince the reader to construct the supersymmetric mechanics within the Hamiltonian framework, or, at least, to combine the superfield approach with the existing methods of Hamiltonian mechanics. The standard approach to construct the supersymmetric mechanics with more than two supercharges is the Lagrangian superfield approach. Surely, superfield formalism is a quite powerful method for the construction of supersymmetric theories. However, all superfield formalisms, being developed á priori for field theory, are convenient for the construction of the field-theoretical models, which are covariant with respect to space-time coordinate transformations. However, the supermultiplets (i.e., the basic ingredients of superfield formalisms) do not respect the transformations mixing field variables. On the other hand, in supersymmetric mechanics these variables appear as spatial coordinates. In other words, the superfield approach, being applied to supersymmetric mechanics, provides us with a local construction of mechanical models. Moreover, the obtained models need to be reformulated in the Hamiltonian framework, for the subsequent quantization. In addition, many of the numerous methods and statements in the Hamiltonian formalism could be easily extended to supersymmetric systems and applied there. Independently from the specific preferences, the "Hamiltonian view" of the existing models of supersymmetric mechanics, which were built within the superfield approach, could establish unexpected links between different supermultiplets

A. Nersessian: *Elements of (Super-)Hamiltonian Formalism*, Lect. Notes Phys. **698**, 139–188 (2006)

DOI 10.1007/3-540-33314-2_4

and models. Finally, the superfield methods seem to be too general in the context of simple mechanical systems.

For this reason, we tried to present some elements of Hamiltonian formalism, which do not usually appear in the standard textbooks on classical mechanics, but appear to be useful in the context of supersymmetric mechanics. We pay much attention to the procedure of Hamiltonian reduction, having in mind that it could be used for the construction of the lower-dimensional supersymmetric models from the existing higher-dimensional ones. Also, we devote a special attention to the Hopf maps and Kähler spaces, which are typical structures in supersymmetric systems. Indeed, to extend the number of supersymmetries (without extension of the fermionic degrees of freedom) we usually equip the configuration/phase space with complex structures and restrict them to be Kähler, hyper-Kähler, quaternionic and so on, often via a choice of the appropriate supermultiplets related to the real, complex, quaternionic structures. We illustrated these matters by examples of Hamiltonian reductions related with Hopf maps, having in mind that they could be straightforwardly applied to supersymmetric systems. Also, we included some less known material related with Hopf fibrations. It concerns the generalization of the oscillator to spheres, complex projective spaces, and quaternionic projective spaces, as well as the reduction of the oscillator systems to Coulomb ones.

Most of the presented constructions are developed only for the zero and first Hopf maps. We tried to present them in the way, which will clearly show, how to extend them to the second Hopf map and the quaternionic case.

The last two sections are devoted to the super-Hamiltonian formalism. We present the superextensions of the Hamiltonian constructions, underlying the specific "super"-properties, and present some examples. Then we provide the list of supersymmetric mechanics constructed within the Hamiltonian approach. Also in this case, we tried to arrange the material in such a way, as to make clear the relation of these constructions to complex structures and their possible extension to quaternionic ones.

The main references to the generic facts about Hamiltonian mechanics are the excellent textbooks [1, 2], and on the supergeometry there exist the monographs [3, 4]. There are numerous reviews on supersymmetric mechanics. In our opinion the best introduction to the subject is given in [5, 6].

4.2 Hamiltonian Formalism

In this section we present some basic facts about the Hamiltonian formalism, which could be straightforwardly extended to the super-Hamiltonian systems.

We restrict ourselves to considering Hamiltonian systems with nondegenerate Poisson brackets. These brackets are defined, locally, by the expressions

$$\{f, g\} = \frac{\partial f}{\partial x^i} \omega^{ij}(x) \frac{\partial g}{\partial x^j}, \quad \det \omega^{ij} \neq 0, \tag{4.1}$$

where

$$\{f, g\} = -\{g, f\}, \Leftrightarrow \omega^{ij} = -\omega^{ji} \tag{4.2}$$

$$\{\{f, g\}, h\} + \text{cycl.perm}(f, g, h) = 0, \Leftrightarrow \omega^{ij}_{,n}\omega^{nk} + \text{cycl.perm}(i, j, k) = 0 . \tag{4.3}$$

Equation (4.2) is known as a "antisymmetricity condition," and (4.3) is called Jacobi identity. Owing to the nondegeneracy of the matrix ω^{ij}, one can construct the nondegenerate two-form, which is closed due to Jacobi identity

$$\omega = \frac{1}{2}\omega_{ij}\mathrm{d}x^i \wedge \mathrm{d}x^j \;\; : \;\; \mathrm{d}\omega = 0 \; \Leftrightarrow \; \omega_{ij,k} + \text{cycl.perm}(i, j, k) = 0 . \tag{4.4}$$

The manifold M equipped with such a form is called symplectic manifold, and denoted by (M, ω). It is clear that M is an even-dimensional manifold, $\dim M = 2N$.

The Hamiltonian system is defined by the triple (M, ω, H), where $H(x)$ is a scalar function called Hamiltonian.

The Hamiltonian equations of motion yield the vector field preserving the symplectic form ω

$$\frac{\mathrm{d}x^i}{\mathrm{d}t} = \{H, x^i\} = V_H^i \;\; : \qquad \mathcal{L}_{V_H}\omega = 0 . \tag{4.5}$$

Here, $\mathcal{L}_\mathbf{V}$ denotes the Lie derivative along vector field $\mathbf{V}$.

Vice versa, any vector field, preserving the symplectic structure, is locally a Hamiltonian one. The easiest way to see it is to use homotopy formula

$$\imath_\mathbf{V}\mathrm{d}\omega + \mathrm{d}\imath_\mathbf{V}\omega = \mathcal{L}_\mathbf{V}\omega \; \Rightarrow \; \mathrm{d}\imath_\mathbf{V}\omega = 0 . \tag{4.6}$$

Hence, $\imath_\mathbf{V}\omega$ is a closed one-form and could be locally presented as follows: $\imath_\mathbf{V}\omega = \mathrm{d}H(x)$. The local function $H(x)$ is precisely the Hamiltonian, generating the vector field $\mathbf{V}$. The transformations preserving the symplectic structure are called *canonical transformations*.

Any symplectic structure could be locally presented in the form (*Darboux theorem*)

$$\omega_\text{can} = \sum_{i=1}^{N} \mathrm{d}p_i \wedge \mathrm{d}q^i , \tag{4.7}$$

where (p_i, q^i) are the local coordinates of the symplectic manifold.

The vector field $\mathbf{V}$ defines a symmetry of the Hamiltonian system, if it preserves both the Hamiltonian $\mathcal{H}$ and the symplectic form ω: $\mathcal{L}_\mathbf{V}\omega = 0$, $\mathbf{V}\mathcal{H} = 0$. Hence,

$$\mathbf{V} = \{\mathcal{J}, \}, \qquad \{\mathcal{J}, \mathcal{H}\} = 0 . \tag{4.8}$$

The $2N$-dimensional Hamiltonian system is called an *integrable system*, when it has N functionally independent constants of motion being in involution (*Liouville theorem*),

$$\{\mathcal{J}_i, \mathcal{J}_j\} = 0, \quad \{\mathcal{H}, \mathcal{J}_i\} = 0, \quad \mathcal{H} = \mathcal{J}_1, \quad i, j = 1, \ldots, N. \qquad (4.9)$$

When the constants of motion are noncommutative, the integrability of the system needs more than N constants of motion. If

$$\{\mathcal{J}_\mu, \mathcal{J}_\nu\} = f_{\mu\nu}(\mathcal{J}), \quad \text{corank } f_{\mu\nu} = K_0, \quad \mu, \nu = 1, \ldots, K \geq K_0, \qquad (4.10)$$

then the system is integrable, if $2N = K + K_0$. The system with $K + K_0 \geq 2N$ constants of motion is sometimes called a *superintegrable system*.

The cotangent bundle T^*M_0 of any manifold M_0 (parameterized by local coordinates q^i) could be equipped with the canonical symplectic structure (4.7).

The dynamics of a free particle moving on M_0 is given by the Hamiltonian system

$$\left(T^*M_0, \quad \omega_{\text{can}}, \quad \mathcal{H}_0 = \frac{1}{2} g^{ij}(q) p_i p_j \right), \qquad (4.11)$$

where $g^{ij} g_{jk} = \delta_k^i$, and $g_{ij} dq^i dq^j$ is a metric on M_0.

The interaction with a potential field could be incorporated in this system by the appropriate change of Hamiltonian,

$$\mathcal{H}_0 \quad \rightarrow \quad \mathcal{H} = \frac{1}{2} g^{ij}(q) p_i p_j + U(q), \qquad (4.12)$$

where $U(q)$ is a scalar function called potential. Hence, the corresponding Hamiltonian system is given by the triplet $(T^*M_0, \quad \omega_{\text{can}}, \quad \mathcal{H})$.

In contrast to the potential field, the interaction with a magnetic field requires a change of symplectic structure. Instead of the canonical symplectic structure ω_{can}, we have to choose

$$\omega_F = \omega_{\text{can}} + F, \quad F = \frac{1}{2} F_{ij}(q) dq^i \wedge dq^j, \quad dF = 0, \qquad (4.13)$$

where F_{ij} are components of the magnetic field strength.

Hence, the resulting system is given by the triplet $(T^*M_0, \quad \omega_F, \quad \mathcal{H})$. Indeed, taking into account that the two-form F is locally exact, $F = dA$, $A = A_a(q) dq^a$, we could pass to the canonical coordinates $(\pi_a = p_a + A_a, \quad q^a)$. In these coordinates the Hamiltonian system assumes the conventional form

$$\left(T^*M_0, \quad \omega_{\text{can}} = d\pi_a \wedge dq^a, \quad \mathcal{H} = \frac{1}{2} g^{ab}(\pi_a - A_a)(\pi_b - A_b) + U(q) \right).$$

Let us also remind, that in the three-dimensional case the magnetic field could be identified with vector, whereas in the two-dimensional case it could be identified with (pseudo)scalar.

The generic Hamiltonian system could be described by the following (phase space) action

$$S = \int dt \left(\mathcal{A}_i(x) \dot{x}^i - \mathcal{H}(x) \right), \qquad (4.14)$$

where $\mathcal{A} = \mathcal{A}_i dx^i$ is a symplectic one-form: $d\mathcal{A} = \omega$. Indeed, varying the action, we get the equations

$$\delta \mathcal{S} = 0 , \quad \Leftrightarrow \quad \dot{x}^i \omega_{ij}(x) = \frac{\partial H}{\partial x^i} , \quad \omega_{ij} = \frac{\partial A_i}{\partial x^j} - \frac{\partial A_j}{\partial x^i} .$$

Though $\mathcal{A}$ is defined up to closed (locally exact) one-form, $\mathcal{A} \to \mathcal{A} + df(x)$, this arbitrariness has no impact in the equations of motion. It change the Lagrangian on the total derivative $f_{,i}\dot{x}^i = df(x)/dt$.

As an example, let us consider the particle in a magnetic field. The symplectic one-form corresponding to the symplectic structure (4.13), could be chosen in the form $\mathcal{A} = (p_a + A_a) dq^a$, $d\mathcal{A} = \omega_F$. Hence, the action (4.14) reads

$$\mathcal{S} = \int dt \left((p_a + A_a)\dot{q}^a - \frac{1}{2}g^{ab}(q)p_a p_b - U(q) \right) . \tag{4.15}$$

Varying this action by p, we get, on the extrema, the conventional second-order action for the system in a magnetic field

$$\mathcal{S}_0 = \int dt \left(\frac{1}{2}g_{ab}\dot{q}^a\dot{q}^b + A_a\dot{q}^a - U(q) \right) . \tag{4.16}$$

The presented manipulations are nothing but the Legendre transformation from the Hamiltonian formalism to the Lagrangian one.

4.2.1 Particle in the Dirac Monopole Field

Let us consider the special case of a system on three-dimensional space moving in the magnetic field of a Dirac monopole. Its symplectic structure is given by the expression

$$\omega_D = dp_i \wedge dq^i + s\frac{q^i}{2|q|^3}\epsilon_{ijk} dq^j \wedge dq^k . \tag{4.17}$$

The corresponding Poisson brackets are given by the relations

$$\{p_i, q^j\} = \delta_i^j , \quad \{q^i, q^j\} = 0 , \quad \{p_i, p_j\} = s\epsilon_{ijk}\frac{q^k}{|q|^3} . \tag{4.18}$$

It is clear that the monopole field does not break the rotational invariance of the system. The vector fields generating $SO(3)$ rotations are given by the expressions

$$\mathbf{V}_i = \epsilon_{ijk}q^j\frac{\partial}{\partial q^k} - \epsilon_{ijk}p_j\frac{\partial}{\partial p_k} , \quad [\mathbf{V}_j, \mathbf{V}_j] = \epsilon_{ijk}\mathbf{V}_k . \tag{4.19}$$

The corresponding Hamiltonian generators could be easily found as well

$$\imath_{\mathbf{V}_i}\omega_D = d\mathcal{J}_i, \quad \{\mathcal{J}_i, \mathcal{J}_j\} = \epsilon_{ijk}\mathcal{J}_k ,$$

where

$$\mathcal{J}_i = \epsilon_{ijk} q^j p_k + s \frac{q^i}{|q|} , \qquad \mathcal{J}_i q^i = s|q| . \tag{4.20}$$

Now, let us consider the system given by the symplectic structure (4.17), and by the $so(3)$-invariant Hamiltonian

$$\mathcal{H} = \frac{p_i p_i}{2g} + U(|q|) , \qquad \{\mathcal{J}_i, \mathcal{H}\} = 0 , \tag{4.21}$$

where $g(|q|)\, dq^i\, dq^i$ is $so(3)$-invariant metric on M_0. In order to find the trajectories of the system, it is convenient to direct the q^3 axis along the vector $\mathbf{J} = (\mathcal{J}_1, \mathcal{J}_2, \mathcal{J}_3)$, i.e., to assume that $\mathbf{J} = \mathcal{J}_3 \equiv J$. Upon this choice of the coordinate system one has

$$\frac{q^3}{|q|} = \frac{s}{J} . \tag{4.22}$$

Then, we introduce the angle

$$\phi = \arctan \frac{q^1}{q^2} , \qquad \frac{d\phi}{dt} = \frac{2J}{g|q|^2}, \tag{4.23}$$

and get, after obvious manipulations

$$\mathcal{E} = \frac{J^2 - s^2}{|q|^2 g} + \frac{J^2}{g} \left(\frac{d|q|}{d\phi} \right)^2 + U(|q|). \tag{4.24}$$

Here, $\mathcal{E}$ denotes the energy of the system.

From the expression (4.23) we find,

$$\phi = J \int \frac{d|q|}{|q|\sqrt{(\mathcal{E} - U)N^2 - J^2 + s^2}} . \tag{4.25}$$

It is seen that, upon the replacement

$$U(q) \to U(q) + \frac{s^2}{g|q|^2} , \tag{4.26}$$

we shall eliminate in (4.25) the dependence on s, i.e., on a monopole field. The only impact of the monopole field on the trajectory will be the shift of the orbital plane given by (4.22).

Let us summarize our considerations. Let us consider the $so(3)$-invariant three-dimensional system

$$\omega_{\text{can}} = d\mathbf{p} \wedge d\mathbf{q} , \qquad \mathcal{H} = \frac{\mathbf{p}^2}{2g} + U(|\mathbf{q}|) , \qquad \{\mathbf{J}_0, \mathcal{H}\} = 0 , \qquad \mathbf{J}_0 = \mathbf{p} \times \mathbf{q} . \tag{4.27}$$

Then, replacing it by the following one:

$$\omega_{\text{can}} + s\frac{\mathbf{q} \times d\mathbf{q} \times d\mathbf{q}}{2|\mathbf{q}|^3} , \quad \mathcal{H} = \frac{1}{2g}\left(\mathbf{p}^2 + \frac{s^2}{|\mathbf{q}|^2}\right) + U(|\mathbf{q}|) , \quad \mathbf{J} = \mathbf{J}_0 + s\frac{\mathbf{q}}{|\mathbf{q}|} ,$$

$$\tag{4.28}$$

we shall preserve the form of the orbit of the initial system, but shift it along $\mathbf{J}$ in accordance with (4.22).

One can expect that, when the initial system has a symmetry, additional with respect to the rotational one, the latter system will also inherit it. For the Coulomb system, $U = \gamma/|\mathbf{q}|$, this is indeed a case. The modified system (which is known as a MIC-Kepler system) possesses the hidden symmetry given by the analog of the Runge–Lenz vector, which is completely similar to the Runge–Lenz vector of the Kepler system [7].

4.2.2 Kähler Manifolds

One of the most important classes of symplectic manifolds is that of Kähler manifolds. The Hermitian manifold $(M, g_{a\bar{b}} \, dz^a \, dz^b)$ is called Kähler manifold, if the imaginary part of the Hermitian structure is a symplectic two-form (see, e.g. [2, 8]):

$$\omega = ig_{a\bar{b}}\, dz^a \wedge d\bar{z}^b \ : \ d\omega = 0 , \quad \det g_{a\bar{b}} \neq 0 .$$

$$\tag{4.29}$$

The Poisson brackets associated with this symplectic structure read

$$\{f, g\}_0 = i\frac{\partial f}{\partial \bar{z}^a} g^{\bar{a}b} \frac{\partial g}{\partial z^b} - i\frac{\partial g}{\partial z^b} g^{\bar{a}b} \frac{\partial f}{\partial \bar{z}^a}, \quad \text{where} \quad g^{\bar{a}b}g_{b\bar{c}} = \delta^{\bar{a}}_{\bar{c}}.$$

$$\tag{4.30}$$

From the closeness of (4.29) it immediately follows, that the Kähler metric can be locally represented in the form

$$g_{a\bar{b}}\, dz^a \, d\bar{z}^b = \frac{\partial^2 K}{\partial z^a \partial \bar{z}^b}\, dz^a \, d\bar{z}^b,$$

$$\tag{4.31}$$

where $K(z, \bar{z})$ is some real function called the Kähler potential. The Kähler potential is defined as moduloholomorphic and antiholomorphic functions

$$K(z, \bar{z}) \to K(z, \bar{z}) + U(z) + \bar{U}(\bar{z}) .$$

$$\tag{4.32}$$

The local expressions for the differential–geometric objects on Kähler manifolds are also very simple. For example, the nonzero components of the metric connections (Cristoffel symbols) look as follows:

$$\Gamma^a_{bc} = g^{\bar{n}a}g_{b\bar{n},c}, \quad \Gamma^{\bar{a}}_{\bar{b}\bar{c}} = \overline{\Gamma}^a_{bc} ,$$

$$\tag{4.33}$$

while the nonzero components of the curvature tensor read

$$R^a_{bc\bar{d}} = -(\Gamma^a_{bc})_{,d}, \quad R^{\bar{a}}_{\bar{b}\bar{c}d} = \overline{R}^a_{bc\bar{d}} .$$

$$\tag{4.34}$$

The isometries of Kähler manifolds are given by the *holomorphic Hamiltonian vector fields*

$$\mathbf{V}_\mu = V_\mu^a(z)\frac{\partial}{\partial z^a} + \bar{V}_\mu^{\bar{a}}(\bar{z})\frac{\partial}{\partial \bar{z}^a}, \qquad \mathbf{V}_\mu = \{\mathbf{h}_\mu, \}_0, \qquad (4.35)$$

where $\mathbf{h}_\mu$ is a real function, $\mathbf{h}_\mu = \bar{\mathbf{h}}_\mu$, called Killing potential. One has

$$[\mathbf{V}_\mu, \mathbf{V}_\nu] = C_{\mu\nu}^\lambda \mathbf{V}_\lambda, \qquad \{\mathbf{h}_\mu, \mathbf{h}_\nu\}_0 = C_{\mu\nu}^\lambda \mathbf{h}_\lambda + \text{const.} ,$$

and

$$\frac{\partial^2 \mathbf{h}_\mu}{\partial z^a \partial z^b} - \Gamma_{ab}^c \frac{\partial \mathbf{h}_\mu}{\partial z^c} = 0 .$$

The dynamics of a particle moving on the Kähler manifold in the presence of a constant magnetic field is described by the Hamiltonian system

$$\Omega_B = dz^a \wedge d\pi_a + d\bar{z}^a \wedge d\bar{\pi}_a + iB g_{a\bar{b}} dz^a \wedge d\bar{z}^b, \quad \mathcal{H}_0 = g^{a\bar{b}}\pi_a\bar{\pi}_b \quad (4.36)$$

The isometries of a Kähler structure define the Noether constants of motion

$$\mathcal{J}_\mu \equiv J_\mu + B\mathbf{h}_\mu = V_\mu^a \pi_a + \bar{V}_\mu^{\bar{a}}\bar{\pi}_{\bar{a}} + B\mathbf{h}_\mu : \begin{cases} \{\mathcal{H}_0, J_\mu\} = 0, \\ \{J_\mu, J_\nu\} = C_{\mu\nu}^\lambda J_\lambda . \end{cases} \quad (4.37)$$

One can easily check that the vector fields generated by $\mathcal{J}_\mu$ are independent of B

$$\mathbf{V} = V^a(z)\frac{\partial}{\partial z^a} - V_{,b}^a \pi_a \frac{\partial}{\partial \pi_a} + \bar{V}^a(\bar{z})\frac{\partial}{\partial \bar{z}^a} - \bar{V}_{,\bar{b}}^a \bar{\pi}_a \frac{\partial}{\partial \bar{\pi}_a} . \quad (4.38)$$

Hence, the inclusion of a constant magnetic field preserves the whole symmetry algebra of a free particle moving on a Kähler manifold.

4.2.3 Complex Projective Space

The most known nontrivial example of a Kähler manifold is the complex projective space $\mathbb{C}P^N$. It is defined as a space of complex lines in $\mathbb{C}^{N+1}$: $u^{\tilde{a}} \sim \lambda u^{\tilde{a}}$, where $u^{\tilde{a}}$, $\tilde{a} = 0, 1, \ldots, N$ are the Euclidean coordinates of $\mathbb{C}^{N+1}$, and $\lambda \in \mathbb{C} - \{0\}$. Equivalently, the complex projective space is the coset space $\mathbb{C}P^N = SU(N+1)/U(N)$.

The complex projective space $\mathbb{C}P^N$ could be covered by $N+1$ charts marked by the indices $\tilde{a} = 0, a$. The zero chart could be parameterized by the functions (coordinates) $z_{(0)}^a = u^a/u^0$, $a = 1, \ldots N$; the first chart by $z_{(1)}^a = z^a/z^1$, $a = 0, 2, 3 \ldots, N$, and so on.

Hence, the transition function from the $\tilde{b}$-th chart to the $\tilde{c}$-th one has the form

$$z_{(\tilde{c})}^{\tilde{a}} = \frac{z_{(\tilde{b})}^{\tilde{a}}}{z_{(\tilde{b})}^{\tilde{c}}}, \quad \text{where} \quad z_{(\tilde{a})}^{\tilde{a}} = 1 . \quad (4.39)$$

One can equip the $\mathbb{C}P^N$ by the Kähler metric, which is known under the name of Fubini-Study metric

$$g_{a\bar{b}}\, dz^a\, dz^b = \frac{dz\, d\bar{z}}{1+z\bar{z}} - \frac{(\bar{z}dz)(zd\bar{z})}{(1+z\bar{z})}.$$

(4.40)

Its Kähler potential is given by the expression

$$K = \log(1 + z\bar{z}).$$

(4.41)

Indeed, it is seen that upon transformation from one chart to the other, given by (4.39), this potential changes by holomorphic and antiholomorphic functions, i.e., the Fubini-Study metric is globally defined on $\mathbb{C}P^N$.

The Poisson brackets on $\mathbb{C}P^N$ are defined by the following relations:

$$\{z^a, \bar{z}^b\} = (1 + z\bar{z})(\delta^{a\bar{b}} + z^a \bar{z}^b), \qquad \{z^a, z^b\} = \{\bar{z}^a, \bar{z}^b\} = 0.$$

(4.42)

It is easy to see that $\mathbb{C}P^N$ is a constant curvature space, with the symmetry algebra $su(N+1)$. This algebra is defined by the Killing potentials

$$h_{\bar{a}b} = \frac{z^a \bar{z}^b - N\delta_{\bar{a}b}}{1+z\bar{z}}, \qquad h_a^- = \frac{z^a}{1+z\bar{z}}, \qquad h_a^+ = \frac{\bar{z}^a}{1+z\bar{z}}.$$

(4.43)

The manifold $\mathbb{C}P^1$ (complex projective plane) is isomorphic to the two-dimensional sphere S^2. Indeed, it is covered by the two charts, with the transition function $z \to 1/z$. The symmetry algebra of $\mathbb{C}P^1$ is $su(2) = so(3)$

$$\{x^i, x^j\} = \epsilon^{ijk}x^k, \quad i, j, k = 1, 2, 3$$

(4.44)

where the Killing potentials x^i look as follows:

$$x^1 + ix^2 = \frac{2z}{1+z\bar{z}}, \qquad x^3 = \frac{1 - z\bar{z}}{1+z\bar{z}}.$$

(4.45)

It is seen that these Killing potentials satisfy the condition

$$x^i x^i = 1,$$

i.e., x^i defines the sphere S^2 in the three-dimensional ambient space $\mathbb{R}^3$. It is straightforwardly checked that z are the coordinates of the sphere in the stereographic projection on $\mathbb{R}^2 = \mathbb{C}$. The real part of the Fubini-Study structure gives the linear element of S^2, and the imaginary part coincides with the volume element of S^2.

On the other hand, these expressions give the embedding of the S^2 in S^3 (with ambient coordinates u^1, u^2) defining the so-called first Hopf map $S^3/S^1 = S^2$. Below we shall describe this map in more detail.

4.2.4 Hopf Maps

The Hopf maps (or Hopf fibrations) are the fibrations of the sphere over a sphere,

$$S^{2p-1}/S^{p-1} = S^p, \quad p = 1, 2, 4, 8 . \tag{4.46}$$

These fibrations reflect the existence of real $(p = 1)$, complex $(p = 2)$, quaternionic $(p = 4)$, and octonionic $(p = 8)$ numbers.

We are interested in the so-called zeroth, first, and second Hopf maps:

$$
\begin{aligned}
S^1/S^0 = S^1 \quad &\text{(zero Hopf map)} \\
S^3/S^1 = S^2 \quad &\text{(first Hopf map)} \\
S^7/S^3 = S^4 \quad &\text{(second Hopf map)} .
\end{aligned}
\tag{4.47}
$$

Let us describe the Hopf maps in explicit terms. For this purpose, we consider the functions $\mathbf{x}(u, \bar{u}), x_0(u, \bar{u})$

$$\mathbf{x} = 2u_1\bar{u}_2, \qquad x_{p+1} = u_1\bar{u}_1 - u_2\bar{u}_2 , \tag{4.48}$$

where u_1, u_2, could be real, complex, or quaternionic numbers. So, one can consider them as a coordinates of the $2p$-dimensional space $\mathbb{R}^{2p}$, where $p = 1$ when $u_{1,2}$ are real numbers; $p = 2$ when $u_{1,2}$ are complex numbers; $p = 4$ when $u_{1,2}$ are quaternionic numbers; $p = 8$ when $u_{1,2}$ are octonionic ones.

In all cases x_{p+1} is a real number, while $\mathbf{x}$ is, respectively, a real number $(p = 1)$, complex number $(p = 2)$, quaternion$(p = 4)$, or octonion $(p = 8)$. Hence, $(x_0, \mathbf{x})$ parameterize the $(p + 1)$-dimensional space $\mathbb{R}^{p+1}$.

The functions $\mathbf{x}, x_{p+1}$ remain invariant under transformations

$$u_a \rightarrow g u_a, \quad \text{where} \quad g\bar{g} = 1 . \tag{4.49}$$

Hence,

$$
\begin{aligned}
g &= \pm 1 \quad \text{for} \quad p = 1 & \tag{4.50} \\
g &= \lambda_1 + i\lambda_2 , \qquad \lambda_1^2 + \lambda_2^2 = 1 \quad \text{for} \quad p = 2 & \tag{4.51} \\
g &= \lambda_1 + i\lambda_2 + j\lambda_3 + k\lambda_4 , \qquad \lambda_1^2 + \cdots + \lambda_4^2 = 1 \quad \text{for} \quad p = 4 . & \tag{4.52}
\end{aligned}
$$

and similarly for the octonionic case $p = 8$.

So, g parameterizes the spheres S^{p-1} of unit radius. Notice that S^1, S^3, S^7 are the only parallelizable spheres. We shall also use the following isomorphisms between these spheres and groups: $S^0 = Z_2$, $S^1 = U(1)$, $S^3 = SU(2)$.

We get that (4.48) defines the fibrations

$$\mathbb{R}^2/S^0 = \mathbb{R}^2, \quad \mathbb{R}^4/S^1 = \mathbb{R}^3, \quad \mathbb{R}^8/S^3 = \mathbb{R}^5, \quad \mathbb{R}^{16}/S^7 = \mathbb{R}^9 . \tag{4.53}$$

One could immediately check that the following equation holds:

$$\mathbf{x}\bar{\mathbf{x}} + x_{p+1}^2 = (u_1\bar{u}_1 + u_2\bar{u}_2)^2 . \tag{4.54}$$

Thus, defining the $(2p-1)$-dimensional sphere in $\mathbb{R}^{2p}$ of the radius r_0: $u_a\bar{u}_a = r_0$, we will get the p-dimensional sphere in $\mathbb{R}^{p+1}$ with radius $R_0 = r_0^2$

$$u_1\bar{u}_1 + u_2\bar{u}_2 = r_0^2 \quad \Rightarrow \quad \mathbf{x}\bar{\mathbf{x}} + x_0^2 = r_0^4 . \tag{4.55}$$

So, we arrive at the Hopf maps given by (4.47). The last, fourth Hopf map, $S^{15}/S^7 = S^8$, corresponding to $p = 8$, is related to octonions in the same manner.

For our purposes it is convenient to describe the expressions (4.48) in a less unified way. For the zero Hopf map it is convenient to consider the initial and resulting ambient spaces $\mathbb{R}^2$ as complex spaces $\mathbb{C}$, parameterized by the single complex coordinates w and z. In this case the map (4.48) could be represented in the form

$$w = z^2 , \tag{4.56}$$

which is known as a Bohlin (or Levi–Civita) transformation relating the Kepler problem with the circular oscillator.

For the first and second Hopf maps it is convenient to represent the transformation (4.48) in the following form:

$$\mathbf{x} = u\gamma\bar{u} . \tag{4.57}$$

Here, for the first Hopf map $\mathbf{x} = (x^1, x^2, x^3)$ parameterizes $\mathbb{R}^3$, and u_1, u_2 parameterize $\mathbb{C}^2$, and $\gamma = (\sigma^1, \sigma^2, \sigma^3)$ are Pauli matrices. This transformation is also known under the name of Kustaanheimo–Stiefel transformation. For the second Hopf map $\mathbf{x} = (x^1, \ldots, x^5)$ parameterizes $\mathbb{R}^5$, and $u_1, \ldots, u_4$ parameterize $\mathbb{C}^4 = \mathbb{H}^2$, and $\gamma = (\gamma^1, \ldots, \gamma^4, \gamma^5 = \gamma^1\gamma^2\gamma^3\gamma^4)$, where $\gamma^1, \ldots, \gamma^4$ are Euclidean four-dimensional gamma-matrices. The latter transformation is sometimes called Hurwitz transformation, or "generalized Kustaanheimo-Stiefel" transformation.

4.3 Hamiltonian Reduction

A Hamiltonian system which has a constant(s) of motion, can be reduced to a lower-dimensional one. The corresponding procedure is called Hamiltonian reduction. Let us explain the meaning of this procedure in the simplest case of the Hamiltonian reduction by a single constant of motion.

Let $(\omega, \mathcal{H})$ be a given $2N$-dimensional Hamiltonian system, with the phase space (local) coordinates x^A, and let $\mathcal{J}$ be its constant of motion, $\{\mathcal{H}, \mathcal{J}\} = 0$. We go from the local coordinates x^A to another set of coordinates, $(\mathcal{H}, y^i, u)$, where $y^i = y^i(x)$ are $2N - 2$ independent functions, which commute with $\mathcal{J}$,

$$\{y^i, \mathcal{J}\} = 0, \quad i = 1, \ldots, 2N - 2 . \tag{4.58}$$

In this case the latter coordinate, $u = u(x)$, necessarily has a nonzero Poisson bracket with $\mathcal{J}$ (because the Poisson brackets are nondegenerate):

$$\{u(x), \mathcal{J}\} \neq 0 . \tag{4.59}$$

Then, we immediately get that in these coordinates the Hamiltonian is independent of u

$$\{\mathcal{J}(\mathcal{H},y,u),\mathcal{H}\} = \frac{\partial \mathcal{H}}{\partial u} \cdot \{u, \mathcal{J}\} \neq 0, \Rightarrow \mathcal{H} = \mathcal{H}(\mathcal{J}, y) . \tag{4.60}$$

On the other hand, from the Jacobi identity we get

$$\{\{y^i, y^j\}, \mathcal{J}\} = \frac{\partial \{y^i, y^j\}}{\partial u} \{u, \mathcal{J}\} = 0 \Rightarrow \{y^i, y^j\} = \omega^{ij}(y, \mathcal{J}) . \tag{4.61}$$

Since $\mathcal{J}$ is a constant of motion, we can fix its value

$$\mathcal{J} = c , \tag{4.62}$$

and describe the system in terms of the local coordinates y^i only

$$(\omega(x), \mathcal{H}(x)) \rightarrow (\omega_{\text{red}}(y,c) = \omega_{ij}(y,c)\,\mathrm{d}y^i \wedge \mathrm{d}y^j, \mathcal{H}_{\text{red}} = \mathcal{H}(y,c)) . \tag{4.63}$$

Hence, we reduced the initial $2N$-dimensional Hamiltonian system to a $(2N - 2)$-dimensional one.

Geometrically, the Hamiltonian reduction by $\mathcal{J}$ means that we fix the $(2N-1)$- dimensional level surface M_c by (4.62), and then factorize it by the action of a vector field $\{\mathcal{J}, \ \}$, which is tangent to M_c. The resulting space $\mathcal{M}_0 = M_c/\{\mathcal{J}, \ \}$ is a phase space of the reduced system.

The Hamiltonian reduction by the K commuting constants of motion $\mathcal{J}$, $\{\mathcal{J}_\alpha, \mathcal{J}_\beta\} = 0$ is completely similar to the above procedure. It reduces the $2N$-dimensional Hamiltonian system to a $2(N - K)$-dimensional one.

When the constants of motion do not commute with each other, the reduction procedure is a bit more complicated.

Let the initial Hamiltonian system have K constants of motion,

$$\{\mathcal{J}_\alpha, \mathcal{H}\} = 0, \qquad \{\mathcal{J}_\alpha, \mathcal{J}_\beta\} = \omega_{\alpha\beta}(\mathcal{J}), \quad \text{corank } \omega_{\alpha\beta}|_{\mathcal{J}_\alpha = c_\alpha} = K_0. \tag{4.64}$$

Hence, one could choose the K_0 functions, which commute with the whole set of the constants of motion

$$\widetilde{\mathcal{J}}_{\tilde{\alpha}}(\mathcal{J}) : \{\widetilde{\mathcal{J}}_{\tilde{\alpha}}, \mathcal{J}_\beta\}|_{\mathcal{J}=c} = 0, \quad \tilde{\alpha} = 1, \ldots K_0 . \tag{4.65}$$

The vector fields $\{\widetilde{\mathcal{J}}_{\tilde{\alpha}}, \ \}$ are tangent to the level surface

$$M_c : \mathcal{J}_\alpha = c_\alpha \quad \dim M_c = 2N - K . \tag{4.66}$$

Factorizing M_c by the action of the commuting vector fields $\{\widetilde{\mathcal{J}}_{\tilde{\alpha}}, \ \}$, we arrive at the phase space of the reduced system, $\mathcal{M}_0 = M_c/\{\mathcal{J}, \ \}$, whose dimension is given by the expression

$$\dim \mathcal{M}_0 = 2N - K - K_0 . \tag{4.67}$$

In contrast to the commuting case, the reduced system could depend on the parameters $\widetilde{c}_{\tilde{\alpha}}$ only.

Notice that the Hamiltonian system could also possess a discrete symmetry. In this case the reduced system has the same dimension as the previous one. To be more precise, the reduction by the discrete symmetry group could be described by a *local* canonical transformation. However, the quantum mechanical counterpart of this canonical transformation could yield a system with nontrivial physical properties.

Below, we shall illustrate the procedure of (Hamiltonian) reduction by discrete, commutative, and noncommutative symmetry generators on examples related to Hopf maps.

4.3.1 Zero Hopf Map: Magnetic Flux Tube

The transformation of the Hamiltonian system associated with the zero Hopf map corresponds to the reduction of the system by the discrete group Z_2. It is a (local) canonical transformation. As a consequence, the resulting system has the same dimension as the initial one.

Let us consider the Hamiltonian system with four-dimensional phase space, parameterized by the pair of canonically conjugated complex coordinates, $(\omega = d\pi \wedge dz + d\bar{\pi} \wedge d\bar{z}, \mathcal{H})$, which is invariant under the following action of Z_2 group:

$$\mathcal{H}(z, \bar{z}, \pi, \bar{\pi}) = \mathcal{H}(-z, -\bar{z}, -\pi, -\bar{\pi}), \quad \omega(\pi, \bar{\pi}, z, \bar{z}) = \omega(-\pi, -\bar{\pi}, -z, -\bar{z}) \; .$$

We can pass now to the coordinates, which are invariant under this transformation (clearly, it is associated with the zero Hopf map)

$$w = z^2, \qquad p = \pi/2z \tag{4.68}$$
$$\omega = d\pi \wedge dz + d\bar{\pi} \wedge d\bar{z} = dp \wedge dw + d\bar{p} \wedge d\bar{w} \; . \tag{4.69}$$

However, one can see that the angular momentum of the initial systems looks as a doubled angular momentum of the transformed one

$$J = i(z\pi - \bar{z}\bar{\pi}) = 2i(wp - \bar{w}\bar{p}). \tag{4.70}$$

This indicates that the global properties of these two systems could be essentially different. This difference has to be reflected in the respective quantum-mechanical systems.

Let us consider the Schrödinger equation

$$\mathcal{H}(\pi, \bar{\pi}, z, \bar{z})\Psi(z, \bar{z}) = E\Psi(z, \bar{z}), \quad \pi = -i\partial_z, \; \bar{\pi} = -i\partial_{\bar{z}} \; , \tag{4.71}$$

with the wave function which obeys the condition

$$\Psi(|z|, \arg z + 2\pi) = \Psi(|z|, \arg z) \; . \tag{4.72}$$

Let us reduce it by the action of Z_2 group, restricting ourselves to even ($\sigma = 0$) or odd ($\sigma = \frac{1}{2}$) solutions of (4.71)

$$\Psi_\sigma(z,\bar z) = \psi_\sigma(z^2, \bar z^2)\, e^{2i\sigma \arg z}, \quad \sigma = 0, 1/2, \tag{4.73}$$

and then perform the Bohlin transformation (4.68). According to (4.73), the wave functions ψ_σ satisfy the condition

$$\psi_\sigma(|w|, \arg w + 2\pi) = \psi_\sigma(|w|, \arg w), \tag{4.74}$$

which implies that the range of definition $\arg w \in [0, 4\pi)$ can be restricted, without loss of generality, to $\arg w \in [0, 2\pi)$. In terms of ψ_σ the Schrödinger equation (4.71) reads

$$\mathcal{H}(\hat p_\sigma, \hat p_\sigma^+, w, \bar w)\psi_\sigma(w, \bar w) = E\psi_\sigma, \quad \hat p_\sigma = -i\partial_w - \frac{i\sigma}{w}. \tag{4.75}$$

Equation (4.75) can be interpreted as the Schrödinger equation of a particle with electric charge e in the static magnetic field given by the potential $A_w = \frac{i\sigma}{ew}$, $\sigma = 0, 1/2$. It is a potential of an infinitely thin solenoid-"magnetic flux tube" (or magnetic vortex, in the two-dimensional interpretation): it has zero strength of the magnetic field $B = \mathrm{rot} A_w = 0$ ($w \in \dot{\mathbb{C}}$) and nonzero magnetic flux $2\pi\sigma/e$.

In accordance with (4.70), the angular momentum transforms as follows:

$$J \to 2J_\sigma, \quad J_\sigma = \frac{i}{\hbar}\left(w\hat p_\sigma - \bar w \hat p_\sigma^+\right), \tag{4.76}$$

where J_σ is the angular momentum operator of the reduced system. Hence, the eigenvalues of the angular momenta of the reduced and initial systems, m_σ and M, are related by the expression $M = 2m_\sigma$, from which it follows that

$$m_\sigma = \pm\sigma, \pm(1+\sigma), \pm(2+\sigma), \dots. \tag{4.77}$$

Hence, the Z_2-reduction related to zero Hopf map transforms the even states of the initial system to the complete basis of the resulting one. The odd states of the initial system yield the wave functions of the resulting system in the presence of magnetic flux generating spin $1/2$. Similarly to the above consideration, one can show that the reduction of the two-dimensional system by the Z_N group yields the N systems with the fractional spin $\sigma = 0, 1/N, 2/N, \dots, (N-1)/N$ (see [9]).

4.3.2 1st Hopf Map: Dirac Monopole

Now, we consider the Hamiltonian reduction by the action of the $U(1)$ group, which is associated with the first Hopf map. It is known under the name of Kustaanheimo–Stiefel transformation.

Let us consider the Hamiltonian system on the four-dimensional Hermitian space $(M_0, g_{a\bar b}\, dz^a\, d\bar z^b)$, $\dim_{\mathbb{C}} M_0 = 2$,

$$T^* M_0, \quad \omega = dz^a \wedge d\pi_a + d\bar z^a \wedge d\bar\pi_a, \quad \mathcal{H} = g^{a\bar b}\pi_a\bar\pi_b + V(z,\bar z). \tag{4.78}$$

We define, on the T^*M_0 space, the Hamiltonian action of the $U(2)$ group given by the generators

$$\mathbf{J} = iz\boldsymbol{\sigma}\boldsymbol{\pi} - i\bar{\boldsymbol{\pi}}\boldsymbol{\sigma}\bar{z}, \quad J_0 = iz\boldsymbol{\pi} - i\bar{z}\bar{\boldsymbol{\pi}} : \tag{4.79}$$

$$\{J_0, J_k\} = 0, \quad \{J_k, J_l\} = 2\epsilon_{klm}J_m, \tag{4.80}$$

where $\boldsymbol{\sigma}$ are Pauli matrices.

Let us consider the Hamiltonian reduction of the phase space (T^*M_0, ω) by the (Hamiltonian) action of the $U(1) = S^1$ group given by the generator J_0. Since J_0 commutes with J_i, the latter will generate the Hamiltonian action of the $su(2) = so(3)$ algebra on the reduced space as well.

To perform the Hamiltonian reduction, we have to fix the level surface

$$J_0 = 2s , \tag{4.81}$$

and then factorize it by the action of the vector field $\{J_0, \}$.

The resulting six-dimensional phase space T^*M^{red} could be parameterized by the following $U(1)$-invariant functions:

$$\mathbf{y} = z\boldsymbol{\sigma}\bar{z}, \quad \boldsymbol{\pi} = \frac{z\boldsymbol{\sigma}\boldsymbol{\pi} + \bar{\boldsymbol{\pi}}\boldsymbol{\sigma}\bar{z}}{2z\bar{z}} : \quad \{\mathbf{y}, J_0\} = \{\boldsymbol{\pi}, J_0\} = 0 . \tag{4.82}$$

In these coordinates the reduced symplectic structure and the generators of the angular momentum are given by the expressions [compare with (4.17), (4.20)]

$$\Omega_{\text{red}} = d\boldsymbol{\pi} \wedge d\mathbf{y} + s\frac{\mathbf{y} \times d\mathbf{y} \times d\mathbf{y}}{2|\mathbf{y}|^3}, \quad \mathbf{J}_{\text{red}} = \mathbf{J}/2 = \boldsymbol{\pi} \times \mathbf{y} + s\frac{\mathbf{y}}{|\mathbf{y}|}.$$

Hence, we get the phase space of the Hamiltonian system describing the motion of a nonrelativistic scalar particle in the magnetic field of the Dirac monopole.

Let M_0 be a $U(2)$-invariant Kähler space with a metric generated by the Kähler potential $K(z\bar{z})$ [10]

$$g_{a\bar{b}} = \frac{\partial^2 K(z\bar{z})}{\partial z^a \partial \bar{z}^b} = a(z\bar{z})\delta_{a\bar{b}} + a'(z\bar{z})\bar{z}^a z^b , \tag{4.83}$$

where

$$a(y) = \frac{dK(y)}{dy}, \quad a'(y) = \frac{d^2K(y)}{dy^2} .$$

Let the potential be also $U(2)$-invariant, $V = V(z\bar{z})$, so that $U(2)$ is a symmetry of the Hamiltonian: $\{J_0, \mathcal{H}\} = \{J_i, \mathcal{H}\} = 0$.

Hence, the Hamiltonian could also be restricted to the reduced six-dimensional phase space. The reduced Hamiltonian looks as follows:

$$\mathcal{H}_{\text{red}} = \frac{1}{a}\left[y\boldsymbol{\pi}^2 - b(\mathbf{y}\boldsymbol{\pi})^2 \right] + s^2\frac{1 - by}{ay} + V(y),$$

where

$$y \equiv |\mathbf{y}|, \quad b = \frac{a'(y)}{a + ya'(y)} .$$

Let us perform the canonical transformation $(\mathbf{y}, \boldsymbol{\pi}) \rightarrow (\mathbf{x}, \mathbf{p})$ to the conformal-flat metric

$$\mathbf{x} = f(y)\mathbf{y}, \qquad \boldsymbol{\pi} = f\mathbf{p} + \frac{df}{dy}\frac{(\mathbf{yp})}{y}\mathbf{y} ,$$

where

$$\left(1 + \frac{yf'(y)}{f}\right)^2 = 1 + \frac{ya'(y)}{a} \quad \Rightarrow \quad \left(\frac{d\log x}{dy}\right)^2 = \frac{d\log ya(y)}{y\,dy}, \quad x < 1 .$$

In the new coordinates the Hamiltonian takes the form

$$\mathcal{H}_{\mathrm{red}} = \frac{x^2(y)}{ya(y)}\mathbf{p}^2 + \frac{s^2}{y(a + ya'(y))} + V\left(y(x)\right) .$$

To express the y, $a(y)$, $a'(y)$ via x, it is convenient to introduce the function

$$\tilde{A}(y) \equiv \int (a + ya'(y))yf(y)\,dy$$

and consider its Legendre transform $A(x)$,

$$A(x) = A(x, y)|_{\partial A(x,y)/\partial y}, \qquad A(x, y) = xa(y)y - \tilde{A}(y) .$$

Then, we immediately get

$$\frac{dA(x)}{dx} = a(y)y, \qquad x\frac{d^2A}{dx^2} = y\sqrt{a(a + ya'(y))} .$$

By the use of these expressions, we can represent the reduced Hamiltonian as follows:

$$\mathcal{H}_{\mathrm{red}} = \frac{x^2}{N^2}\mathbf{p}^2 + \frac{s^2}{(2xN'(x))^2} + V\left(y(x)\right), \quad N^2(x) \equiv \frac{dA}{dx}. \tag{4.84}$$

The Kähler potential of the initial system is connected with N via the equations

$$\frac{dK}{dx} = \frac{N^3(x)}{2x^2N'(x)}, \qquad \frac{d\log y}{dx} = \frac{N}{2x^2N'(x)} . \tag{4.85}$$

Hence, for $s = 0$ we shall get the system (4.27). However, when $s \neq 0$, by comparing the reduced system with (4.28), we conclude that the only Kähler space which yields a "well-defined system with monopole" is flat space.

4.3.3 $\mathbb{C}^{N+1} \to \mathbb{C}\mathbf{P}^N$ and $T^*\mathbb{C}^{N+1} \to T^*\mathbb{C}\mathbf{P}^N$

Now, we consider the Hamiltonian reduction of the space $(\mathbb{C}^{N+1}, \omega = du^0 d\bar{u}^0 + du^a d\bar{u}^a)$, to the complex projective space $\mathbb{C}\mathbf{P}^N$.

The $U(N+1) = U(1) \times SU(N)$ isometries of this space are defined by the following Killing potentials:

$$J_0 = u\bar{u}, \quad J_{su(N+1)} = u\hat{T}\bar{u}, \quad \{J_0, J_{su(N+1)}\} = 0 ,$$

where $T = T^\dagger$, $\mathrm{Tr}T = 0$ are $(N+1) \times (N+1)$-dimensional traceless matrices defining the $su(N+1)$ algebra. The Poisson brackets, corresponding to the Kähler structure, are defined by the relations $\{u^0, \bar{u}^0\} = i$, $\{u, \bar{u}^b\} = i\delta^{ab}$.

Let us perform the Hamiltonian reduction by the action of J_0. The reduced phase space is a $2N$-dimensional one. Let us choose for this space the following local complex coordinates:

$$z^a = \frac{u^a}{u^0} \quad : \{z^a, J_0\} = 0, \quad a = 1, \dots, N \tag{4.86}$$

and fix the level surface

$$J_0 = r_0^2 \Rightarrow |u^0|^2 = \frac{r_0^2}{1 + z\bar{z}} . \tag{4.87}$$

Then, we immediately get the Poisson brackets for the reduced space

$$\{z^a, \bar{z}^b\} = \frac{i}{r_0^2}(1 + z\bar{z})(\delta^{ab} + z^a\bar{z}^b) , \qquad \{z^a, z^b\} = \{\bar{z}^a, \bar{z}^b\} = 0 . \tag{4.88}$$

Hence, the reduced Poisson bracket are associated with the Kähler structure. It could be easily seen, that this Kähler structure is given by the Fubini-Study metric (4.40) multiplied on r_0^2. The restriction of the generators $J_{su(N+1)}$ on the level surface (4.87) yields the expressions (4.43).

In the above example $\mathbb{C}^{N+1}$ and $\mathbb{C}\mathbf{P}^N$ appeared as the phase spaces. Now, let us show, how to reduce the $T^*\mathbb{C}^{N+1}$ to $T^*\mathbb{C}\mathbf{P}^N$, i.e., let us consider the case when $\mathbb{C}^{N+1}$ and $\mathbb{C}\mathbf{P}^N$ play the role of the configuration spaces of the mechanical systems. Since the dimension of $T^*\mathbb{C}^{N+1}$ is $4(N+1)$, and the dimension of $T^*\mathbb{C}^N$ is $4N$, the reduction has to be performed by two commuting generators.

Let us equip the initial space with the canonical symplectic structure (4.78), and perform the reduction of this phase space by the action of the generators

$$J_0 = i\pi u - \bar{\pi}\bar{u}, \qquad h_0 = u\bar{u} \quad : \quad \{J_0, h_0\} = 0 . \tag{4.89}$$

We choose the following local coordinates of the reduced space:

$$z^a = \frac{u^a}{u^0}, \quad p_a = g_{a\bar{b}}(z, \bar{z})\left(\frac{\bar{\pi}^a}{\bar{u}^0} - \bar{z}^a\frac{\bar{\pi}^0}{\bar{z}^0}\right) :$$

$$\{z^a, J_0\} = \{z^a, h_0\} = \{p_a, J_0\} = \{p_a, h_0\} = 0 \,,$$

where $g_{a\bar{b}}$ is defined by the expression (4.40). Then, calculating the Poisson brackets between these functions, and fixing the value of the generators J_0, h_0,

$$h_0 = r_0^2, \qquad J_0 = s \,, \tag{4.90}$$

we get

$$\{p_a, z^b\} = \delta_a^b, \qquad \{p_a, \bar{p}_b\} = i\frac{s}{r_0^2}g_{a\bar{b}}(z, \bar{z}) \,. \tag{4.91}$$

Hence, we arrive at the phase space structure of the particle moving on $\mathbb{C}\mathrm{P}^N$ in the presence of a constant magnetic field with $B_0 = s/r_0^2$ strength.

4.3.4 2nd Hopf Map: $SU(2)$ Instanton

In the above examples we have shown that the zero Hopf map is related to the canonical transformation corresponding to the reduction of the two-dimensional system by the discrete group $Z_2 = S^0$, and transforms the system with two-dimensional configuration space to the system of the same dimension, which has a spin $\sigma = 0, 1/2$. The first Hopf map corresponds to the reduction of the system with four-dimensional configuration space by the Hamiltonian action of $U(1) = S^1$ group, and yields the system moving on the three-dimensional space in the presence of the magnetic field of the Dirac monopole. Similarly, with the second Hopf map one can relate the Hamiltonian reduction of the cotangent bundle of eight-dimensional space (say, $T^*\mathbb{C}^4 = T^*\mathbb{IH}^2$) by the action of $SU(2) = S^3$ group. When the $SU(2)$ generators I_i have nonzero values, $I_i = c_i, \sum_i |c_i| \neq 0$, the reduced space is a $(2 \cdot 8 - 3 - 1 =)$12-dimensional one, $T^*\mathbb{R}^5 \times S^2$. It is the phase space of a colored particle moving on $\mathbb{R}^5$ in the presence of the $SU(2)$ Yang monopole [11] (here S^2 appears as a isospin space).

When $c_1 = c_2 = c_3 = 0$, the J_i generators commute with each other, and the reduced space is a $(2 \cdot 8 - 2 \cdot 3 =)$10-dimensional one, $T^*\mathbb{R}^5$. Such a reduction is also known under the name of Hurwitz transformation relating the eight-dimensional oscillator with the five-dimensional Coulomb problem.

We shall describe a little bit different reduction, associated with the fibration $\mathbb{C}\mathrm{P}^3/\mathbb{C}\mathrm{P}^1 = S^4$ [12]. This fibration could be immediately obtained by factorization of the second Hopf map $S^7/S^3 = S^4$ by $U(1)$. Indeed, the second Hopf map is described by the formulae (4.48), (4.49), where S^7 is embedded in the two-dimensional quaternionic space $\mathbb{IH}^2 = \mathbb{C}^4$, parameterized by four complex (two quaternionic) Euclidean coordinates

$$u_i = v_i + jv_{i+1}, \quad i = 1, 2, \quad u_1, \mathbf{u}_2 \in \mathbb{IH}, \quad v_1, v_2, v_3, v_4 \in \mathbb{C} \,. \tag{4.92}$$

Here, S^4 is embedded in $\mathbb{R}^5$ parameterized by the Eucludean coordinates $(\mathbf{x}, x_5)$ given by (4.48). This embedding is invariant under the right action of a $SU(2)$ group given by (4.49), so that $\mathbf{g}$ defines a three-sphere (4.52). The complex projective space $\mathbb{C}P^3$ is defined as $S^7/U(1)$, while the inhomogeneous coordinates z_a appearing in the Fubini-Study metric of $\mathbb{C}P^3$, are related to the coordinates of $\mathbb{C}^4$ as follows: $z_a = v_a/v_4$, $a = 1, 2, 3$. The expressions (4.48) defining S^4 are invariant under $U(1)$-factorization, while $S^3/U(1) = S^2$. Thus, we arrive to the conclusion that $\mathbb{C}P^3$ is the S^2-fibration over $S^4 = \mathbb{IHP}^1$. The expressions for z_a yield the following definition of the coordinates of S^4:

$$w_1 = \frac{\bar{z}_2 + z_1 \bar{z}_3}{1 + z_3 \bar{z}_3}, \qquad w_2 = \frac{z_2 \bar{z}_3 - \bar{z}_1}{1 + z_3 \bar{z}_3} . \tag{4.93}$$

Choosing z_3 as a local coordinate of $S^2 = \mathbb{C}P^1$,

$$u = z_3 , \tag{4.94}$$

we get the expressions

$$z_1 = w_1 u - \bar{w}_2, \quad z_2 = w_2 u + \bar{w}_1, \quad z_3 = u . \tag{4.95}$$

In these coordinates the Fubini-Study metric on $\mathbb{C}P^3$ looks as follows:

$$g_{a\bar{b}} \, dz_a \, d\bar{z}_b = \frac{dz \, d\bar{z}}{1 + z\bar{z}} - \frac{(\bar{z} \, dz)(z \, d\bar{z})}{(1 + z\bar{z})^2} = \frac{dw_i \, d\bar{w}_i}{(1 + w\bar{w})^2} + \frac{(du + \mathcal{A})(d\bar{u} + \bar{\mathcal{A}})}{(1 + u\bar{u})^2}, \tag{4.96}$$

where

$$\mathcal{A} = \frac{(\bar{w}_1 + w_2 u)(u \, dw_1 - d\bar{w}_2) + (\bar{w}_2 - w_1 u)(u \, dw_2 + d\bar{w}_1)}{1 + w\bar{w}} . \tag{4.97}$$

Hence, w_1, w_2, and u are the conformal-flat complex coordinates of $S^4 = \mathbb{IHP}^1$ and $S^2 = \mathbb{C}P^1$, while the connection $\mathcal{A}$ defines the $SU(2)$ gauge field.

Now, let us consider the Hamiltonian system describing the motion of a free particle on $\mathbb{C}P^3$

$$\mathcal{H}_{\mathbb{C}P^3} = g^{a\bar{b}} \pi_a \bar{\pi}_b , \qquad \{z_a, \pi_b\} = i\delta_{ab} \tag{4.98}$$

Let us extend the coordinate transformation (4.95) to the $T^*\mathbb{C}P^3$, by the following transformation of momenta:

$$\pi_1 = \frac{\bar{u}p_1 - \bar{p}_2}{1 + u\bar{u}}, \qquad \pi_2 = \frac{\bar{u}p_2 + \bar{p}_1}{1 + u\bar{u}},$$

$$\pi_3 = p_u + \frac{\bar{p}_2 w_1 - \bar{p}_1 w_2 - \bar{u}(w_1 p_1 + w_2 p_2)}{1 + u\bar{u}} . \tag{4.99}$$

This extended transformation is a canonical transformation,

$$\{w_i, p_j\} = \delta_{ij}, \qquad \{u, p_u\} = 1 . \tag{4.100}$$

In the new coordinates the Hamiltonian reads

$$\mathcal{H}_{\mathbb{CP}^3} = (1 + w\bar{w})^2 P_i \bar{P}_i + (1 + u\bar{u})^2 p_u \bar{p}_u .\qquad(4.101)$$

Here, we introduced the covariant momenta

$$P_1 = p_1 - i\frac{\bar{w}_1}{1 + w\bar{w}}I_1 - \frac{w_2}{1 + w\bar{w}}I_+, \quad P_2 = p_2 - i\frac{\bar{w}_2}{1 + w\bar{w}}I_1 + \frac{w_1}{1 + w\bar{w}}I_+,$$
$$(4.102)$$

and the $su(2)$ generators $I_\pm, I_1$ defining the isometries of S^2

$$I_1 = -i(p_u u - \bar{p}_u \bar{u}), \quad I_- = p_u + \bar{u}^2 \bar{p}_{\bar{u}}, \quad I_+ = \bar{p}_{\bar{u}} + u^2 p_u \qquad(4.103)$$
$$\{I_\pm, I_1\} = \mp i I_\pm, \quad \{I_+, I_-\} = 2i I_1 .$$

The nonvanishing Poisson brackets between P_i and w_i are given by the following relations (and their complex conjugates):

$$\{w_i, P_j\} = \delta_{ij}, \quad \{P_1, P_2\} = -\frac{2I_+}{(1 + w\bar{w})^2}, \quad \{P_i, \bar{P}_j\} = -i\frac{2I_1 \delta_{ij}}{(1 + w\bar{w})^2} .$$
$$(4.104)$$

The expressions in the r.h.s. define the strength of a homogeneous $SU(2)$ instanton (the "angular part" of the $SU(2)$ Yang monopole), written in terms of conformal–flat coordinates of $S^4 = \mathbb{HP}^1$. Hence, the first part of the Hamiltonian, i.e., $\mathcal{D}_4 = (1 + w\bar{w})^2 P_i \bar{P}_i$, describes a particle on the four-dimensional sphere in the field of a $SU(2)$ instanton.

The Poisson brackets between P_i and $u, \bar{u}, p_u, \bar{p}_u$ are defined by the following nonzero relations and their complex conjugates:

$$\{P_i, p_u\} = -\frac{\bar{w}_i + 2\epsilon_{ij} w_j u}{1 + \bar{w}w}p_u, \quad \{P_i, \bar{p}_u\} = \frac{\bar{w}_i \bar{p}_u}{1 + \bar{w}w} ,$$

$$\{P_i, u\} = \frac{(\bar{w}_i + \epsilon_{ij} w_j u) u}{1 + \bar{w}w}, \quad \{\bar{P}_i, u\} = \frac{\epsilon_{ij}\bar{w}_j - w_i u}{1 + \bar{w}w} .$$

The second part of the Hamiltonian defines the motion of a free particle on the two-sphere. It could be represented as a Casimir of $SU(2)$

$$\mathcal{D}_{S^2} = (1 + u\bar{u})^2 p_u \bar{p}_u = I_+ I_- + I_1^2 \equiv I^2 .\qquad(4.105)$$

It commutes with the Hamiltonian $\mathcal{D}_0$, as well as with $I_1, I_\pm$ and P_i, w_i

$$\{\mathcal{D}_{\mathbb{CP}^3}, I^2\} = \{P_i, I^2\}_B = \{w_i, I^2\}_B = \{I_1, I^2\}_B = \{I_\pm, I^2\}_B = 0 .\quad(4.106)$$

Hence, we can perform a Hamiltonian reduction by the action of the generator $\mathcal{D}_2$, which reduces the initial 12-dimensional phase space $T_*\mathbb{CP}^3 = T^*(S^4 \times S^2)$ to a 10-dimensional one. The relations (4.106) allow us to parameterize the reduced 10-dimensional phase space in terms of the coordinates $P_i, w_i, I_\pm, I_1$, where the latter obey the relation

$$I_+ I_- + I_1^2 \equiv I^2 = \text{const} .\qquad(4.107)$$

Thus, the reduced phase space is nothing but $T^*S^4 \times S^2$, where S^2 is the internal space of the instanton.

Let us collect the whole set of nonzero expressions defining the Poisson brackets on $T_*S^4 \times S^2$

$$\{w_i, P_j\} = \delta_{ij},$$

$$\{P_1, P_2\} = -\frac{2I_+}{(1 + w\bar{w})^2},$$

$$\{P_i, \bar{P}_j\} = -i\frac{2I_1\delta_{ij}}{(1 + w\bar{w})^2},$$

$$\{P_i, I_1\} = i\frac{\epsilon_{ij}w_j I_+}{1 + w\bar{w}} \qquad (4.108)$$

$$\{P_i, I_+\} = \frac{\bar{w}_i I_+}{1 + \bar{w}w},$$

$$\{P_i, I_-\} = -\frac{\bar{w}_i I_- + 2i\epsilon_{ij}w_j I_1}{1 + \bar{w}w}$$

$$\{I_+, I_-\} = 2iI_1, \quad \{I_\pm, I_1\} = \mp iI_\pm \ .$$

The reduced Hamiltonian is $\mathcal{H}_{\mathbb{CP}^3}^{\mathrm{red}} = (1 + w\bar{w})^2 P\bar{P} + I^2$. So, the Hamiltonian of the colored particle on S^4 interacting with the $SU(2)$ instanton is connected with the Hamiltonian of a particle on $\mathbb{CP}^3$ as follows:

$$\mathcal{D}_{S^4} = \mathcal{D}_{\mathbb{CP}^3}^{\mathrm{red}} - I^2 \quad (> 0). \qquad (4.109)$$

This yields an intuitive explanation of the degeneracy in the ground state in the corresponding quantum system on S^4. Indeed, since the l.h.s. is positive, the ground state of the quantum system on S^4 corresponds to the excited state of a particle on $\mathbb{CP}^3$, which is a degenerate one. On the other hand, the ground state of a particle on $\mathbb{CP}^3$ can be reduced to the free particle on S^4, when $I = 0$.

Now, let us consider a similar reduction for the particle on $\mathbb{CP}^3$, in the presence of constant magnetic field (4.36).

Passing to the coordinates (4.95) and momenta (4.102) we get the Poisson brackets defined by the nonzero relations given by (4.105) and

$$\{p_u, \bar{p}_u\}_B = \frac{iB}{(1 + u\bar{u})^2}, \qquad (4.110)$$

$$\{w_i, P_j\}_B = \delta_{ij}, \quad \{P_1, P_2\}_B = -\frac{2I_+}{(1 + w\bar{w})^2}, \qquad (4.111)$$

$$\{P_i, \bar{P}_j\}_B = -i\frac{2I_1\delta_{ij}}{(1 + w\bar{w})^2}. \qquad (4.112)$$

where $\mathcal{I}_\pm, \mathcal{I}_1$ are defined by the expressions

$$\mathcal{I}_1 = I_1 + \frac{B}{2}\frac{1-u\bar{u}}{1+u\bar{u}}, \quad \mathcal{I}_- = I_- - B\frac{i\bar{u}}{1+u\bar{u}}, \quad \mathcal{I}_+ = I_+ + B\frac{iu}{1+u\bar{u}} \quad (4.113)$$

Notice that the expressions (4.112) are similar to (4.104) and the generators (4.113) form, with respect to the new Poisson brackets, the $su(2)$ algebra

$$\{\mathcal{I}_\pm, \mathcal{I}_1\}_B = \mp i\mathcal{I}_\pm, \qquad \{\mathcal{I}_+, \mathcal{I}_-\} = 2i\mathcal{I}_1. \qquad (4.114)$$

It is clear that these generators define the isometries of the "internal" two-dimensional sphere with a magnetic monopole located at the center.

Once again, as in the absence of a magnetic field, we can reduce the initial system by the Casimir of the $SU(2)$ group

$$\mathcal{I}^2 \equiv \mathcal{I}_1^2 + \mathcal{I}_+\mathcal{I}_- = \mathcal{D}_{S^2} + B^2/4, \quad \Rightarrow \quad \mathcal{I} \geq B/2 \ . \qquad (4.115)$$

To perform the Hamiltonian reduction, we have to fix the value of $\mathcal{I}^2$, and then factorize by the action of the vector field $\{\mathcal{I}^2, \}_B$.

The coordinates (4.93), (4.102) commute with the Casimir (4.115),

$$\{P_i, \mathcal{I}^2\}_B = \{w_i, \mathcal{I}^2\}_B = \{\mathcal{I}_1, \mathcal{I}^2\}_B = \{\mathcal{I}_\pm, \mathcal{I}^2\}_B = 0 \ . \qquad (4.116)$$

Hence, as we did above, we can choose P_i, w_i, and $\mathcal{I}_\pm$ as the coordinates of the reduced, 10-dimensional phase space.

The coordinates $\mathcal{I}_\pm$, $\mathcal{I}_1$ obey the condition

$$\mathcal{I}_1^2 + \mathcal{I}_-\mathcal{I}_+ = \mathcal{I}^2 = \text{const.} \qquad (4.117)$$

The resulting Poisson brackets are defined by the expressions (4.108), with $I_1, I_\pm$ replaced by $\mathcal{I}_\pm, \mathcal{I}_1$.

Hence, the particle on $\mathbb{C}P^3$ moving in the presence of a constant magnetic field reduces to a colored particle on S^4 interacting with the instanton field. The Hamiltonians of these two systems are related as follows:

$$\mathcal{D}_{S^4} = \mathcal{D}_{\mathbb{C}P^3}^{\text{red}} - \mathcal{I}^2 + B^2/4, \quad \mathcal{I} \geq B/2 \qquad (4.118)$$

Notice that, upon quantization, we must replace $\mathcal{I}^2$ by $\mathcal{I}(\mathcal{I}+1)$ and require that both $\mathcal{I}$ and B take (half)integer values (since we assume unit radii for the spheres, this means that the "monopole number" obeys a Dirac quantization rule). The extension of this reduction to quantum mechanics relates the theories of the quantum Hall effect on S^4 [13] and $\mathbb{C}P^3$ [14].

Notice that the third Hopf map could also be related with the generalized quantum Hall effect theory [15].

4.4 Generalized Oscillators

Among the integrable systems with hidden symmetries the oscillator is the simplest one. In contrast to other systems with hidden symmetries (e.g., Coulomb systems), its symmetries form a Lie algebra. The N-dimensional oscillator on $T^*\mathbb{R}^N$,

$$\mathcal{H} = \frac{1}{2}\left(p_a p_a + \alpha^2 q^a q^a\right), \quad \omega_{\mathrm{can}} = dp_a \wedge dq^a, \quad a = 1,\dots,N \qquad (4.119)$$

besides the rotational symmetry $so(N)$, has also hidden ones, so that the whole symmetry algebra is $su(N)$. The symmetries of the oscillator are given by the generators

$$J_{ab} = p_a q^b - p_b q^a, \qquad I_{ab} = p_a p_b + \alpha^2 q^a q^b. \qquad (4.120)$$

The huge number of hidden symmetries allows us to construct generalizations of the oscillator on curved spaces, which inherit many properties of the initial system.

The generalization of the oscillator on the sphere was suggested by Higgs [16]. It is given by the following Hamiltonian system:

$$\mathcal{H} = \frac{1}{2}\left(g^{ab} p_a p_b + \alpha^2 q^2\right), \quad \omega = dp_a \wedge dq^a, \quad q^a = \frac{x_a}{x_0}, \qquad (4.121)$$

where x^a, x_0 are the Euclidean coordinates of the ambient space $\mathbb{R}^{N+1}$: $x_0^2 + x^a x^a = 1$, and $g_{ab}\, dq^a\, dq^b$ is the metric on S^N. This system inherits the rotational symmetries of the flat oscillator given by (4.120), and possesses the hidden symmetries given by the following constants of motion [compare with (4.120)]:

$$I_{ab} = J_a J_b + \alpha^2 q^a q^b, \qquad (4.122)$$

where J_a are the translation generators on S^N.

In contrast to the flat oscillator, whose symmetry algebra is $su(N)$, the spherical (Higgs) oscillator has a nonlinear symmetry algebra.

This construction has been extended to the complex projective spaces in [17], where the oscillator on $\mathbb{C}P^N$ was defined by the Hamiltonian

$$\mathcal{H} = \left(g^{\bar{a}b}\bar{\pi}_a \pi_b + \alpha^2 z\bar{z}\right), \qquad (4.123)$$

with $z^a = u^a/u^0$ denoting inhomogeneous coordinates of $\mathbb{C}P^N$ and $g_{a\bar{b}}\, dz^a\, d\bar{z}^b$ being Fubini-Study metric (4.40).

It is easy to see that this system has constants of motion given by the expressions

$$J_{a\bar{b}} = i(z^b \pi_a - \bar{\pi}_b \bar{z}^a), \qquad I_{a\bar{b}} = J_a^+ J_{\bar{b}}^- + \omega^2 \bar{z}^a z^b, \qquad (4.124)$$

where $J_a^+ = \pi_a + (\bar{z}\bar{\pi})\bar{z}^a$, $J_a^- = \bar{J}_a^+$ are the translation generators on $\mathbb{C}P^N$. The generators $J_{a\bar{b}}$ define the kinematical symmetries of the system and form a $su(N)$ algebra. When $N > 1$, the generators $I_{a\bar{b}}$ are functionally independent of $\mathcal{H}$, $J_{a\bar{b}}$ and define hidden symmetries. As in the spherical case, their algebra is a nonlinear one

$$\{J_{\bar{a}b}, J_{\bar{c}d}\} = i\delta_{\bar{a}d}J_{\bar{b}c} - i\delta_{\bar{c}b}J_{\bar{a}d},$$
$$\{I_{a\bar{b}}, J_{cd}\} = i\delta_{c\bar{b}}I_{ad} - i\delta_{ad}I_{c\bar{b}}$$
$$\{I_{a\bar{b}}, I_{c\bar{d}}\} = i\alpha^2\delta_{c\bar{b}}J_{a\bar{d}} - i\alpha^2\delta_{ad}J_{c\bar{b}} +$$
$$+iI_{c\bar{b}}(J_{a\bar{d}} + J_0\delta_{a\bar{d}}) - iI_{a\bar{d}}(J_{c\bar{b}} + J_0\delta_{c\bar{b}}) \ . \tag{4.125}$$

Hence, it is seen that for $N = 1$, i.e., in the case of the two-dimensional sphere $S^2 = \mathbb{C}\mathrm{P}^1$, the suggested system has no hidden symmetries, as opposed to the Higgs oscillator on S^2. Nevertheless, this model is exactly solvable both for $N = 1$ and $N > 1$ [18]. Moreover, it remains exactly solvable, even after inclusion of a constant magnetic field, for any N (including $N = 1$, when it has no hidden symmetries). The magnetic field does not break the symmetry algebra of the system! As opposed to the described model, the constant magnetic field breaks the hidden symmetries, as well as the exact solvability, of the Higgs oscillator on $S^2 = \mathbb{C}\mathrm{P}^1$.

Remark. The Hamiltonian (4.123) could be represented as follows:

$$\mathcal{H} = g^{a\bar{b}}(\pi_a\bar{\pi}_b + \alpha^2\partial_a K\bar{\partial}_b K) \ , \tag{4.126}$$

where $K(z, \bar{z}) = \log(1 + z\bar{z})$ is the Kähler potential of the Fubini-Study metric.

Although this potential is not uniquely defined, it provides the system with some properties, which are general for the few oscillator models on Kähler spaces. By this reason we postulate it as an oscillator potential on arbitrary Kähler manifolds.

Now, let us compare these systems with the sequence which we like: real, complex, quaternionic numbers (and zeroth, first, second Hopf map). Let us observe, that the S^N-oscillator potential is defined, in terms of the ambient space $\mathbb{R}^{N+1}$, in complete similarity to the $\mathbb{C}\mathrm{P}^N$-oscillator potential in terms of the "ambient" space $\mathbb{C}^{N+1}$. The latter system preserves its exact solvability in the presence of a constant magnetic ($U(1)$ gauge) field.

Hence, continuing this sequence, one can define on the quaternionic projective spaces $\mathbb{H}\mathrm{P}^N$ the oscillator-like system given by the potential

$$V_{\mathbb{H}\mathrm{P}^N} = \alpha^2 w^a \bar{w}^a = \alpha^2 \frac{u_1^a \bar{u}_1^a + u_2^a \bar{u}_2^a}{u_1^0 \bar{u}_1^0 + u_1^0 \bar{u}_1^0} \ , \tag{4.127}$$

where

$$w^a = \frac{u_1^a + ju_2^a}{u_1^0 + ju_2^0}, \quad u_1^a \bar{u}_1^a + u_2^a \bar{u}_2^a + u_1^0 \bar{u}_1^0 + u_2^0 \bar{u}_2^0 = 1 \ .$$

Here, w^a are inhomogeneous (quaternionic) coordinates of the quaternionic projective space $\mathbb{H}\mathrm{P}^N$, and $u_0^a + ju_1^a, u_1^0 + ju_2^0$ are the Euclidean coordinates of the "ambient" quaternionic space $\mathbb{H}^{N+1} = \mathbb{C}^{2N+2}$.

One can expect that this system will be a superintegrable one and will be exactly solvable also in the presence of an $SU(2)$ instanton field.

In the simplest case of $\mathbb{IHP}^1 = S^4$ we shall get the alternative (with respect to the Higgs) model of the oscillator on the four-dimensional sphere. In terms of the ambient space $\mathbb{R}^5$, its potential will be given by the expression

$$V_{S^4} = \alpha^2 \frac{1 - x^0/x}{1 + x^0/x} = \alpha^2 \frac{1 - \cos\theta}{1 + \cos\theta} . \tag{4.128}$$

Checking this system for this simplest case, we found, that it is indeed exactly solvable in the presence of the instanton field [19].

Let us mention that the Higgs (spherical) oscillator could be straightforwardly extended to (one- and two-sheet) hyperboloids, and the $\mathbb{CP}^N$-oscillator to the Lobachevsky spaces $\mathcal{L}_N = SU(N+1)/U(N)$. In both cases these systems have hidden symmetries.

Notice also that, on the spheres S^N, there exists the analog of the Coulomb system suggested by Schrödinger [20]. It is given by the potential

$$V_{\text{Coulomb}} = -\frac{\gamma}{r_0} \frac{y_{N+1}}{|\mathbf{y}|}, \quad y_{N+1}^2 + |\mathbf{y}|^2 = r_0^2 . \tag{4.129}$$

This system inherits the hidden symmetry of the conventional Coulomb system on $\mathbb{R}^N$.

Probably, as in the case of the oscillator, one can define superintegrable analogs of the Coulomb system on the complex projective spaces $\mathbb{CP}^N$ and on the quaternionic projective spaces $\mathbb{IHP}^N$. However, up to now, this question has not been analyzed.

4.4.1 Relation of the (Pseudo)Spherical Oscillator and Coulomb Systems

The oscillator and Coulomb systems, being the best known among the superintegrable mechanical systems, possess many similarities both at the classical and quantum mechanical levels. Writing down these systems in spherical coordinates, one can observe that the radial Schrödinger equation of the $(p+1)$ dimensional Coulomb system could be transformed in the Shrödinger equation of the $2p$-dimensional oscillator by the transformation (see, e.g. [21])

$$r = R^2,$$

where r and R are the radial coordinates of the Coulomb and oscillator systems, respectively.

Due to the existence of the Hopf maps, in the cases of $p = 1, 2, 4$ one can establish a complete correspondence between these systems. Indeed, their angular parts are, respectively, p- and $(2p-1)$-dimensional spheres, while the above relation follows immediately from (4.54). Considering the Hamiltonian reductions related to the Hopf maps (as it was done in the previous section), one can deduce, that the $(p+1)$-dimensional Coulomb systems could be obtained from the $2p$-dimensional oscillator, by a reduction under the $G = S^{(p-1)}$

group. Moreover, for nonzero values of those generators we shall get generalizations of the Coulomb systems, specified by the presence of a magnetic flux ($p = 1$), a Dirac monopole ($p = 2$), a Yang monopole ($p = 4$) [9, 22, 23]. However, this procedure assumes a change in the roles of the coupling constants and the energy. To be more precise, these reductions convert the energy surface of the oscillator in the energy surface of the Coulomb-like system, while there is no one-to-one correspondence between their Hamiltonians.

As we have seen above, there exists well-defined generalizations of the oscillator systems on the spheres, hyperboloids, complex projective spaces, and Lobachevsky spaces. The Coulomb system could also be generalized on the spheres and hyperboloids. Hence, the following natural question arises. Is it possible to relate the oscillator and Coulomb systems on the spheres and hyperboloids, similarly to those in the flat cases? The answer is positive, but it is rather strange. The oscillators on the $2p$-dimensional sphere and two-sheet hyperboloid (pseudosphere) result in the Coulomb-like systems on the $(p + 1)$-dimensional pseudosphere, for $p = 1, 2, 4$ [24].

Below, following [24], we shall show how to relate the oscillator and Coulomb systems on the spheres and two-sheet hyperboloids. In the planar limit this relation results in the standard correspondence between the conventional (flat) oscillator and the Coulomb-like system. We shall discuss mainly the $p = 1$ case, since the treatment could be straightforwardly extended to the $p = 2, 4$ cases.

Let us introduce the complex coordinate z parameterizing the sphere by the complex projective plane $\mathbb{C}P^1$ and the two-sheeted hyperboloid by the Poincaré disk (Lobachevsky plane, pseudosphere) $\mathcal{L}$)

$$\mathbf{x} \equiv x_1 + ix_2 = R_0 \frac{2z}{1 + \epsilon z\bar{z}}, \qquad x_3 = R_0 \frac{1 - \epsilon z\bar{z}}{1 + \epsilon z\bar{z}} . \tag{4.130}$$

In these coordinates the metric becomes conformally-flat

$$ds^2 = R_0^2 \frac{4 dz \, d\bar{z}}{(1 + \epsilon z\bar{z})^2}. \tag{4.131}$$

Here $\epsilon = 1$ corresponds to the system on the sphere, and $\epsilon = -1$ to that on the pseudosphere. The lower hemisphere and the lower sheet of the hyperboloid are parameterized by the unit disk $|z| < 1$, while the upper hemisphere and the upper sheet of the hyperboloid are specified by $|z| > 1$, and transform one into another by the inversion $z \to 1/z$. In the limit $R_0 \to \infty$ the lower hemisphere (the lower sheet of the hyperboloid) turns into the whole two-dimensional plane. In these terms the oscillator and Coulomb potentials read

$$V_{\text{osc}} = \frac{2\alpha^2 R_0^2 z\bar{z}}{(1 - \epsilon z\bar{z})^2}, \qquad V_C = -\frac{\gamma}{R_0} \frac{1 - \epsilon z\bar{z}}{2|z|} , \tag{4.132}$$

Let us equip the oscillator phase space $T^*\mathbb{C}P^1$ ($T^*\mathcal{L}$) with the symplectic structure

$$\omega = d\pi \wedge dz + d\bar\pi \wedge d\bar z \tag{4.133}$$

and introduce the rotation generators defining the $su(2)$ algebra for $\epsilon = 1$ and the $su(1.1)$ algebra for $\epsilon = -1$

$$\mathbf{J} \equiv \frac{iJ_1 - J_2}{2} = \pi + \epsilon \bar z^2 \bar\pi, \ J \equiv \frac{\epsilon J_3}{2} = i(z\pi - \bar z \bar\pi) \,. \tag{4.134}$$

These generators, together with $\mathbf{x}/R_0, x_3/R_0$, define the algebra of motion of the (pseudo)sphere via the following nonvanishing Poisson brackets:

$$\{\mathbf{J}, \mathbf{x}\} = 2x_3, \ \{\mathbf{J}, x_3\} = -\epsilon \bar{\mathbf{x}}, \ \{J, \mathbf{x}\} = i\mathbf{x}, \\ \{\mathbf{J}, \bar{\mathbf{J}}\} = -2i\epsilon J, \ \{J, \mathbf{J}\} = i\mathbf{J} \,. \tag{4.135}$$

In these terms, the Hamiltonian of a free particle on the (pseudo)sphere reads

$$H_0^\epsilon = \frac{\mathbf{J}\bar{\mathbf{J}} + \epsilon J^2}{2R_0^2} = \frac{(1 + \epsilon z\bar z)^2 \pi\bar\pi}{2R_0^2} \,, \tag{4.136}$$

whereas the oscillator Hamiltonian is given by the expression

$$H_{\rm osc}^\epsilon(\alpha, R_0|\pi, \bar\pi, z, \bar z) = \frac{(1 + \epsilon z\bar z)^2 \pi\bar\pi}{2R_0^2} + \frac{2\alpha^2 R_0^2 z\bar z}{(1 - \epsilon z\bar z)^2} \,. \tag{4.137}$$

It can be easily verified that the latter system possesses the hidden symmetry given by the complex (or vectorial) constant of motion [16]

$$\mathbf{I} = I_1 + iI_2 = \frac{\mathbf{J}^2}{2R_0^2} + \frac{\alpha^2 R_0^2}{2} \frac{\bar{\mathbf{x}}^2}{x_3^2} \,, \tag{4.138}$$

which defines, together with J and $H_{\rm osc}$, the cubic algebra

$$\{\mathbf{I}, J\} = 2i\mathbf{I}, \ \{\bar{\mathbf{I}}, \mathbf{I}\} = 4i\left(\alpha^2 J + \frac{\epsilon J H_{\rm osc}}{R_0^2} - \frac{J^3}{2R_0^4}\right) \,. \tag{4.139}$$

The energy surface of the oscillator on the (pseudo)sphere $H_{\rm osc}^\epsilon = E$ reads

$$\frac{(1 - (z\bar z)^2)^2 \pi\bar\pi}{2R_0^4} + 2\left(\alpha^2 + \epsilon \frac{E}{R_0^2}\right)z\bar z = \frac{E}{R_0^2}(1 + (z\bar z)^2) \,. \tag{4.140}$$

Now, performing the canonical Bohlin transformation (4.68) one can rewrite the expression (4.140) as follows:

$$\frac{(1 - w\bar w)^2 p\bar p}{2r_0^2} - \frac{\gamma}{r_0} \frac{1 + w\bar w}{2|w|} = \mathcal{E}_C \,, \tag{4.141}$$

where we introduced the notation

$$r_0 = R_0^2, \quad \gamma = \frac{E}{2}, \quad -2\mathcal{E}_C = \alpha^2 + \epsilon \frac{E}{r_0} \,. \tag{4.142}$$

Comparing the l.h.s. of (4.141) with the expressions (4.132), (4.136) we conclude that (4.141) defines the energy surface of the Coulomb system on the pseudosphere with "radius" r_0, where w, p denote the complex stereographic coordinate and its conjugated momentum, respectively. In the above, r_0 is the "radius" of the pseudosphere, while $\mathcal{E}_C$ is the energy of the system. Hence, we related classical isotropic oscillators on the sphere and pseudosphere with the classical Coulomb problem on the pseudosphere.

The constants of motion of the oscillators, J and $\mathbf{I}$ (which coincide on the energy surfaces (4.140)) are converted, respectively, into the doubled angular momentum and the doubled Runge–Lenz vector of the Coulomb system

$$J \to 2J_C, \quad \mathbf{I} \to 2\mathbf{A}, \quad \mathbf{A} = -\frac{iJ_C\mathbf{J}_C}{r_0} + \gamma\frac{\bar{\mathbf{x}}_C}{|\mathbf{x}_C|}, \tag{4.143}$$

where $\mathbf{J}_C$, J_C, $\mathbf{x}_C$ denote the rotation generators and the pseudo-Euclidean coordinates of the Coulomb system.

We have shown above that, for establishing the quantum–mechanical correspondence, we have to supplement the quantum–mechanical Bohlin transformation with the reduction by the Z_2 group action, choosing either even ($\sigma = 0$) or odd ($\sigma = 1/2$) wave functions (4.73). The resulting Coulomb system is spinless for $\sigma = 0$, and it possesses spin $1/2$ for $\sigma = 1/2$.

The presented construction could be straightforwardly extended to higher dimensions, concerning the $2p$-dimensional oscillator on the (pseudo)sphere and the $(p + 1)$-dimensional Coulomb-like systems, $p = 2, 4$. It is clear, that the $p = 2$ case corresponds to the Hamiltonian reduction, associated with the first Hopf map, and the $p = 4$ case is related to the second Hopf map. Indeed, the oscillator on the $2p$-dimensional (pseudo)sphere is also described by the Hamiltonian (4.137), where the following replacement is performed: $(z, \pi) \to (z^a, \pi_a)$, $a = 1, \ldots, p$, with the summation over these indices understood. Consequently, the oscillator energy surfaces are again given by (4.140). Then, performing the Hamiltonian reduction, associated with the pth Hopf maps (see the previous section) we shall get the Coulomb-like system on the $(p + 1)$-dimensional pseudosphere.

For example, if $p = 2$, we reduce the system under consideration by the Hamiltonian action of the $U(1)$ group given by the generator $J = i(z\pi - \bar{z}\bar{\pi})$. This reduction was described in detail in Sect. 4.4. For this purpose, we have to fix the level surface $J = 2s$ and choose the $U(1)$-invariant stereographic coordinates in the form of the conventional Kustaanheimo–Stiefel transformation (4.82). The resulting symplectic structure takes the form (4.17). The oscillator energy surface reads

$$\frac{(1 - \mathbf{q}^2)^2}{8r_0^2}\left(\mathbf{p}^2 + \frac{s^2}{\mathbf{q}^2}\right) - \frac{\gamma}{r_0}\frac{1 + \mathbf{q}^2}{2|\mathbf{q}|} = \mathcal{E}_C, \tag{4.144}$$

where r_0, γ, $\mathcal{E}_C$ are defined by the expressions (4.142).

Interpreting $\mathbf{q}$ as the (real) stereographic coordinates of the three-dimensional pseudosphere

$$\mathbf{x} = r_0 \frac{2\mathbf{q}}{1 - \mathbf{q}^2}, \qquad x_4 = r_0 \frac{1 + \mathbf{q}^2}{1 - \mathbf{q}^2}, \qquad (4.145)$$

we conclude that (4.144) defines the energy surface of the pseudospherical analog of a Coulomb-like system proposed in [7], which is also known under the name of "MIC-Kepler" system.

In the $p = 4$ case, we have to reduce the system by the action of the $SU(2)$ group and choose the $SU(2)$-invariant stereographic coordinates and momenta in the form corresponding to the standard Hurwitz transformation, which yields a pseudospherical analog of the so-called $SU(2)$-Kepler (or Yang–Coulomb) system [23]. The potential term of the resulting system will be given by the expression

$$V_{SU(2)-\text{Kepler}} = \frac{I^2}{r_0^2} \left(\frac{x_5^2}{2\mathbf{x}^2} - 2 \right) - \frac{\gamma}{2r_0} \frac{x_5}{|\mathbf{x}|}, \qquad (4.146)$$

where $(\mathbf{x}, x_5)$ are the (pseudo)-Euclidean coordinates of the ambient space $\mathbb{R}^{1.5}$ of the five-dimensional hyperboloid, $|\mathbf{x}|^2 - x_5^2$; I^2 is the value of the generator $\mathcal{J}_i^2$, under which the $SU(2)$ reduction has been performed. The constants r_0, γ are defined by the expressions (4.142).

It is interesting to clarify, which systems will the $\mathbb{C}P^N$-oscillators, after similar reductions, result in. We have checked it only for the first Hopf map, corresponding to the case $p = 2$ [17, 25].To our surprise, we found that the oscillators on $\mathbb{C}P^2$ and $\mathcal{L}_2$ also resulted, after reduction, in the pseudospherical MIC-Kepler system!

4.5 Supersymplectic Structures

In the previous section, we presented some elements of Hamiltonian formalism which, in our belief, could be useful in the study of supersymmetric mechanics.

In the present section, we shall briefly discuss the Hamiltonian formalism on superspaces (super-Hamiltonian formalism). The super-Hamiltonian formalism, in its main lines, is a straightforward extension of the ordinary Hamiltonian formalism to superspace, with a more or less obvious placement of sign factors. Probably, from the supergeometrical viewpoint, the only qualitative difference appears in the existence of the odd Poisson brackets (antibrackets), which have no analogs in ordinary spaces, and in the respect of the differential forms to integration. Fortunately, these aspects are inessential for our purposes.

The Poisson brackets of the functions $f(x)$ and $g(x)$ on superspaces are defined by the expression

$$\{f, g\}_\kappa = \frac{\partial_r f}{\partial x^A} \Omega_\kappa^{AB}(x) \frac{\partial_l g}{\partial x^B}, \qquad \kappa = 0, 1. \qquad (4.147)$$

They obey the conditions

$$p(\{f,g\}_\kappa) = p(f) + p(g) + \kappa \quad \text{(grading)},$$
$$\{f,g\}_\kappa = -(-1)^{(p(f)+\kappa)(p(g)+\kappa)}\{g,f\}_\kappa \quad (\text{"antisymmetricity"}), \quad (4.148)$$
$$(-1)^{(p(f)+1)(p(h)+\kappa)}\{f,\{g,h\}_\kappa\}_\kappa + \text{cycl.perm.(f, g, h)} = 0 \quad (\text{Jacobi id.}) .$$
$$(4.149)$$

Here, x^A are local coordinates of superspace, while $\frac{\partial_r}{\partial x^A}$ and $\frac{\partial_l}{\partial x^A}$ denote right and left derivatives, respectively.

It is seen that the nondegenerate odd Poisson brackets can be defined on the $(N.N)$-dimensional superspaces, and the nondegenerate even Poisson brackets could be defined on the $(2N.M)$-dimensional ones. In this case the Poisson brackets are associated with the supersymplectic structure

$$\Omega_\kappa = dz^A \Omega_{(\kappa)AB} \, dz^B, \quad d\Omega_\kappa = 0 , \quad (4.150)$$

where $\Omega_{(\kappa)AB}\Omega_\kappa^{BC} = \delta_A^C$.

The generalization of the Darboux theorem states that locally, the nondegenerate Poisson brackets could be transformed to the canonical form. The canonical odd Poisson brackets look as follows:

$$\{f,g\}_1^{\text{can}} = \sum_{i=1}^N \left(\frac{\partial_r f}{\partial x^i} \frac{\partial_l g}{\partial \theta_i} - \frac{\partial_r f}{\partial \theta_i} \frac{\partial_l g}{\partial x^i} \right) , \quad (4.151)$$

where $p(\theta_i) = p(x^i) + 1 = 1$. The canonical even Poisson brackets read

$$\{f,g\}_0 = \sum_{i=1}^N \left(\frac{\partial f}{\partial x^i} \frac{\partial g}{\partial x^{i+N}} - \frac{\partial f}{\partial x^{i+N}} \frac{\partial g}{\partial x^i} \right) + \sum_{\alpha=1}^M \epsilon_\alpha \frac{\partial_r f}{\partial \theta^\alpha} \frac{\partial_l Lg}{\partial \theta^\alpha}, \quad \epsilon_\alpha = \pm 1 .$$
$$(4.152)$$

Here, x^i, x^{i+N} denote even coordinates, $p(x) = 0$, and θ^α are the odd ones $p(\theta) = 1$.

In a completely similar way to the ordinary (non-"super") space, one can show that the vector field preserving the supersymplectic structure is a locally Hamiltonian one. Hence, both types of supersymplectic structures can be related with the Hamiltonian systems, which have the following equations of motion:

$$\frac{dx^A}{dt} = \{\mathcal{H}_\kappa, x^A\}_\kappa, \quad p(\mathcal{H}_\kappa) = \kappa . \quad (4.153)$$

Any supermanifold $\mathcal{M}$ underlied by the bosonic manifold M_0 can be associated with some vector bundle VM_0 of M_0 [3], in the following sense. One can choose on $\mathcal{M}$ local coordinates (x^i, θ^μ), such that the transition functions from one chart (parameterized by (x^i, θ^μ)) to the other chart (parameterized by $(\tilde{x}^i, \tilde{\theta}^\mu)$) look as follows:

$$\tilde{x}^i = \tilde{x}^i(x), \quad \tilde{\theta}^\mu = A_\nu^\mu(x)\theta^\nu . \quad (4.154)$$

Changing the parity of θ: $p(\theta^\mu) = 1 \rightarrow p(\theta^\mu) = 0$, we shall get the vector bundle VM_0 of M_0.

Any supermanifold equipped with the odd symplectic structure, is associated with the cotangent bundle of M_0 [26], so that the odd symplectic structure could be globally transformed to the canonical form, with the odd Poisson bracket given by the expression (4.151). Hence, the functions on the odd symplectic manifold could be interpreted as contravariant antisymmetric tensors on M_0.

The structure of the even symplectic manifold is not so rigid: there is a variety of ways to extend the given symplectic manifold (M_0, ω) to the supersymplectic ones, associated with the vector bundle VM_0. On these supermanifolds one can (globally) define the even symplectic structure

$$\Omega = \omega + \mathrm{d}\left(\theta^\mu g_{\mu\nu}(x)\mathcal{D}\theta^\nu\right)$$
$$= \omega + \frac{1}{2}R_{\nu\mu ki}\theta^\nu\theta^\mu\,\mathrm{d}x^i \wedge \mathrm{d}x^k + g_{\mu\nu}\mathcal{D}\theta^\nu \wedge \mathcal{D}\theta^\mu\,, \qquad (4.155)$$

Here x^i are local coordinates of $\mathcal{M}_0$ and θ^μ are the (odd) coordinates in the bundle; $g_{\mu\nu} = g_{\mu\nu}(x)$ are the components of the metrics in the bundle, while $\mathcal{D}\theta^\mu = d\theta^\mu + \Gamma^\mu_{\nu i}\theta^\nu dx^i$, where $\Gamma^\mu_{i\nu}$ are the connection components respecting the metric in the bundle

$$g_{\mu\nu;k} = g_{\mu\nu,k} - g_{\mu\alpha}\Gamma^\alpha_{k\nu} - g_{\alpha\nu}\Gamma^\alpha_{k\mu} = 0\,. \qquad (4.156)$$

We used the following notation as well: $R_{\mu\nu ki} = g_{\mu\alpha}R^\alpha_{\nu ki}$, where $R^\mu_{\nu ki}$ are the components of connection's curvature

$$R^\nu_{\alpha ki} = -\Gamma^\nu_{k\alpha,i} + \Gamma^\nu_{i\alpha,k} + \Gamma^\nu_{k\beta}\Gamma^\beta_{i\alpha} - \Gamma^\nu_{i\beta}\Gamma^\beta_{k\alpha}\,; \quad R^\nu_{\alpha ik} = -R^\nu_{\alpha ki}\,.$$

Let us consider the coordinate transformation (4.154). With respect to this transformation, the connection components transform as follows:

$$\bar{\Gamma}^\mu_{i\nu} = A^\mu{}_\lambda\Gamma^\lambda_{k\alpha}\frac{\partial_r x^k}{\partial\bar{x}^i}B^\alpha{}_\nu - A^\mu{}_{\alpha,k}B^\alpha{}_\nu\frac{\partial_r x^k}{\partial\bar{x}^i}\,, \quad A_\mu{}^\nu B_\nu{}^\lambda = \delta^\lambda_\mu\,. \quad (4.157)$$

Since $\mathcal{D}\theta^\nu$ transforms homogeneously under (4.154), $\mathcal{D}\bar{\theta}^\nu = \mathcal{D}\theta^\mu A_\mu{}^\nu(x)$, we conclude that the supersymplectic structure (4.155) is covariant under (4.154) as well.

The corresponding Poisson brackets look as follows:

$$\{f,g\} = (\nabla_i f)\,\tilde{\omega}^{ij}(\nabla_j g) + \alpha\frac{\partial_r f}{\partial\theta^\mu}\,g^{\mu\nu}\,\frac{\partial_l g}{\partial\theta^\nu}\,; \qquad (4.158)$$

where

$$\tilde{\omega}^{im}\left(\omega_{mj} + \frac{1}{2}R_{\nu\mu mj}\theta^\nu\theta^\mu\right) = \delta^i_j\,, \qquad \nabla_i = \frac{\partial}{\partial x^i} - \Gamma^k_{ij}(x)\,\theta^{ja}\frac{\partial}{\partial\theta^{ka}}\,.$$

On the supermanifolds one can define also the analog of the Kähler structures. We shall call the complex symplectic supermanifold an even (odd)

Kähler one, when the even (odd) symplectic structure is defined by the expression

$$\Omega_\kappa = i(-1)^{p_A(p_B+\kappa+1)}g_{(\kappa)A\bar{B}}\,dz^A \wedge d\bar{z}^B\,, \qquad (4.159)$$

where

$$g_{(\kappa)A\bar{B}} = (-1)^{(p_A+\kappa+1)(p_B+\kappa+1)+\kappa+1}\overline{g_{(\kappa)B\bar{A}}}, \quad p(g_{(\kappa)A\bar{B}}) = p_A + p_B + \kappa.$$

Here and in the following, the index $\kappa = 0(1)$ denotes the even (odd) case.

The Kähler potential on the supermanifold is a local real even (odd) function $K_\kappa(z,\bar{z})$ defining the Kähler structure

$$g_{(\kappa)A\bar{B}} = \frac{\partial_l}{\partial z^A}\frac{\partial_r}{\partial \bar{z}^B}K_\kappa(z,\bar{z}). \qquad (4.160)$$

As in the usual case, K_κ is defined up to arbitrary holomorphic and antiholomorphic functions.

With the even (odd) form Ω_κ one can associate the even (odd) Poisson bracket

$$\{f,g\}_\kappa = i\left(\frac{\partial_r f}{\partial \bar{z}^A}g^{(\kappa)\bar{A}B}\frac{\partial_l g}{\partial z^B} - (-1)^{(p_A+\kappa)(p_B+\kappa)}\frac{\partial_r f}{\partial z^A}g^{(\kappa)\bar{A}B}\frac{\partial_l g}{\partial \bar{z}^B}\right), \qquad (4.161)$$

where

$$g^{(\kappa)\bar{A}B}g_{(\kappa)B\bar{C}} = \delta^{\bar{A}}_{\bar{C}}\,, \quad \overline{g^{(\kappa)\bar{A}B}} = (-1)^{(p_A+\kappa)(p_B+\kappa)}g^{(\kappa)\bar{B}A}\,.$$

Example. Let us consider the supermanifold ΛM associated with the tangent bundle of the Kähler manifold M_0. On this supermanifold one can define the even and odd Kähler potentials [27]

$$K_0 = K(z,\bar{z}) + F(ig_{a\bar{b}}\sigma^a\bar{\sigma}^b), \qquad K_1 = \frac{\partial K(z,\bar{z})}{\partial z^a}\sigma^a + \frac{\partial K(z,\bar{z})}{\partial \bar{z}^a}\bar{\sigma}^a\,, \quad (4.162)$$

where $K(z,\bar{z})$ is a Kähler potential on M_0, $g_{a\bar{b}} = \partial^2 K/\partial z^a\partial\bar{z}^b$, and $F(x)$ is a real function which obeys the condition $F'(0) \neq 0$. It is clear that these functions define even and odd Kähler structures on ΛM_0, respectively.

Finally, let us notice that the analog of the Liouville measure for the even supersymplectic symplectic structure Ω_0 reads

$$\rho = \sqrt{\mathrm{Ber}\Omega_{(0)AB}}\,, \qquad (4.163)$$

while the odd symplectic structure has no similar invariant [28]. Indeed, one can verify that the even super-Hamiltonian vector field is always divergenceless, $\mathrm{str}\{H,\ \}_0 = 0$ (similarly to the non-super-Hamiltonian vector field), while in the case of the odd super-Hamiltonian vector field this property of the Hamiltonian vector field fails. As a consequence, in the latter case the so-called Δ-operator can be defined [29], which plays a crucial role in the Batalin–Vilkovisky formalism (Lagrangian BRST quantization formalism) [30].

4.5.1 Odd Super-Hamiltonian Mechanics

Let us consider the supermanifold ΛM, associated with the tangent bundle of the symplectic manifold (M, ω), i.e., the external algebra of (M, ω). In other words, the odd coordinates θ^i transform from one chart to another like dx^i, and they can be interpreted as the basis of the 1-forms on M. By the use of the ω we can equip ΛM with the odd symplectic structure

$$\Omega_1 = d\left(\omega_{ij}\theta^j \, dx^i\right) = \omega_{ij} \, dx^i \wedge d\theta^j + \frac{1}{2}\omega_{ki,j}\theta^j \, dx^k \wedge dx^i. \qquad (4.164)$$

The corresponding odd Poisson brackets are defined by the following relations:

$$\{x^i, x^j\}_1 = 0, \quad \{x^i, \theta^j\}_1 = \omega^{ij}, \quad \{\theta^i, \theta^j\}_1 = \frac{\partial \omega^{ij}}{\partial x^k}\theta^k \,, \qquad (4.165)$$

where $\omega^{ij}\omega_{jk} = \delta_k^i$.

Let us define, on ΛM, the even function

$$F = -\frac{1}{2}\theta^i\omega_{ij}\theta^j, \; : \quad \{F, F\}_1 = 0 \,, \qquad (4.166)$$

where the latter equation holds due to the closeness of ω. By making use of this function, one can define the map of any function on M in the odd function on ΛM

$$f(x) \to Q_f(x, \theta) = \{f(x), F(x, \theta)\}_1 \,, \qquad (4.167)$$

which possesses the following important property:

$$\{f(x), g(x)\} = \{f(x), Q_g(x, \theta)\}_1 \quad \text{for} \quad \text{any} \quad f(x), g(x) \,. \qquad (4.168)$$

In particular (4.167) maps the Hamiltonian mechanics $(M, \omega, H(x))$ in the following super-Hamiltonian one: $(\Lambda M, \Omega_1, Q_H = \{H, F\}_1)$, where Q_H plays the role of the odd Hamiltonian on ΛM.

The functions H, F, Q_H form the superalgebra

$$\{H \pm F, H \pm F\}_1 = \pm 2Q_H,$$
$$\{H + F, H - F\}_1 = \{H \pm F, Q_H\}_1 = \{Q_H, Q_H\}_1 = 0 \,, \qquad (4.169)$$

i.e., the resulting mechanics possesses the supersymmetry transformation defined by the "supercharge" $H + F$. This superalgebra has a transparent interpretation in terms of base manifold (M, ω)

$$\{H, \quad\}_1 = \xi_H^i \frac{\partial}{\partial \theta^i} \quad \to \quad \hat{\imath}_H - \text{contraction with } \xi_H,$$

$$\{F, \quad\}_1 = \theta^i \frac{\partial}{\partial x^i} \quad \to \quad \hat{d} - \text{exterior} \quad \text{differential},$$

$$\{Q, \quad\}_1 = \xi_H^i \frac{\partial}{\partial x^i} + \xi_{H,k}^i \theta^k \frac{\partial}{\partial \theta^i} \quad \to \quad \hat{\mathcal{L}}_H - \text{Lie} \quad \text{derivative} \quad \text{along} \quad \xi_H,$$

while, using the Jacobi identity (4.149), we get

$$\{H, F\}_1 = Q_H \rightarrow \hat{d}\hat{\imath}_H + \hat{\imath}_H\hat{d} = \hat{\mathcal{L}}_H - \text{homotopy   formula}.$$

Hence, the above dynamics could be useful for the description of the differential calculus on the symplectic (and Poisson) manifolds. Particularly, it has a nice application in equivariant cohomology and related localization formulae (see [31] and refs. therein).

However, the presented supersymmetric model has no deep dynamical meaning, since the odd Poisson brackets do not admit any consistent quantization scheme. Naively, this is reflected in the fact that conjugated operators should have opposite Grassmann grading, so that the Planck constant must be a Grassmann-odd number.

Moreover, the presented supersymmetric mechanics is not interesting even from the classical viewpoint. Its equations of motion read

$$\frac{\mathrm{d}x^i}{\mathrm{d}t} = \{x^i, Q_H\}_1 = \xi_H^i, \qquad \frac{\mathrm{d}\theta^i}{\mathrm{d}t} = \{\theta^i, Q_H\}_1 = \frac{\partial \xi_H^i}{\partial x^j}\theta^j,$$

i.e., the "fermionic" degrees of freedom have no impact in the dynamics of the "bosonic" degrees of freedom.

Nevertheless, the odd Poisson brackets are widely known, since 1981, in the theoretical physics community under the name of "antibrackets." That was the year, when Batalin and Vilkovisky suggested their *Covariant Lagrangian BRST quantization formalism* (which is known presently as the Batalin–Vilkovisky formalism) [30], where the antibrackets (odd Poisson brackets) play the key role. However, only decades after, this elegant formalism was understood in terms of conventional supergeometrical constructions [26, 29]. It seems that the Batalin–Vilkovisky formalism could also be useful for the geometrical (covariant) formulation of the superfield approach to the construction of supersymmetric Lagrangian field-theoretical and mechanical models [32].

We shall not touch upon these aspects of super-Hamiltonian systems, and will restrict ourselves to the consideration of supersymmetric Hamiltonian systems with even symplectic structure.

4.5.2 Hamiltonian Reduction: $\mathbb{C}^{N+1.M} \rightarrow \mathbb{CP}^{N.M}$, $\Lambda\mathbb{C}^{N+1} \rightarrow \Lambda\mathbb{CP}^N$

The procedure of super-Hamiltonian reduction is very similar to the Hamiltonian one. The main difference is in the counting of the dimensionality of the phase superspace. Namely, we should separately count the number of "fermionic" and "bosonic" degrees of freedom, which were eliminated during the reduction.

Instead of describing the extension of the Hamiltonian reduction to the supercase, we shall illustrate it by considering superextensions of the reduction

$\mathbb{C}^{N+1} \to \mathbb{C}P^N$ presented in Sect. 4.5. These examples were considered in details in [33].

Let us consider the complex superspace $\mathbb{C}^{N+1,M}$ parameterized by the complex coordinates $(u^{\tilde{a}}, \eta^n)$, $\tilde{a} = 0, 1, \ldots, N$, $n = 1, \ldots, M$. Let us equip it with the canonical symplectic structure

$$\Omega^0 = i(\mathrm{d}u^{\tilde{a}} \wedge \bar{\mathrm{d}}u^{\tilde{a}} - i\mathrm{d}\eta^n \wedge \mathrm{d}\bar{\eta}^n)$$

and with the corresponding even Poisson bracket

$$\{f, g\}_0 = i\left(\frac{\partial f}{\partial u^{\tilde{a}}} \frac{\partial g}{\partial \bar{u}^{\tilde{a}}} - \frac{\partial f}{\partial \bar{u}^{\tilde{a}}} \frac{\partial g}{\partial u^{\tilde{a}}}\right) + \frac{\partial_r f}{\partial \eta^n} \frac{\partial_l g}{\partial \bar{\eta}^n} + \frac{\partial_r f}{\partial \bar{\eta}^n} \frac{\partial_l g}{\partial \eta^n}. \qquad (4.170)$$

The (super-)Hamiltonian action of the $U(1)$ group is given, on this space, by the generator

$$\mathcal{J}_0 = u^{\tilde{a}} \bar{u}^{\tilde{a}} - i\eta^n \bar{\eta}^n . \qquad (4.171)$$

For the reduction of $\mathbb{C}^{N+1,M}$ by this generator, we have to factorize the $(2N+1.2M)_{\mathbb{R}}$-dimensional level supersurface

$$\mathcal{J}_0 = r_0^2 \qquad (4.172)$$

by the even super-Hamiltonian vector field $\{\mathcal{J}_0, \ \}$ (which is tangent to that surface). Hence, the resulting phase superspace is a $(2N.2M)_{\mathbb{R}}$-dimensional one.

Hence, for the role of local coordinates of the reduced phase space, we have to choose the N even and M odd complex functions commuting with $\mathcal{J}_0$. On the chart $u^{\tilde{a}} \neq 0$, appropriate functions are the following ones:

$$z_{(\tilde{a})}^A = \left(z_{(\tilde{a})}^a = \frac{u^a}{u^{\tilde{a}}}, \ \theta_{(\tilde{a})}^k = \frac{\eta^k}{u^{\tilde{a}}}, \quad a \neq \tilde{a}\right) : \{z_{(\tilde{a})}^A, \mathcal{J}_0\}_0 = 0. \qquad (4.173)$$

The reduced Poisson brackets could be defined by the expression $\{f, g\}_0^{\mathrm{red}} = \{f, g\}_0|_{\mathcal{J}_0 = r_0^2}$, where f, g are functions depending on the coordinates $z_{(\tilde{a})}^A, \bar{z}_{(\tilde{b})}^A$. Straightforward calculations yield the result

$$\{z^A, z^B\}_0^{\mathrm{red}} = \{\bar{z}^A, \bar{w}^B\}_0^{\mathrm{red}} = 0,$$

$$\{z^A, \bar{z}^B\}_0^{\mathrm{red}} = (i)^{p_A p_B + 1} \frac{1 + (-i)^{p_C} z^C \bar{z}^C}{r_0^2} \left(\delta^{AB} + (-i)^{p_A p_B} z^A \bar{w}^B\right).$$

It is seen that these Poisson brackets are associated with a Kähler structure. This Kähler structure is defined by the potential

$$K = r_0^2 \log(1 + (-i)^{p_C} z^C \bar{z}^{\tilde{C}}) . \qquad (4.174)$$

The transition functions from the $\tilde{a}$th chart to the $\tilde{b}$th one look as follows:

$$z_{(\tilde{a})}^{\tilde{c}} = \frac{z_{(\tilde{b})}^{\tilde{c}}}{z_{(\tilde{b})}^{\tilde{a}}}, \qquad \theta_{(\tilde{a})}^{k} = \frac{\theta_{(\tilde{b})}^{k}}{z_{(\tilde{b})}^{\tilde{a}}}, \qquad \text{where} \quad z_{(\tilde{b})}^{\tilde{a}} = \left(w_{(\tilde{b})}^{a}, \; w_{(\tilde{a})}^{\tilde{a}} = 1\right). \tag{4.175}$$

Upon these transformations the Kähler potential changes on the holomorphic and antiholomorphic functions, i.e., the reduced phase space is indeed a Kähler supermanifold. We shall refer to it as $\mathbb{C}P^{N.M}$. The quantization of this supermanifold is considered in [34].

Now, let us consider the Hamiltonian reduction of the superspace $\mathbb{C}^{N+1,N+1}$ by the action of the $\mathcal{N} = 2$ superalgebra, given by the generators

$$\mathcal{J}_0 = u^{\tilde{a}} \bar{u}^{\tilde{a}} - i \eta^{\tilde{a}} \bar{\eta}^{\tilde{a}}, \qquad \Theta^+ = u^{\tilde{a}} \bar{\eta}^{\tilde{a}}, \qquad \Theta^- = \bar{u}^{\tilde{a}} \eta^{\tilde{a}} :$$
$$\{\Theta^+, \Theta^-\} = \mathcal{J}_0, \quad \{\Theta^\pm, \Theta^\pm\} = \{\Theta^\pm, \mathcal{J}_0\} = 0. \tag{4.176}$$

The equations

$$\mathcal{J}_0 = r_0^2, \qquad \Theta^\pm = 0 \tag{4.177}$$

define the $(2N + 1.2N)$-dimensional level surface $M_{r_0^2, 0, 0}$. The reduced phase superspace can be defined by the factorization of $M_{r_0^2, 0, 0}$ by the action of the tangent vector field $\{\mathcal{J}, \}_0$. Hence, the reduced phase superspace is a $(2N.2N)_\mathbb{R}$-dimensional one. The conventional local coordinates of the reduced phase superspace could be chosen as follows (on the chart $u^0 \neq 0$):

$$\sigma^a = -i\{z^a, \Theta^+\} = \theta^a - \theta^0 z^a, \qquad w^a = z^a + i\frac{\Theta^-}{\mathcal{J}_0}\sigma^a, \tag{4.178}$$

where z^a, θ^0, θ^a are defined by (4.173). The reduced Poisson brackets are defined as follows:

$$\{f, g\}_0^{\text{red}} = \{f, g\}_0 \mid_{\mathcal{J} = r_0^2, \Theta^\pm = 0},$$

where f, g are the functions on (w^a, σ^a). Straightforward calculations result in the following relations:

$$\{w^A, w^B\}_0^{\text{red}} = \{\bar{w}^A, \bar{w}^B\}_0^{\text{red}} = 0, \quad \text{where} \quad w^A = (w^a, \sigma^a)$$

$$\{w^a, \bar{w}^b\}_0^{\text{red}} = i\frac{A}{r_0^2}(\delta^{ab} + w^a \bar{w}^b) - \frac{\sigma^a \bar{\sigma}^b}{r_0^2},$$

$$\{w^a, \bar{\sigma}^b\}_0^{\text{red}} = i\frac{A}{r_0^2}\left(w^a \bar{\sigma}^b + \mu(\delta^{ab} + w^a \bar{w}^b)\right) \tag{4.179}$$

$$\{\sigma^a, \bar{\sigma}^b\}_0^{\text{red}} = \frac{A}{r_0^2}\left((1 + i\mu\bar{\mu})\delta^{ab} + w^a \bar{w}^b + i(\sigma^a + \mu w^a)(\bar{\sigma}^b + \bar{\mu}\bar{w}^b)\right),$$

and

$$A = 1 + w^a \bar{w}^a - i\sigma^a \bar{\sigma}^a + \frac{i\sigma^a \bar{w}^a \bar{\sigma}^b w^b}{1 + w^c \bar{w}^c}, \qquad \mu = \frac{\bar{w}^a \sigma^a}{1 + w^b \bar{w}^b}.$$

These Poisson brackets are associated with the Kähler structure defined by the potential

$$K = r_0^2 \log A(w, \bar{w}, \sigma, \bar{\sigma}) = r_0^2 \log(1 + w^a \bar{w}^a) + r_0^2 \log(1 - i g_{a\bar{b}} \sigma^a \bar{\sigma}^b) .$$

$$(4.180)$$

where $g_{a\bar{b}}(w, \bar{w})$ is the Fubini-Study metric on $\mathbb{C}P^N$.

The transition functions from the $\tilde{a}$th chart to the $\tilde{b}$th one reads

$$w_{(\tilde{b})}^{\tilde{c}} = \frac{w_{(\tilde{a})}^{\tilde{c}}}{w_{(\tilde{a})}^{\tilde{b}}}, \qquad \sigma_{(\tilde{b})}^{\tilde{c}} = \frac{\sigma_{(\tilde{a})}^{\tilde{c}} x_{(\tilde{a})}^{\tilde{b}} - w_{(\tilde{a})}^{\tilde{c}} \sigma_{(\tilde{a})}^{\tilde{b}}}{(w_{(\tilde{a})}^{\tilde{b}})^2} ,$$

where $(w_{(\tilde{a})}^{\tilde{a}} = 1, \sigma_{(\tilde{b})}^{\tilde{b}} = 0)$. Hence, σ^a transforms like dw^a, i.e., the reduced phase superspace is $\Lambda \mathbb{C}P^N$, the external algebra of the complex projective space $\mathbb{C}P^N$.

Remark 1. On $\mathbb{C}^{N+1,N+1}$ one can define the odd Kähler structure as well, $\Omega^1 = du^n \wedge d\bar{\eta}^n + d\bar{u}^n \wedge d\eta^n$. It could be reduced to the odd Kähler structure on $\Lambda \mathbb{C}P^N$ by the action of the generators

$$J_0 = z\bar{z}, \quad Q = z\bar{\eta} + \bar{z}\eta .$$

Remark 2. The generalization of the reduction $T^* \mathbb{C}^2 \to T^* \mathbb{R}^3$, where the latter is specified by the presence of a Dirac monopole, is also straightforward. One should consider the $(4.M)_{\mathbb{C}}$-dimensional superspace equipped with the canonical even symplectic structure $\Omega_0 = d\pi \wedge dz + d\bar{\pi} \wedge d\bar{z} + d\eta \wedge d\bar{\eta}$, and reduce it by the Hamiltonian action of the $U(1)$ group given by the generator $\mathcal{J} = i\pi z - i\bar{\pi}\bar{z} - i\eta\bar{\eta}$. The resulting space is a $(6.2M)_{\mathbb{R}}$-dimensional one. Its even local coordinates could be defined by the same expressions, as in the bosonic case (4.82), while the odd coordinates could be chosen as follows: $\theta^m = f(z\bar{z})\bar{z}_0\eta^m$.

4.6 Supersymmetric Mechanics

In the previous sections, we presented some basic elements of the Hamiltonian and super-Hamiltonian formalism. We paid special attention to the examples, related with Kähler geometry, keeping in mind that the latter is of a special importance in supersymmetric mechanics. Indeed, the incorporation of the Kähler structure(s) is one of the standard ways to increase the number of supersymmetries of the system.

Our goal is to construct the supersymmetric mechanics with $\mathcal{N} \geq 2$ supersymmetries. This means that, on the given phase superspace equipped with even symplectic structure, we should construct the Hamiltonian $\mathcal{H}$ which has $\mathcal{N} = N$ odd constants of motion Q_i forming the superalgebra

$$\{Q_i, Q_j\} = 2\delta_{ij}\mathcal{H}, \qquad \{Q_i, \mathcal{H}\} = 0 . \qquad (4.181)$$

This kind of mechanics is referred to as "$\mathcal{N} = N$ supersymmetric mechanics".

It is very easy to construct the $\mathcal{N} = 1$ supersymmetric mechanics with single supercharges: we should simply take the square (under a given nondegenerate even Poisson bracket) of the arbitrary odd function Q_1, and consider the resulting even function as the Hamiltonian

$$\{Q_1, Q_1\} \equiv 2\mathcal{H}_{SUSY} :\Rightarrow \{Q_1, \mathcal{H}_{SUSY}\} = 0 . \tag{4.182}$$

However, the case of $\mathcal{N} = 1$ supersymmetric mechanics is not an interesting system, both from the dynamical and field-theoretical viewpoints.

If we want to construct the $\mathcal{N} > 1$ supersymmetric mechanics, we must specify both the underlying system and the structure of phase superspace.

Let us illustrate it on the simplest examples of $\mathcal{N} = 2$ supersymmetric mechanics. For this purpose, it is convenient to present the $\mathcal{N} = 2$ superalgebra as follows:

$$\{Q^+, Q^-\} = \mathcal{H}, \qquad \{Q^\pm, Q^\pm\} = 0 , \tag{4.183}$$

where $Q^\pm = (Q_1 \pm iQ_2)/\sqrt{2}$. Hence, we have to find the odd complex function, which is nilpotent with respect to the given nondegenerate Poisson bracket, in order to construct the appropriate system.

Let us consider a particular example, when the underlying system is defined on the cotangent bundle T^*M_0, and it is given by (4.12).

To supersymmetrize this system, we extend the canonical symplectic structure as follows:

$$\Omega = dp_a \wedge dx^a + \frac{1}{2}R_{abcd}\theta_+^a\theta_-^b \, dx^c \wedge dx^d + g_{ab}D\theta_+^a \wedge D\theta_-^b , \tag{4.184}$$

where $D\theta_\pm^a \equiv d\theta_\pm^a + \Gamma_{bc}^a\theta_\pm^b dx^c$, and Γ_{bc}^a, R_{abcd} are the components of the connection and curvature of the metrics $g_{ab}dx^a dx^b$ on M_0.

We choose the following candidate for a complex supercharge:

$$Q_\pm = (p_a \pm iW_{,a})\theta^a{}_\pm \ : \{Q_\pm, Q_\pm\} = 0 . \tag{4.185}$$

Hence, the supersymmetric Hamiltonian could be constructed by the calculation of the Poisson brackets of these supercharges.

$$\mathcal{H} \equiv \{Q_+, Q_-\} = \frac{1}{2}g^{ab}(p_a p_b + W_{,a}W_{,b}) + W_{a;b}\theta_+^a\theta_-^b + R_{abcd}\theta_-^a\theta_+^b\theta_-^c\theta_+^d . \tag{4.186}$$

The "minimal" coupling of the magnetic field, $\Omega \to \Omega + F_{ab}dx^a \wedge dx^b$, breaks the $\mathcal{N} = 2$ supersymmetry of the system

$$\{Q_\pm, Q_\pm\} = F_{ab}\theta_\pm^a\theta_\pm^b, \qquad \{Q_+, Q_-\} = \mathcal{H} + iF_{ab}\theta_+^a\theta_-^b .$$

Notice that the Higgs oscillator on the sphere S^N, considered in Sect. 4.5, could be supersymmetrized in this way, choosing $W = \frac{\alpha}{2}\log\frac{2+q^2}{2-q^2}$, with $\mathbf{q}$ being the conformal coordinates of the sphere.

One of the ways to extend this construction to $\mathcal{N} = 4$ supersymmetric mechanics is the doubling of the number of odd degrees of freedom. It was considered, within the (Lagrangian) superfield approach in [35]. In this paper, the authors considered the $(2N.2N)_{\mathbb{R}}$-dimensional superspace and the supercharges containing term cubic on odd variables. Calculating the Poisson brackets, the authors found that the admissible metrics of the configuration space of that system should have the following local form:

$$g_{ab} = \frac{\partial^2 A(x)}{\partial x^a \partial x^b} .$$
(4.187)

The admissible set of potentials looks, in this local coordinates, as follows: $V = g_{abc}c^{ab} + g^{ab}d_{af}$, where c^{ab} and d_{ab} are constant matrices.

So, considering the Hamiltonian system with generic phase spaces, we found that without any efforts it could be extended to $\mathcal{N} = 1$ supersymmetric mechanics. For the construction of $\mathcal{N} = 2$ supersymmetric mechanics we were forced to restrict ourselves to systems on the cotangent bundle of Riemann manifolds. Even after this strong restriction, we found that the inclusion of a magnetic field breaks the supersymmetry of the system. On the other hand, in trying to construct $\mathcal{N} = 4$ supersymmetric mechanics, we found that in this case even the metric of the configuration space and the admissible set of potentials are strongly restricted.

In further examples we shall show that the transition to Kähler geometry makes these restrictions much weaker.

4.6.1 $\mathcal{N} = 2$ Supersymmetric Mechanics with Kähler Phase Space

Let us consider a supersymmetric mechanics whose phase superspace is the external algebra of the Kähler manifold ΛM, where $(M, g_{a\bar{b}}(z,\bar{z})dz^a d\bar{z}^{\bar{b}})$ is the phase space of the underlying Hamiltonian mechanics [36]. The phase superspace is $(D|D)_{\mathbb{C}}$-dimensional supermanifold equipped with the Kähler structure

$$\Omega = i\partial\bar{\partial}\left(K - ig_{a\bar{b}}\theta^a\bar{\theta}^{\bar{b}} \right) = i(g_{a\bar{b}} + iR_{a\bar{b}c\bar{d}}\theta^c\bar{\theta}^{\bar{d}})\,dz^a \wedge d\bar{z}^{\bar{b}} + g_{a\bar{b}}D\theta^a \wedge D\bar{\theta}^{\bar{b}},$$

where $D\theta^a = d\theta^a + \Gamma^a_{bc}\theta^c\,dz^c$, and Γ^a_{bc}, $R_{a\bar{b}c\bar{d}}$ are the Cristoffel symbols and curvature tensor of the underlying Kähler metrics $g_{a\bar{b}} = \partial_a\partial_{\bar{b}}K(z,\bar{z})$, respectively.

The corresponding Poisson bracket can be presented in the form

$$\{\ \ ,\ \ \} = i\tilde{g}^{a\bar{b}}\nabla_a \wedge \bar{\nabla}_{\bar{b}} + g^{a\bar{b}}\frac{\partial}{\partial\theta^a} \wedge \frac{\partial}{\partial\bar{\theta}^{\bar{b}}}$$
(4.188)

where

$$\nabla_a = \frac{\partial}{\partial z^a} - \Gamma^c_{ab}\theta^b\frac{\partial}{\partial\theta^c}, \qquad \tilde{g}^{-1}_{a\bar{b}} = (g_{a\bar{b}} + iR_{a\bar{b}c\bar{d}}\theta^c\bar{\theta}^{\bar{d}}) .$$

On this phase superspace one can immediately construct $\mathcal{N} = 2$ supersymmetric mechanics, defined by the supercharges

$$Q^0_+ = \partial_a K(z, \bar{z}) \theta^a, \qquad Q^0_- = \partial_{\bar{a}} K(z, \bar{z}) \bar{\theta}^{\bar{a}} \qquad (4.189)$$

where $K(z, \bar{z})$ is the Kähler potential of M, defined up to holomorphic and antiholomorphic functions, $K(z, \bar{z}) \rightarrow K(z, \bar{z}) + U(z) + \bar{U}(\bar{z})$.

The Hamiltonian of the system reads

$$\mathcal{H}_0 = g^{a\bar{b}} \partial_a K \partial_{\bar{b}} K - i g_{a\bar{b}} \theta^a \bar{\theta}^{\bar{b}} + i \theta^c K_{c;a} \tilde{g}^{a\bar{b}} K_{\bar{b};\bar{d}} \bar{\theta}^{\bar{d}} \qquad (4.190)$$

where $K_{a;b} = \partial_a \partial_b K - \Gamma^c_{ab} \partial_c K$.

Another example of $\mathcal{N} = 2$ supersymmetric mechanics is defined by the supercharges

$$Q^c_+ = \partial_a G(z, \bar{z}) \theta^a, \qquad Q^c_- = \partial_{\bar{a}} G(z, \bar{z}) \bar{\theta}^{\bar{a}}, \qquad (4.191)$$

where the real function $G(z, \bar{z})$ is the Killing potential of the underlying Kähler structure

$$\partial_a \partial_b G - \Gamma^c_{ab} \partial_c G = 0, \qquad G^a(z) = g^{a\bar{b}} \partial_{\bar{b}} G(z, \bar{z}) . \qquad (4.192)$$

In this case the Hamiltonian of system reads

$$\mathcal{H}^c = g_{a\bar{b}} G^a G^{\bar{b}} + i \bar{\theta}^{\bar{d}} G_{a\bar{d}} \tilde{g}^{a\bar{b}} G_{\bar{c}b} \theta^c , \qquad (4.193)$$

where $G_{a\bar{b}} = \partial_a \partial_{\bar{b}} G(z, \bar{z})$.

The commutators of the supercharges in these particular examples read

$$\{Q^c_\pm, Q^0_\pm\} = \mathcal{R}_\pm, \qquad \{Q^c_\pm, Q^0_\mp\} = \mathcal{Z} , \qquad (4.194)$$

where

$$\tilde{\mathcal{Z}} \equiv G(z, \bar{z}) + i G_{a\bar{b}}(z, \bar{z}) \theta^a \bar{\theta}^{\bar{b}}, \quad \mathcal{R}_+ = i \theta^c K_{c;a} \tilde{g}^{a\bar{b}} G_{\bar{b};d} \theta^d, \quad \mathcal{R}_- = \bar{\mathcal{R}}_+ . \qquad (4.195)$$

Hence, introducing the supercharges

$$\Theta_\pm = Q^0_\pm \pm i Q^c_\mp , \qquad (4.196)$$

we can define $N = 2$ SUSY mechanics specified by the presence of the central charge $\mathcal{Z}$

$$\begin{aligned} \{\Theta_+, \Theta_-\} = \tilde{\mathcal{H}}, \quad \{\Theta_\pm, \Theta_\pm\} = \pm i \mathcal{Z} \\ \{\mathcal{Z}, \Theta_\pm\} = 0, \quad -\{\tilde{\mathcal{H}}, \Theta_\mp\} = 0, \quad \{\mathcal{Z}, \tilde{\mathcal{H}}\} = 0 . \end{aligned} \qquad (4.197)$$

The Hamiltonian of this generalized mechanics is defined by the expression

$$\tilde{\mathcal{H}} = \mathcal{H}_0 + \mathcal{H}_c + i \mathcal{R}_+ - i \mathcal{R}_- . \qquad (4.198)$$

A "fermionic number" is of the form

$$\tilde{\mathcal{F}} = i g_{a\bar{b}} \theta^a \bar{\theta}^{\bar{b}} : \{\tilde{\mathcal{F}}, \Theta_\pm,\} = \pm i \Theta_\pm . \qquad (4.199)$$

It seems that, on the external algebra of the hyper-Kähler manifold, in the same manner one could construct $\mathcal{N} = 4$ supersymmetric mechanics. On the other hand, the hyper-Kähler manifolds are the cotangent bundle of the Kähler manifolds equipped with Ricci-flat metrics.

We shall demonstrate, in the next examples, that these restrictions can be too strong. Namely, choosing the underlying phase space to be the cotangent bundle of the Kähler manifold, we will double the number of supercharges and get the $\mathcal{N} = 4$ supersymmetric mechanics on the cotangent bundles of generic Kähler manifolds and the $\mathcal{N} = 8$ ones on the cotangent bundles of the special Kähler manifolds.

4.6.2 $\mathcal{N} = 4$ Supersymmetric Mechanics

Let us show that the Hamiltonian mechanics (4.12) could be easily extended to the $\mathcal{N} = 4$ supersymmetric mechanics, when the configuration space M_0 is the Kähler manifold $(M_0, g_{a\bar{b}} \, dz^a \, d\bar{z}^{\bar{b}})$, $g_{a\bar{b}} = \partial^2 K(z,\bar{z})/\partial z^a \partial \bar{z}^{\bar{b}}$, and the potential term has the form

$$V(z,\bar{z}) = \frac{\partial \bar{U}(\bar{z})}{\partial \bar{z}^a} g^{\bar{a}b} \frac{\partial U(z)}{\partial z^b} \ .$$

For this purpose, let us define the supersymplectic structure

$$\begin{aligned}
\Omega &= \omega_0 - i \partial \bar{\partial} \mathbf{g} \\
&= d\pi_a \wedge dz^a + d\bar{\pi}_a \wedge d\bar{z}^{\bar{a}} + R_{a\bar{b}c\bar{d}} \eta_i^a \bar{\eta}_i^{\bar{b}} \, dz^a \wedge d\bar{z}^b + g_{a\bar{b}} D\eta_i^a \wedge D\bar{\eta}_i^{\bar{b}}
\end{aligned} \quad (4.200)$$

where

$$\mathbf{g} = i g_{a\bar{b}} \eta^a \sigma_0 \bar{\eta}^{\bar{b}}, \quad D\eta_i^a = d\eta_i^a + \Gamma_{bc}^a \eta_i^a \, dz^a, \quad i = 1, 2$$

Γ_{bc}^a, $R_{a\bar{b}c\bar{d}}$ are the connection and curvature of the Kähler structure, respectively, and the odd coordinates η_i^a belong to the external algebra ΛM_0, i.e., they transform as dz^a. This symplectic structure becomes canonical in the coordinates (p_a, χ^k)

$$\begin{aligned}
p_a &= \pi_a - \tfrac{i}{2} \partial_a \mathbf{g}, \quad \chi_i^m = e_b^m \eta_i^b : \\
\Omega &= dp_a \wedge dz^a + d\bar{p}_{\bar{a}} \wedge d\bar{z}^{\bar{a}} + d\chi_i^m \wedge d\bar{\chi}_i^{\bar{m}} \ ,
\end{aligned} \quad (4.201)$$

where e_a^m are the einbeins of the Kähler structure: $e_a^m \delta_{m\bar{m}} \bar{e}_b^{\bar{m}} = g_{a\bar{b}}$. The corresponding Poisson brackets are defined by the following nonzero relations (and their complex-conjugates):

$$\begin{aligned}
\{\pi_a, z^b\} &= \delta_a^b, \quad \{\pi_a, \eta_i^b\} = -\Gamma_{ac}^b \eta_i^c \ , \\
\{\pi_a, \bar{\pi}_b\} &= -R_{a\bar{b}c\bar{d}} \eta_k^c \bar{\eta}_k^{\bar{d}}, \quad \{\eta_i^a, \bar{\eta}_j^{\bar{b}}\} = g^{a\bar{b}} \delta_{ij} \ .
\end{aligned}$$

Let us represent the $\mathcal{N} = 4$ supersymmetry algebra as follows:

$$\{Q_i^+, Q_j^-\} = \delta_{ij} \mathcal{H}, \quad \{Q_i^\pm, Q_j^\pm\} = \{Q_i^\pm, \mathcal{H}\} = 0, \quad i = 1, 2 \ , \quad (4.202)$$

and choose the supercharges given by the functions

$$Q_1^+ = \pi_a \eta_1^a + i U_{\bar a} \bar\eta_2^{\bar a}, \quad Q_2^+ = \pi_a \eta_2^a - i U_{\bar a} \bar\eta_1^{\bar a}. \tag{4.203}$$

Then, calculating the commutators (Poisson brackets) of these functions, we get that the supercharges (4.203) belong to the superalgebra (4.202), when the functions $U_a, \bar U_{\bar a}$ are of the form

$$U_a(z) = \frac{\partial U(z)}{\partial z^a}, \qquad \bar U_{\bar a}(\bar z) = \frac{\partial \bar U(\bar z)}{\partial \bar z^a}, \tag{4.204}$$

while the Hamiltonian reads

$$\mathcal{H} = g^{a\bar b}(\pi_a \bar\pi_b + U_a \bar U_{\bar b}) - i U_{a;b}\eta_1^a \eta_2^b + i \bar U_{\bar a;\bar b}\bar\eta_1^{\bar a}\bar\eta_2^{\bar b} - R_{a\bar b c\bar d}\eta_1^a \bar\eta_1^b \eta_2^a \bar\eta_2^d, \tag{4.205}$$

where $U_{a;b} \equiv \partial_a \partial_b U - \Gamma^c_{ab}\partial_c U$.

Now, following [37], let us extend this system to $\mathcal{N} = 4$ supersymmetric mechanics with central charge

$$\{\Theta_i^+, \Theta_j^-\} = \delta_{ij}\mathcal{H} + \mathcal{Z}\sigma_{ij}^3, \quad \{\Theta_i^\pm, \Theta_j^\pm\} = 0, \{\mathcal{Z}, \mathcal{H}\} = \{\mathcal{Z}, \Theta_k^\pm\} = 0. \tag{4.206}$$

For this purpose one introduces the supercharges

$$\begin{aligned}
\Theta_1^+ &= (\pi_a + iG_{,a}(z, \bar z))\, \eta_1^a + i\bar U_{,\bar a}(\bar z)\bar\eta_2^{\bar a}, \\
\Theta_2^+ &= (\pi_a - iG_{,a}(z, \bar z))\, \eta_2^a - i\bar U_{,\bar a}(\bar z)\bar\eta_1^{\bar a},
\end{aligned} \tag{4.207}$$

where the real function $G(z, \bar z)$ obeys the conditions (4.192) and $\partial_{\bar a}G g^{\bar a b}U_b = 0$. So, G is a Killing potential defining the isometry of the underlying Kähler manifold (given by the vector $\mathbf{G} = G^a(z)\partial_a + \bar G^a(\bar z)\bar\partial_a, \quad G^a = ig^{a\bar b}\bar\partial_b G$) which leaves the holomorphic function $U(z)$ invariant

$$\mathcal{L}_{\mathbf{G}}U = 0 \Rightarrow G^a(z)U_a(z) = 0.$$

Calculating the Poisson brackets of these supercharges, we get explicit expressions for the Hamiltonian

$$\begin{aligned}
\mathcal{H} \equiv g^{a\bar b}\left(\pi_a \bar\pi_{\bar b} + G_{,a}G_{\bar b} + U_{,a}\bar U_{,\bar b}\right) \\
-i U_{a;b}\eta_1^a \eta_2^b + i\bar U_{\bar a;\bar b}\bar\eta_1^{\bar a}\bar\eta_2^{\bar b} + \tfrac{1}{2}G_{a\bar b}(\eta_k^a \bar\eta_k^b) - R_{a\bar b c\bar d}\eta_1^a \bar\eta_1^b \eta_2^c \bar\eta_2^d
\end{aligned} \tag{4.208}$$

and for the central charge

$$\mathcal{Z} = i(G^a \pi_a + G^{\bar a}\bar\pi_{\bar a}) + \frac{i}{2}\partial_a \bar\partial_{\bar b}G(\eta^a \sigma_3 \bar\eta^{\bar b}). \tag{4.209}$$

It can be checked by a straightforward calculation that the function $\mathcal{Z}$ indeed belongs to the center of the superalgebra (4.206). The scalar part of each phase with standard $\mathcal{N} = 2$ supersymmetry can be interpreted as a particle moving on the Kähler manifold in the presence of an external magnetic field, with strength $F = iG_{a\bar b}dz^a \wedge d\bar z^{\bar b}$, and in the potential field $U_{,a}(z)g^{a\bar b}\bar U_{,\bar b}(\bar z)$.

Assuming that $(M_0, g_{a\bar b}\, dz^a\, d\bar z^{\bar b})$ is the hyper-Kähler metric, $U(z) + \bar U(\bar z)$ is a triholomorphic function and $G(z, \bar z)$ defines a triholomorphic Killing vector, one should get $\mathcal{N} = 8$ supersymmetric mechanics. In this case, instead of the phase with standard $\mathcal{N} = 2$ supersymmetry arising in the Kähler case, we shall get the phase with standard $\mathcal{N} = 4$ supersymmetry. This system could be straightforwardly constructed by the dimensional reduction of the $\mathcal{N} = 2$ supersymmetric $(1 + 1)$-dimensional sigma-model by Alvarez-Gaumé and Freedman [38].

$\mathcal{N} = 8$ Mechanics

We have seen that the transition from the generic Riemann space to the generic Kähler space allows one to double the number of supersymmetries from $\mathcal{N} = 2$ to $\mathcal{N} = 4$, with the appropriate restriction of the admissible set of potentials.

On the other hand, we mentioned that the doubling of the number of odd variables and the restriction the Riemann metric allow one to construct the $\mathcal{N} = 4$ supersymmetric mechanics [35]. Now, following the paper [39], we shall show that a similar procedure, applied to the systems on Kähler manifolds, permits to construct the $\mathcal{N} = 8$ supersymmetric mechanics, with the supersymmetry algebra[1]

$$\{Q_{i\alpha}, Q_{j\beta}\} = \{\overline{Q}_{i\alpha}, \overline{Q}_{j\beta}\} = 0, \quad \{Q_{i\alpha}, \overline{Q}_{j\beta}\} = \epsilon_{\alpha\beta}\epsilon_{ij}\mathcal{H}_{SUSY}, \quad (4.210)$$

where $i, j = 1, 2, \alpha, \beta = 1, 2$.

We present the results for the mechanics without (bosonic) potential term. The respective systems with potential terms are constructed in [40].

To construct the $\mathcal{N} = 8$ supersymmetric mechanics, let us define the $(2d.4d)_{\mathbb{C}}$-dimensional symplectic structure

$$\Omega = \mathrm{d}\mathcal{A} = \mathrm{d}\pi_a \wedge \mathrm{d}z^a + \mathrm{d}\bar{\pi}_a \wedge \mathrm{d}\bar{z}^a - R_{a\bar{b}c\bar{d}}\eta_{i\alpha}^c \bar{\eta}^{d|i\alpha}\,\mathrm{d}z^a \wedge \mathrm{d}\bar{z}^b + g_{a\bar{b}}D\eta_{i\alpha}^a \wedge D\bar{\eta}^{b|i\alpha}\,,$$

where

$$\mathcal{A} = \pi_a \mathrm{d}z^a + \bar{\pi}_a\,\mathrm{d}\bar{z}^a + \frac{1}{2}\eta_{i\alpha}^a g_{a\bar{b}}D\bar{\eta}^{b|i\alpha} + \frac{1}{2}\bar{\eta}_{i\alpha}^b g_{a\bar{b}}D\eta^{a|i\alpha}\,, \quad (4.211)$$

and $D\eta_{i\alpha}^a = \mathrm{d}\eta_{i\alpha}^a + \Gamma_{bc}^a\eta_{i\alpha}^b\mathrm{d}z^c$. The corresponding Poisson brackets are given by the following nonzero relations (and their complex-conjugates):

$$\{\pi_a, z^b\} = \delta_a^b, \quad \{\pi_a, \eta_{i\alpha}^b\} = -\Gamma_{ac}^b\eta_{i\alpha}^c\,,$$
$$\{\pi_a, \bar{\pi}_b\} = R_{a\bar{b}c\bar{d}}\eta_{i\alpha}^c\bar{\eta}^{d|i\alpha}, \quad \{\eta_{i\alpha}^a, \bar{\eta}^{b|j\beta}\} = g^{ab}\delta_i^j\delta_\alpha^\beta\,. \quad (4.212)$$

Let us search the supercharges among the functions

$$Q_{i\alpha} = \pi_a\eta_{i\alpha}^a + \frac{1}{3}\bar{f}_{abc}\overline{T}_{i\alpha}^{abc}, \quad \overline{Q}_{i\alpha} = \bar{\pi}_a\bar{\eta}_{i\alpha}^a + \frac{1}{3}f_{abc}T_{i\alpha}^{abc}, \quad (4.213)$$

where $T_{i\alpha}^{abc} \equiv \eta_{i\beta}^a\eta^{bj\beta}\eta_{j\alpha}^c$.

---

[1] We use the following convention: $\epsilon_{ij}\epsilon^{jk} = \delta_i^k$, $\epsilon_{12} = \epsilon^{21} = 1$.

Calculating the mutual Poisson brackets of $Q_{i\alpha}, \overline{Q}_{i\alpha}$ one can get, that they obey the $\mathcal{N} = 8$ supersymmetry algebra, provided the following relations hold:

$$\frac{\partial}{\partial \bar{z}^d} f_{abc} = 0 , \qquad R_{a\bar{b}c\bar{d}} = -f_{ace} g^{e e'} \bar{f}_{e'bd} . \qquad (4.214)$$

The above equations guarantee, respectively, that the first and second equations in (4.210) are fulfilled. Then we could immediately get the $\mathcal{N} = 8$ supersymmetric Hamiltonian

$$\mathcal{H}_{SUSY} = \pi_a g^{ab} \bar{\pi}_b + \frac{1}{3} f_{abc;d} \Lambda^{abcd} + \frac{1}{3} \bar{f}_{abc;d} \bar{\Lambda}^{abcd} + f_{abc} g^{c\bar{c}'} \bar{f}_{c'de} \Lambda_0^{ab\bar{d}\bar{e}}, \qquad (4.215)$$

where

$$\Lambda^{abcd} \equiv -\frac{1}{4} \eta_{i\alpha}^a \eta^{bi\beta} \eta_{k\beta}^c \eta^{dk\alpha}, \qquad \Lambda_0^{ab\bar{c}\bar{d}} \equiv \frac{1}{2}(\eta_i^{a\alpha} \eta_{j\alpha}^b \bar{\eta}^{c\beta i} \bar{\eta}_\beta^{dj} + \eta_\alpha^{ai} \eta_{i\beta}^b \bar{\eta}^{cj\alpha} \bar{\eta}_j^{d\beta}),$$

and $f_{abc;d} = f_{abc,d} - \Gamma_{da}^e f_{ebc} - \Gamma_{db}^e f_{aec} - \Gamma_{dc}^e f_{abe}$ is the covariant derivative of the third rank covariant symmetric tensor.

Equation (4.214) precisely mean that the configuration space M_0 is a special Kähler manifold of the rigid type [41]. Taking into account the symmetrizing of f_{abc} over spatial indices and the explicit expression of $R_{a\bar{b}c\bar{d}}$ in terms of the metric g_{ab}, we can immediately find the local solution for (4.214)

$$f_{abc} = \frac{\partial^3 f(z)}{\partial z^a \partial z^b \partial z^c}, \qquad g_{a\bar{b}} = e^{i\nu} \frac{\partial^2 f(z)}{\partial z^a \partial z^b} + e^{-i\nu} \frac{\partial^2 \bar{f}(\bar{z})}{\partial \bar{z}^a \partial \bar{z}^b} , \qquad (4.216)$$

where $\nu = \text{const.} \in \mathbb{R}$.

Redefining the local function, $f \to i e^{-i\nu} f$, we shall get the ν-parametric family of supersymmetric mechanics, whose metric is defined by the Kähler potential of a special Kähler manifold of the rigid type. Surely, this local solution is not covariant under arbitrary holomorphic transformation, and it assumes the choice of a distinguished coordinate frame.

The special Kähler manifolds of the rigid type became widely known during last decade due to the so-called "T-duality symmetry": in the context of $\mathcal{N} = 2, d = 4$ super-Yang–Mills theory, it connects the UV and IR limit of the theory [42]. The "T-duality symmetry" is expressed in the line below

$$(z^a, f(z)) \Rightarrow \left(u_a = \frac{\partial f(z)}{\partial z^a}, \tilde{f}(u) \right), \qquad \text{where} \qquad \frac{\partial^2 \tilde{f}(u)}{\partial u_a \partial u_c} \frac{\partial f}{\partial z^c \partial z^b} = -\delta_b^a . \qquad (4.217)$$

It is clear that the symplectic structure is *covariant* under the following holomorphic transformations:

$$\tilde{z}^a = \tilde{z}^a(z), \qquad \tilde{\eta}_{i\alpha}^a = \frac{\partial \tilde{z}^a(z)}{\partial z^b} \eta_{i\alpha}^b, \qquad \tilde{\pi}_a = \frac{\partial z^b}{\partial \tilde{z}^a} \pi_b , \qquad (4.218)$$

By the use of (4.218), we can extend the duality transformation (4.217) to the whole phase superspace $(\pi_a, z^a, \eta_{i\alpha}^a) \to (p^a, u_a, \psi_{a|i\alpha})$

$$u_a = \partial_a f(z), \quad p^a \frac{\partial^2 f}{\partial z^a \partial z^b} = -\pi_b, \quad \psi_a^{i\alpha} = \frac{\partial^2 f}{\partial z^a \partial z^b} \eta^{b|i\alpha} . \tag{4.219}$$

Taking into account the expression of the symplectic structure in terms of the presymplectic one-form (4.211), we can easily perform the Legendre transformation of the Hamiltonian to the (second-order) Lagrangian

$$\mathcal{L} = \mathcal{A}(d/dt) - \mathcal{H}_{SUSY}|_{\pi_a = g_{a\bar{b}} \dot{\bar{z}}^b}$$

$$= g_{a\bar{b}} \dot{z}^a \dot{\bar{z}}^b + \frac{1}{2} \eta_{i\alpha}^a g_{ab} \frac{D\bar{\eta}^{b|i\alpha}}{dt} + \frac{1}{2} \bar{\eta}_{i\alpha}^b g_{ab} \frac{D\eta^{a|i\alpha}}{dt} \tag{4.220}$$

$$- \frac{1}{3} f_{abc;d} \Lambda^{abcd} - \frac{1}{3} \bar{f}_{abc;d} \bar{\Lambda}^{abcd} - f_{abc} g^{c\bar{c}'} \bar{f}_{c'de} \Lambda_0^{abd\bar{e}} .$$

Here, we denoted $d/dt = \dot{z}^a \partial/\partial z^a + \dot{\eta}_{i\alpha}^a \partial/\partial \eta_{i\alpha}^a + c.c.$ Clearly, the Lagrangian (4.220) is covariant under holomorphic transformations (4.218), and the duality transformation as well. The prepotential $\tilde{f}(u)$ is connected with $f(z)$ by the Legendre transformation

$$\tilde{f}(u) = \tilde{f}(u, z)|_{u_a = \partial_a f(z)}, \quad \tilde{f}(u, z) = u_a z^a - f(z) .$$

4.6.3 Supersymmetric Kähler Oscillator

So far, the Kähler structure allowed us to double the number of supersymmetries in the system. One can hope that in some cases this could be preserved after inclusion of constant magnetic field, since this field usually respects the Kähler structure. We shall show, on the example of the Kähler oscillator (4.126), that it is indeed a case.

Let us consider, following [17, 43], the supersymmetrization of a specific model of Hamiltonian mechanics on the Kähler manifold $(M_0, g_{a\bar{b}} dz^a d\bar{z}^b)$ interacting with the constant magnetic field B, viz

$$\mathcal{H} = g^{a\bar{b}}(\pi_a \bar{\pi}_b + \alpha^2 \partial_a K \bar{\partial}_b K), \quad \Omega_0 = d\pi_a \wedge dz^a + d\bar{\pi}_a \wedge d\bar{z}^a + iB g_{a\bar{b}} dz^a \wedge d\bar{z}^b , \tag{4.221}$$

where $K(z, \bar{z})$ is a Kähler potential of configuration space.

Remind, that the Kähler potential is defined up to holomorphic and antiholomorphic terms, $K \to K + U(z) + \bar{U}(\bar{z})$. Hence, in the limit $\omega \to 0$ the above Hamiltonian takes the form

$$\mathcal{H} = g^{a\bar{b}}(\pi_a \bar{\pi}_b + \partial_a U(z) \bar{\partial}_b \bar{U}(\bar{z})) , \tag{4.222}$$

i.e., it admits, in the absence of magnetic field, a $\mathcal{N} = 4$ superextension.

Notice, also, that in the "large mass limit," $\pi_a \to 0$, this system results in the following one:

$$\mathcal{H}_0 = \omega^2 g^{a\bar{b}} \partial_a K \bar{\partial}_b K, \quad \Omega_0 = iB g_{a\bar{b}} dz^a \wedge d\bar{z}^b,$$

which could be easily extended to $\mathcal{N} = 2$ supersymmetric mechanics.

We shall show that, although the system under consideration does not possess a standard $\mathcal{N} = 4$ superextension, it admits a superextension in terms of a nonstandard superalgebra with four fermionic generators, including, as subalgebras, two copies of the $\mathcal{N} = 2$ superalgebra. This nonstandard superextension respects the inclusion of a constant magnetic field.

We use the following strategy. At first, we extend the initial phase space to a $(2N.2N)_{\mathbb{C}}$-dimensional superspace equipped with the symplectic structure

$$\Omega = \Omega_B - iR_{a\bar{b}c\bar{d}}\eta_i^c\bar{\eta}_i^d)\,\mathrm{d}z^a \wedge \mathrm{d}\bar{z}^b + g_{a\bar{b}}D\eta_i^a \wedge D\bar{\eta}_i^b \,, \qquad (4.223)$$

where Ω_B is given by (4.36). The corresponding Poisson brackets are defined by the following nonzero relations (and their complex-conjugates):

$$\{\pi_a, z^b\} = \delta_a^b, \quad \{\pi_a, \eta_i^b\} = -\Gamma_{ac}^b\eta_i^c \,,$$
$$\{\pi_a, \bar{\pi}_b\} = i(Bg_{a\bar{b}} + iR_{a\bar{b}c\bar{d}}\eta_i^c\bar{\eta}_i^d), \quad \{\eta_i^a, \bar{\eta}_j^b\} = g^{a\bar{b}}\delta_{ij} \,. \qquad (4.224)$$

Then, in order to construct the system with the exact $\mathcal{N} = 2$ supersymmetry (4.183), we shall search for the odd functions $Q^\pm$, which obey the equations $\{Q^\pm, Q^\pm\} = 0$ (we restrict ourselves to the supersymmetric mechanics whose supercharges are *linear* in the Grassmann variables η_i^a, $\bar{\eta}_i^{\bar{a}}$).

Let us search for the realization of supercharges among the functions

$$Q^\pm = \cos\lambda\,\Theta_1^\pm + \sin\lambda\,\Theta_2^\pm \,, \qquad (4.225)$$

where

$$\Theta_1^+ = \pi_a\eta_1^a + i\bar{\partial}_a W\bar{\eta}_2^a, \quad \Theta_2^+ = \bar{\pi}_a\bar{\eta}_2^a + i\,\partial_a W\eta_1^a, \quad \Theta_{1,2}^- = \bar{\Theta}_{1,2}^+ \,, \quad (4.226)$$

and λ is some parameter.

Calculating the Poisson brackets of the functions, we get

$$\{Q^\pm, Q^\pm\} = i(B\sin 2\lambda + 2\alpha\cos 2\lambda)\mathcal{F}_\pm \,, \qquad (4.227)$$
$$\{Q^+, Q^-\} = \mathcal{H}_{SUSY}^0 + (B\cos 2\lambda - 2\alpha\sin 2\lambda)\,\mathcal{F}_3/2 \,. \qquad (4.228)$$

Here and further, we use the notation

$$\mathcal{H}_{SUSY}^0 = \mathcal{H} - R_{a\bar{b}c\bar{d}}\eta_1^a\bar{\eta}_1^b\eta_2^c\bar{\eta}_2^d - iW_{a;b}\eta_1^a\eta_2^b + iW_{\bar{a};\bar{b}}\bar{\eta}_1^a\bar{\eta}_2^b + B\frac{ig_{a\bar{b}}\eta_i^a\bar{\eta}_i^b}{2}, \quad (4.229)$$

where $\mathcal{H}$ denotes the oscillator Hamiltonian (4.126), and

$$\mathbf{F} = \frac{i}{2}g_{a\bar{b}}\eta_i^a\bar{\eta}_j^b\sigma_{ij}, \qquad \mathcal{F}_\pm = F_1 \pm F_2. \qquad (4.230)$$

One has then

$$\{Q^\pm, Q^\pm\} = 0 \Leftrightarrow B\sin 2\lambda + 2\alpha\cos 2\lambda = 0 \,, \qquad (4.231)$$

so that $\lambda = \lambda_0 + (i - 1)\pi/2$, $i = 1, 2$.

Here, the parameter λ_0 is defined by the expressions

$$\cos 2\lambda_0 = \frac{B/2}{\sqrt{\alpha^2 + (B/2)^2}}, \qquad \sin 2\lambda_0 = -\frac{\alpha}{\sqrt{\alpha^2 + (B/2)^2}}. \qquad (4.232)$$

Hence, we get the following supercharges:

$$Q_\nu^\pm = \cos \lambda_0 \Theta_1^\pm + (-1)^\nu \sin \lambda_0 \Theta_2^\pm, \qquad (4.233)$$

and the pair of $\mathcal{N} = 2$ supersymmetric Hamiltonians

$$\mathcal{H}_{SUSY}^i = \{Q_\nu^+, Q_\nu^-\} = \mathcal{H}_{SUSY}^0 - (-1)^i \sqrt{\alpha^2 + (B/2)^2} \mathcal{F}_3 . \qquad (4.234)$$

Notice that the supersymmetry invariance is preserved in the presence of a constant magnetic field.

Calculating the commutators of $Q_1^\pm$ and $Q_2^\pm$, we get

$$\{Q_1^\pm, Q_2^\pm\} = 2\sqrt{\alpha^2 + (B/2)^2} \mathcal{F}_\pm, \quad \{Q_1^+, Q_2^-\} = 0 . \qquad (4.235)$$

The Poisson brackets between $\mathcal{F}_\pm$ and $Q_\nu^\pm$ look as follows:

$$\{Q_i^\pm, \mathcal{F}_\pm\} = 0, \quad \{Q_i^\pm, \mathcal{F}_\mp\} = \pm \epsilon_{ij} Q_j^\pm, \quad \{Q_i^\pm, \mathcal{F}_3\} = \pm i Q_i^\pm . \qquad (4.236)$$

In the notation $S_1^\pm \equiv Q_1^\pm, \quad S_2^\pm = Q_2^\mp$ the whole superalgebra reads

$$\begin{aligned}
\{S_i^\pm, S_j^\mp\} &= \delta_{ij}\mathcal{H}_{SUSY}^0 + \Lambda \sigma_{ij}^\mu \mathcal{F}_\mu , \\
\{S_i^\pm, \mathcal{F}_\mu\} &= \pm i\sigma_{ij}^\mu S_j^\pm, \quad \{\mathcal{F}_\mu, \mathcal{F}_\nu\} = \epsilon_{\mu\nu\rho}\mathcal{F}_\rho ,
\end{aligned} \qquad (4.237)$$

where

$$\Lambda = \sqrt{\omega^2 + (B/2)^2} . \qquad (4.238)$$

This is precisely the weak supersymmetry algebra considered by Smilga [44]. In the particular case $\omega = 0$ it yields the $\mathcal{N} = 4$ supersymmetric mechanics broken by the presence of a constant magnetic field.

Let us notice the α and B appear in this superalgebra in a symmetric way, via the factor $\sqrt{\alpha^2 + (B/2)^2}$.

Remark. In the case of the oscillator on $\mathbb{C}^N$ we can smoothly relate the above supersymmetric oscillator with a $N = 4$ oscillator, provided we choose

$$K = \cos \gamma \, z\bar{z} + \sin \gamma \, (z^2 + \bar{z}^2)/2 , \qquad \gamma \in [0, \pi/2] .$$

Hence,

$$\mathcal{H} = \pi\bar{\pi} + \alpha_0^2 z\bar{z} + \sin 2\gamma \, \alpha_0^2 (z^2 + \bar{z}^2)/2 ,$$

i.e., for $\gamma = 0, \pi/2$ we have a standard harmonic oscillator, while for $\gamma \neq 0, \pi/2$ we get the anisotropic one, which is equivalent to two sets of N one-dimensional oscillators with frequencies $\alpha_0 \sqrt{1 \pm \sin 2\gamma}$. The frequency α appearing in the superalgebra, is of the form: $\alpha = \alpha_0 \cos \gamma$.

4.7 Conclusion

We presented some constructions of the Hamiltonian formalism related with Hopf maps and Kähler geometry, and a few models of supersymmetric mechanics on Kähler manifolds. One can hope that the former constructions could be useful in supersymmetric mechanics along the following lines. Firstly, one could try to extend the number of supersymmetries, passing from the Kähler manifolds to quaternionic ones. The model suggested in [45] indicates that this could indeed work. One could also expect that the latter system will respect the inclusion of an instanton field. Secondly, one can try to construct the supersymmetric mechanics, performing the Hamiltonian reduction of the existing systems, related with the Hopf maps. In this way one could get new supersymmetric models, specified by the presence of Dirac and Yang monopoles, as well as with constant magnetic and instanton fields.

Acknowldgments

I would like to thank Stefano Bellucci for organizing the Winter School on *Modern trends in supersymmetric mechanics* and inviting me to deliver (and to write down) these lectures, which are the result of extensive discussions with him, and with S. Krivonos and V.P. Nair. Most of the examples included in the paper were obtained in collaboration. I am indebted to all of my coauthors, especially to A. Yeranian, O. Khudaverdian, V. Ter-Antonyan, P.-Y. Casteill. Special thanks to Stefano Bellucci for careful reading of manuscript and substantial improving of the English.

References

1. V.I. Arnold: *Mathematical Methods of Classical Mechanics* (Nauka, Moscow, 1989)
2. A.M. Perelomov: *Integrable Systems of Classical Mechanics and Lie Algebras*, (Birkhauser, Basel, 1990)
3. F.A. Berezin: *Introduction to Superanalysis* (D. Reidel, Dordrecht, 1986)
4. T. Voronov: *Geometric integration theory over supermanifolds*, Sov. Sci. Rev. C, Math. Phys. **9**, 1–138 (1992)
5. L.E. Gendenshtein, I.V. Krive: Sov. Phys. Uspekhi **146** (4), 553 (1985)
6. F. Cooper, A. Khare, U. Sukhatme: Phys. Rep. **251**, 267 (1995)
7. D. Zwanziger: Phys. Rev. **176**, 1480 (1968)
8. S. Kobayashi, K. Nomizu: *Foundations of Differential Geometry*, vol. 2 (Interscience Publishers, New York, 1969)
9. A. Nersessian, V. Ter-Antonian, M.M. Tsulaia: Mod. Phys. Lett. A **11**, 1605 (1996) [arXiv:hep-th/9604197]
10. A. Nersessian, A. Yeranyan: J. Phys. A **37**, 2791 (2004) [arXiv:hep-th/0309196]
11. C.N. Yang: J. Math. Phys. **19**, 320; (1978); C.N. Yang: J. Math. Phys. **19**, 2622 (1978)

12. S. Bellucci, P.Y. Casteill, A. Nersessian: Phys. Lett. B **574**, 121 (2003) [arXiv:hep-th/0306277]
13. S.C. Zhang, J.P. Hu: Science **294**, 823 (2001)
14. D. Karabali, V.P. Nair: Nucl. Phys. B **641**, 533 (2002)
15. B.A. Bernevig, J.P. Hu, N. Toumbas, S.C. Zhang: Phys. Rev. Lett. **91**, 236803 (2003)
16. P.W. Higgs: J. Phys. A **12**, 309 (1979); H.I. Leemon: J. Phys. A **12**, 489 (1979)
17. S. Bellucci, A. Nersessian: Phys. Rev. D **67**, 065013 (2003) [arXiv:hep-th/0211070]
18. S. Bellucci, A. Nersessian, A. Yeranyan: Phys. Rev. D **70**, 085013 (2004) [arXiv:hep-th/0406184]
19. L. Mardoyan, A. Nersessian: Phys. Rev. B **72**, 233303 (2005) [arXiv:hep-th/0508062]
20. E. Schrödinger: Proc. R. Irish Soc. **46**, 9 (1941); E. Schrödinger: Proc. R. Irish Soc. **46**, 183 (1941); E. Schrödinger, Proc. R. Irish Soc. **47**, 53 (1941)
21. V.M. Ter-Antonyan: quant-ph/0003106
22. A. Nersessian, V. Ter-Antonian: Mod. Phys. Lett. A **9**, 2431 (1994) [arXiv:hep-th/9406130]; A. Nersessian, V. Ter-Antonian: Mod. Phys. Lett. A **10**, 2633 (1995)
23. T. Iwai: J. Geom. Phys. **7**, 507 (1990); L.G. Mardoyan, A.N. Sissakian, V.M. Ter-Antonyan: Phys. Atom. Nucl. **61**, 1746 (1998) [hep-th/9803010]
24. A. Nersessian, G. Pogosyan: Phys. Rev. A **63**, 020103 (2001) [arXiv:quant-ph/0006118]
25. S. Bellucci, A. Nersessian, A. Yeranyan: Phys. Rev. D **70**, 045006 (2004) [arXiv:hep-th/0312323]
26. A. Schwarz: Commun. Math. Phys. **155**, 249 (1993)
27. A.P. Nersessian: Theor. Math. Phys. **96**, 866 (1993)
28. O.M. Khudaverdian, A.S. Schwarz, Y.S. Tyupkin: Lett. Math. Phys. **5**, 517 (1981); O.M. Khudaverdian, R.L. Mkrtchian: Lett. Math. Phys. **18**, 229 (1989)
29. O.M. Khudaverdian: J. Math. Phys. **32**, 1934 (1991); O.M. Khudaverdian: Commun. Math. Phys. **247**, 353 (2004)
30. I.A. Batalin, G.A. Vilkovisky: Phys. Lett. B **102**, 27 (1981); I.A. Batalin, G.A. Vilkovisky: Phys. Rev. D **28**, 2567 (1983); [Erratum; I.A. Batalin, G.A. Vilkovisky: Phys. Lett. D **30**, 508 (1984)]; I.A. Batalin, G.A. Vilkovisky: Nucl. Phys. B **234**, 106 (1984)
31. A. Nersessian: Lect. Notes Phys. **524**, 90 (1997) [arXiv:hep-th/9811110]
32. M. Alexandrov, M. Kontsevich, A. Schwartz, O. Zaboronsky: Int. J. Mod. Phys. A **12**, 1405 (1997) [arXiv:hep-th/9502010]
33. O.M. Khudaverdian, A.P. Nersessian: J. Math. Phys. **34**, 5533 (1993) [arXiv:hep-th/9210091]
34. E. Ivanov, L. Mezincescu, P.K. Townsend: arXiv:hep-th/0404108; E. Ivanov, L. Mezincescu, P.K. Townsend: arXiv:hep-th/0311159
35. E.E. Donets, A. Pashnev, J.J. Rosales, M. Tsulaia: Phys. Rev. D **61**, 043512 (2000)
36. S. Bellucci, A. Nersessian: Nucl. Phys. Proc. Suppl. **102**, 227 (2001) [arXiv:hep-th/0103005]
37. S. Bellucci, A. Nersessian: Phys. Rev. D **64**, 021702 (2001) [arXiv:hep-th/0101065]
38. L. Alvarez-Gaumé, D. Freedman: Commun. Math. Phys. **91**, 87 (1983)

39. S. Bellucci, S. Krivonos, A. Nersessian: Phys. Lett. B **605**, 181 (2005) [arXiv:hep-th/0410029]
40. S. Bellucci, S. Krivonos, A. Shcherbakov: Phys. Lett. B **612**, 283 (2005) [arXiv:hep-th/0502245]
41. P. Fre: Nucl. Phys. Proc. Suppl. BC **45**, 59 (1996) [arXiv:hep-th/9512043]
42. N. Seiberg, E. Witten: Nucl. Phys. B **426**, 19 (1994); [Erratum; N. Seiberg, E. Witten: Nucl. Phys. B **430**, 485 (1994)]
43. S. Bellucci, A. Nersessian: arXiv:hep-th/0401232
44. A.V. Smilga: Phys. Lett. B **585**, 173 (2004) [arXiv:hep-th/0311023]
45. S. Bellucci, S. Krivonos, A. Sutulin: Phys. Lett. B **605**, 406 (2005) [arXiv:hep-th/0410276]

5

Matrix Models

C. Sochichiu[1,2,3]

[1] Max-Planck-Institut für Physik (Werner-Heisenberg-Institut), Föhringer Ring 6, 80805 München, Germany
[2] Institutul de Fizică Aplicată AŞ, str. Academiei, nr. 5, Chişinău MD2028, Moldova
[3] Bogoliubov Laboratory of Theoretical Physics, Joint Institute for Nuclear Research, 141980 Dubna, Moscow Reg., Russia
sochichi@mppmu.mpg.de

Matrix models and their connections to string theory and noncommutative geometry are discussed. Various types of matrix models are reviewed. The most interesting are IKKT and BFSS models. They are introduced as 0+0 and 1+0-dimensional reduction of Yang–Mills model respectively. They are obtained via the deformations of string/membrane worldsheet/worldvolume. Classical solutions leading to noncommutative gauge models are considered.

5.1 Introduction

At the beginning let us define the topic of the present lectures. As follows from the title, "Matrix Models" are theories in which the fundamental variable is matrix. The matrix variable can be just a constant or a function of time or even be defined as a function over some space-time manifold. With this definition almost any model existing in modern physics, e.g., Yang–Mills theory, theories of gravity etc., will be a "matrix theory." Therefore, when speaking on the matrix theory usually a simple structure is assumed, e.g., when fundamental variables are constant or at most time dependent. In the first case, i.e. in one of the models of random matrices, one has no time and therefore these models are not dynamical. This is a statistical theory describing random matrix distributions. These models are popular in many areas, e.g., in the context of description of integrable systems in gauge theories, or nuclear systems as well as in the study of the lattice Dirac operators (for a review see e.g. [1–7] and references therein). The special case of interest for us is the Yang–Mills type matrix models arising in string theory as such in [8].

Another case of interest is the so-called matrix mechanics, i.e., theories of time-evoluting matrices. These models along with the random matrix models are of special interest in string theory. Thus, the Yang–Mills type matrix models appear to nonperturbatively describe collective degrees of freedom in string

C. Sochichiu: *Matrix Models*, Lect. Notes Phys. **698**, 189–225 (2006)
DOI 10.1007/3-540-33314-2_5

theory called *branes*. Branes are extended objects on which the "normal" fundamental strings can end. It was conjectured that including the brane degrees of freedom in the "conventional" superstring theories leads to their unification into the *M-theory*, a model giving in its different perturbative regimes all known superstring models. The M-theory is believed to be related to the 12-dimensional membrane. In the light-cone frame it was conjectured to be described by an Yang–Mills type matrix mechanics (BFSS matrix model) [9]. As we will see in the next section, this as well as the IKKT matrix model can be obtained by quantization/deformation of, respectively, the worldvolume of the membrane and the worldsheet of the string.

As it is by now clear, in this lecture note we are considering mainly these two models, which sometimes are called "matrix theories" to underline their fundamental role in string theory.

The plan of this lecture notes is as follows. In the next section, we give the string motivation and introduce the matrix models as dimensional reductions of supersymmetric Yang–Mills model. Next, we consider Nambu–Goto description of the string and membrane and show that the noncommutative deformation of the respectively, worldsheet or worldvolume leads to IKKT or BFSS matrix models. In the following section, we analyze the classical solutions to these matrix models and interpret them as noncommutative gauge models. The fact that these models have a common description in terms of the original matrix model allows one to establish the equivalence relations among them.

5.2 Matrix Models of String Theory

5.2.1 Branes and Matrices

A breakthrough in the development of string theory, "the second string revolution" happened when it was observed that in the dynamics of fundamental string one has additional degrees of freedom corresponding to the dynamics at the string ends [10] (see [11] for a review).

In the open string mode expansion the dynamics at the edge is described by an Abelian gauge field (particle) (for a modern introduction to string theory see e.g. [12, 13]). The corresponding charge of the end of the string is called *Chan–Patton factor*. Allowing a superposition of several, say N such factors, which correspond to an "N-valent" string end, gives rise to a non-abelian $U(N)$ super Yang–Mills gauge field in the effective Lagrangian of the open string. This is the so-called nine-brane. As it was shown in [10], string theory allows brane configurations of different other dimensions p, $0 \leq p + 1 \leq 10$. Depending on the type of the string model they preserve parts of supersymmetry.

So, descending down to the lower-dimensional p-banes one gets the $p + 1$-dimensional reductions of the 10-dimensional super-Yang–Mills model.

It appears that out of all possibilities only two cases are fundamental, namely, this of $p = 0$ and $p = -1$. All other cases can be obtained from either $p = -1$ or $p = 0$ by condensation of -1- or 0-branes into higher-dimensional objects.

5.2.2 The IKKT Matrix Model Family

As it follows from the space-time picture, the -1 branes are nondynamical and, therefore, should be described by a random matrix model which is the reduction of the 10-dimensional Super Yang-Mills theory (SYM) down to zero dimensions:

$$S_{-1} = -\frac{1}{4g^2}\,\text{tr}[X_\mu, X_\nu]^2 - \text{tr}\,\bar{\psi}\gamma^\mu[X_\mu, \psi]\,, \qquad (5.1)$$

where g is some coupling constant depending on SYM coupling g_{YM} and the volume of compactification. Matrices X_μ, $\mu = 1, \ldots, 10$ are Hermitian $N \times N$ matrices, ψ is a 10-dimensional spinor which has $N \times N$ matrix index, γ^μ are 10-dimensional Dirac γ-matrices.

From the 10-dimensional SYM the matrix model (5.1) inherits the following symmetries:

- Shifts:

$$X_\mu \to X_\mu + a_\mu \cdot \mathbb{I}\,, \qquad (5.2)$$

where a_μ is a c-number.
- $SO(10)$ rotation symmetry

$$X_\mu \to \Lambda_\mu{}^\nu X_\nu\,, \qquad (5.3)$$

where $\Lambda \in SO(10)$. This is the consequence of the (euclideanized) Lorentz invariance of the 10-dimensional SYM model.
- $SU(N)$ gauge symmetry

$$X_\mu \to U^{-1}X_\mu U\,, \qquad (5.4)$$

where $U \in SU(N)$, and this is the remnant of the SYM gauge symmetry invariance.
- Also, one has left from the SYM model the supersymmetry invariance:

$$\delta_1 X_\mu = \bar{\epsilon}\gamma_\mu\psi\,, \qquad (5.5)$$
$$\delta_1 \psi = [X_\mu, X_\nu]\gamma^{\mu\nu}\epsilon\,, \qquad (5.6)$$

as well as the second one which is simply the shift of the fermion,

$$\delta_2 X_\mu = 0\,, \qquad (5.7)$$
$$\delta_2 \psi = \eta\,, \qquad (5.8)$$

where ϵ and η are the supersymmetry transformation parameters.

Exercise 1. Find the relation between the coupling g in (5.1) on one side and SYM coupling g_{YM} and the size/geometry of compactification on the other side.

<u>Hint:</u>   Use an appropriate gauge fixing.

Exercise 2. Show that (5.3)–(5.5) are indeed the symmetries of the action (5.1).

The purely bosonic version of IKKT matrix model can be interpreted as the algebraic version of much older Eguchi–Kawai model [14]. The last is formulated in terms of $SU(N)$ group-valued fields U_μ (in contrast to the algebra-valued X_μ). The action for the Eguchi–Kawai model reads as,

$$S_{\text{EK}} = -\frac{1}{4g_{\text{EK}}^2} \sum_{\mu,\nu} \text{tr}(U_\mu U_\nu U_\mu^{-1} U_\nu^{-1} - \mathbb{I}) . \tag{5.9}$$

By the substitution, $U_\mu = \exp a X_\mu$, $g_{\text{EK}}^2 = g^2 a^{4-d}$ and taking the limit $a \to 0$ one formally comes to the bosonic part of the IKKT action (5.1).

Note: From the string interpretation we will discuss in the next section it is worth to add an extra term to the IKKT action (5.1) and, namely, the chemical potential term,

$$\Delta S_{\text{chem}} = -\beta \, \text{tr} \, \mathbb{I} , \tag{5.10}$$

which "controls" the statistical behavior of N. In the string/brane picture β plays the role of the chemical potential for the number of branes. This produces the relative weights for the distributions with different N, which cannot be catched from the arguments we used to write down the action (5.1).

5.2.3 The BFSS Model Family

Let us consider another important model which describes the dynamics of zero branes [9]. Basic ingredients of this model are roughly the same as for the previous one, the IKKT model, except that now the matrices depend on time. The action for this model is the dimensional reduction of the 10-dimensional SYM model down to the only time dimension:

$$S_{\text{BFSS}} = \frac{1}{g_{\text{BFSS}}} \int dt \, \text{tr} \left\{ \frac{1}{2}(\nabla_0 X_i)^2 + \bar\psi \nabla_0 \psi - \frac{1}{4}[X_i, X_j]^2 - \bar\psi \gamma_i [X_i, \psi] \right\} , \tag{5.11}$$

where, now, the index i runs from one to nine.

The action (5.11) describes the dynamics of zero branes in IIA string theory, but it was also proposed as the action for the M-theory membrane in the light-cone approach. As we are going to see in the next section, this model along with the IKKT model can be obtained by worldvolume quantization of the membrane action.

Another known modification of this action for the pp-wave background was proposed by Berenstein–Maldacena–Nastase (BMN) [15–17]. It differs from the BFSS model additional terms which are introduced in order to respect the pp-wave supersymmetry. The action of the BMN matrix model reads:

$$S_{\text{BMN}} = \int dt \, \text{tr} \left[\frac{1}{2(2R)} (\nabla_0 X_i)^2 + \bar{\psi} \nabla_0 \psi \right.$$
$$\left. + \frac{(2R)}{4} [X_i, X_j]^2 - i(2R)\bar{\psi}\gamma^i[X_i, \psi] \right] + S_{\text{mass}}, \quad (5.12)$$

where S_{mass} is given by

$$S_{\text{mass}} = \int dt \, \text{tr} \left[\frac{1}{2(2R)} \left(-\left(\frac{\mu}{3}\right) \sum_{i=1,2,3} X_i^2 - \left(\frac{\mu}{6}\right) \sum_{i=4,\dots,9} X_i^2 \right) \right.$$
$$\left. - \frac{\mu}{4} \bar{\psi} \gamma_{123} \psi - \frac{\mu i}{3} \sum_{jkl=1,\dots,3} \epsilon_{ijk} X_i X_j X_k \right]. \quad (5.13)$$

The essential difference of this model from the standard BFSS one is that due to the mass and the Chern–Simons terms this matrix model allow stable vacuum solutions which can be interpreted as spherical branes (see e.g. [18, 19]. Such vacuum configurations cannot exist in the original BFSS model.

5.3 Matrix Models from the Noncommutativity

In this section, we show that the matrix models which we introduced in the previous section arise when one allows the worldsheet of the string/worldvolume of the membrane to possess noncommutativity. It is interesting to note from the beginning that the "quantization" of the string worldsheet leads to the IKKT matrix model, while the space noncommutative membrane is described by the BFSS model. Let us remind that the above matrix models were introduced to describe, respectively, the −1- and 0-branes, while the string and the membrane are respectively 1- and 2-brane objects. In the shed of the next section this can be interpreted as deconstruction of the 1- and 2-branes into their basic components, namely −1- and 0-brane objects.

In this section, we consider only the bosonic parts. The extension to the fermionic part is not difficult, so this is left to the reader as an exercise.

5.3.1 Noncommutative String and the IKKT Matrix Model

In trying to make the fundamental string noncommutative one immediately meets the following problem: The noncommutativity parameter is a dimensional parameter and, therefore, hardly compatible with the worldsheet conformal symmetry which plays a fundamental role in the string theory. Beyond this there is no theoretical reason to think that the worldsheet of the fundamental string should be noncommutative. On the other hand, these are other string-like objects in the nonperturbative string theory: D1-branes or D-strings. As it was realized, in the presence of the constant nonzero Neveu–Schwarz B-field the brane can be described by a noncommutative gauge models [20–23]. Then, in contrast to the fundamental string, it is natural to make the D-string noncommutative.

Let us start with the Euclidean Nambu–Goto action for the string,

$$S_{\mathrm{NG}} = T \int d^2\sigma \sqrt{\det_{ab} \partial_a X^\mu \partial_b X_\mu} \,, \tag{5.14}$$

where T is the D-string tension and $X^\mu = X^\mu(\sigma)$ are the embedding coordinates. The expression under the square root of the r.h.s. of (5.14) can equivalently rewritten as follows,

$$\det \partial X \cdot \partial X = \frac{1}{2}\Sigma^2 \,, \tag{5.15}$$

where

$$\Sigma^{\mu\nu} = \epsilon^{ab}\partial_a X^\mu \partial_b X^\nu \,, \tag{5.16}$$

which is the induced worldsheet volume form of the embedding $X^\mu(\sigma)$.

The Nambu–Goto action then becomes:

$$S_{\mathrm{NG}} = T \int d^2\sigma \sqrt{\frac{1}{2}\Sigma^2} \,. \tag{5.17}$$

This action is nonlinear and still quite complicate. A much simple form can be obtained using the Polyakov trick. To illustrate the idea of the trick which is widely used in the string theory consider first the example of a particle.

Polyakov's Trick

The relativistic particle is described by the following reparameterization invariant action,

$$S_p = m \int d\tau \sqrt{\dot{x}^2} \,, \tag{5.18}$$

where m is the mass and x is the particle coordinate. The dynamics of the particle (5.18) is equivalent, at least classically to one described by the following action,

$$S_{pp} = \int d\tau \left(\frac{1}{2}e^{-1}\dot{x}^2 + m^2 e\right) \,. \tag{5.19}$$

In this form one has a new variable e which plays the role of the line einbein function, or better to say of the one-dimensional volume form.

To see the classical equivalence between (5.19) and (5.18) one should write down the equations of motion arising from the variation of e,

$$e^2 = \frac{\dot{x}^2}{m^2} \,, \tag{5.20}$$

and use it to substitute e in the action (5.19) which should give exactly (5.18).

Exercise 3. Show this!

As one can see, both actions (5.18) and (5.19) are reparameterization invariant, the difference being that the Polyakov action (5.19) is quadratic in the particle velocity $\dot{x}$. This trick is widely used in the analysis of nonlinear systems with gauge symmetry. In what follows we will apply it too.

Let us turn back to our string and the action (5.17). Applying the Polyakov trick, one can rewrite the action (5.17) in the following (classically) equivalent form,

$$S_{\mathrm{NGP}} = \int d^2\sigma \left(\frac{1}{4} \eta^{-1} \{X_\mu, X_\nu\}^2 + \eta T^2 \right) \,, \tag{5.21}$$

where η is the string "area" density and we introduced the Poisson bracket notation,

$$\{X, Y\} = \epsilon^{ab} \partial_a X \partial_b Y \,. \tag{5.22}$$

It is not very hard to see that the bracket defined by (5.22) satisfies to all properties a Poisson bracket is supposed to satisfy.

Exercise 4. Do it!

Let us note, that the Poisson bracket (5.22) is not a worldsheet reparameterization invariant quantity. Under the reparameterizations $\sigma \mapsto \sigma'(\sigma)$ it transforms like density rather than scalar the same way as η is:

$$\{X, Y\} \mapsto \det\left(\frac{\partial\sigma'}{\partial\sigma}\right)\{X, Y\}' \tag{5.23a}$$

$$\eta \mapsto \det\left(\frac{\partial\sigma'}{\partial\sigma}\right)\eta(\sigma') \,. \tag{5.23b}$$

Having two densities one can master a scalar,

$$\{X, Y\}_s = \eta^{-1}\{X, Y\} \,, \tag{5.24}$$

which is invariant. Actually, these two definitions coincide in the gauge $\eta = 1$, which in some cases may be possible only locally. In terms of the scalar Poisson bracket the action is rewritten in the form as follows

$$S_{\mathrm{NGP}} = \int d^2\sigma\eta \left(\frac{1}{4}\{X_\mu, X_\nu\}^2 + T^2 \right) \,, \tag{5.25}$$

where $d^2\sigma\eta$ is the invariant worldsheet area form.

"Quantization"

Consider the naive quantization procedure we know from the quantum mechanics. The classical mechanics is described by the canonical classical Poisson bracket,

$$\{p, q\} = 1 , \tag{5.26}$$

and the quantization procedure consists, roughly speaking, in the replacement of the canonical variables (p, q) by the operators $\hat{p}, \hat{q}$. At the same time the $-i\hbar \times$(Poisson bracket) is replaced by the commutator of the corresponding operators. In particular,

$$\{p, q\} \mapsto [\hat{p}, \hat{q}] = -i\hbar . \tag{5.27}$$

Afterwards, main task consists in finding the irreducible representation(s) of the obtained algebra[1]. From the undergraduate course of quantum mechanics we know that there are many unitary equivalent ways to do this, e.g., the oscillator basis representation is a good choice.

Under the quantization procedure functions on the phase space are replaced by operators acting on the irreducible representation space of the algebra (5.27). For these functions and operators one have the correspondence between the tracing and the integration over the phase space with the Liouville measure

$$\int \frac{\mathrm{d}p \, \mathrm{d}q}{2\pi\hbar} \ldots \mapsto \mathrm{tr} \ldots \tag{5.28}$$

Let us turn to our string model. As in the case of quantum mechanics, under the quantization we mean the replacing the fundamental worldsheet variables σ^1 and σ^2 by corresponding operators: $\hat{\sigma}^1$ and $\hat{\sigma}^2$, such that the *invariant* Poisson bracket is replaced by the commutator according to the rule:

$$\{\cdot, \cdot\}_{\mathrm{PB}} = i/\theta[\cdot, \cdot] , \tag{5.29}$$

where θ is the deformation parameter (noncommutativity). The worldsheet functions are replaced by the operators on the Hilbert space on which $\hat{\sigma}^a$ act irreducibly. As we have two forms of the Poisson bracket the question is whether one should use the density form of the Poisson bracket (5.22) or the invariant form (5.24)? The correct choice is the invariant form (5.24). This is imposed by the fact that the operator commutator is invariant with respect of the choice of basic operator set (in our case it is given by operators $\hat{\sigma}^a$).

Let us note that with the choice of invariant Poisson bracket in (5.29) the operators $\hat{\sigma}^a$, generally, do not have standard Heisenberg commutation relations. Rather than that, they commute to a nontrivial operator,

$$[\hat{\sigma}^1, \hat{\sigma}^2] = i\theta\widehat{\eta^{-1}} , \tag{5.30}$$

---

[1] In fact, the enveloping algebra rather the Lie algebra itself

where the operator $\widehat{\eta^{-1}}$ corresponds to the inverse density of the string world-sheet area (i.e., its classical limit gives this density). At the same time the trace in the quantum case corresponds to the worldsheet integration with the invariant measure

$$\int d^2\sigma\eta\,[\ldots] \mapsto 2\pi\theta\,\mathrm{tr}[\ldots]\,. \tag{5.31}$$

Having the "quantization rules" (5.29) and (5.31) one is able to write down the noncommutative analog of the Nambu–Goto–Polyakov string action (5.21). It looks as follows,

$$S = \alpha\,\mathrm{tr}\,\frac{1}{4}[X_\mu, X_\nu]^2 + \beta\,\mathrm{tr}\,\mathbb{I}\,, \tag{5.32}$$

where α and β are the couplings of the matrix model. In terms of the string and the deformation parameters they read,

$$\alpha = \frac{2\pi}{\theta}\,, \tag{5.33}$$

$$\beta = \frac{2\pi T^2}{\theta}\,. \tag{5.34}$$

After the identification of couplings the model (5.32) is identical with the IKKT model (5.1). As a bonus we have obtained the chemical potential (5.10). As we see from the construction, the dimensionality of matrices depend on the irreducibility representation of the noncommutative algebra. As one can expect from what is familiar in quantum mechanics, the compact worldsheets should lead to *finite-dimensional* representations and thus are described, respectively, by matrices of finite dimensions. There is no exact equivalence between the worldsheet geometry and the matrix description. However, the consistency requires that one should recover the worldsheet geometry in the semiclassical limit ($\theta \to 0$).

Another interesting remark is that in this picture the Heisenberg operator basis corresponds to the worldsheet parameterization for which η is constant; as it is well known such parameterization can exist globally only for the topologically trivial worldsheets. On the other hand, in the algebra of operators acting on a separable infinite-dimensional Hilbert space one can always find a Heisenberg operator basis.

Example I: Torus

To illustrate the above consider the example of quantization of toric worldsheet. The torus can be described by one complex modulus (or two real moduli). We are not interested here in the possible form of the toric metric, so we can choose the parameterization of the torus for which $\eta = 1$ and the flat worldsheet coordinates span the range

$$0 \le \sigma^1 < l_1, \qquad 0 \le \sigma^2 < l_2\,. \tag{5.35}$$

The first problem arises when one tries to quantize variables with the range (5.35). In spite of the fact that the (invariant) Poisson bracket is canonical the operators $\hat{\sigma}^{1,2}$ cannot satisfy the Heisenberg algebra,

$$[\hat{\sigma}^1, \hat{\sigma}^2] = i\theta , \tag{5.36}$$

and have bounded values such as in (5.35) at the same time.

Exercise 5. Prove this!

To conciliate the compactness and noncommutativity one should use the compact coordinates U_a instead,

$$U_a = \exp 2\pi i \hat{\sigma}_a / l_a, \quad a = 1, 2 . \tag{5.37}$$

The compact coordinates U_a satisfy the following (Weyl) commutation relations

$$U_1 U_2 = q U_2 U_1 , \tag{5.38}$$

where q is the toric deformation parameter,

$$q = e^{2\pi^2 i \theta / l_1 l_2} . \tag{5.39}$$

If $q^N = 1$ for some $N \in \mathbb{Z}_+$, then U_a generate an irreducible representation of dimension N. In this case an arbitrary $N \times N$ matrix $\mathcal{M}$ can be expanded in powers of U_a, e.g.,

$$\mathcal{M} = \sum_{m,n=0}^{N-1} M_{mn} U_1^m U_2^n . \tag{5.40}$$

Expansion (5.40) is in terms of monomials in U_1 and U_2 ordered in such a way that all U_1's are to the left of all U_2 one can alternatively use the Weyl functions W_{mn} defined as

$$W_{mn} = \exp \left(2\pi i m \hat{\sigma}_1 / l_1 + 2\pi i n \hat{\sigma}_2 / l_2\right) , \tag{5.41}$$

which differs from the product $U_1^m U_2^n$ by a polynomial of lower degree, but is symmetrized in $\hat{\sigma}^1$ and $\hat{\sigma}^2$. Using this expansion in terms of the Weyl functions leads one to the description of matrices in terms of the *Weyl symbols* – ordinary functions subject to the *star product* algebra. Weyl symbols as well as the star product algebras we are going to consider in the next sections.

As a result we have that quantization of the torus surface leads to the description in terms of $N \times N$ matrices where the dimensionality N of the matrices depends on the torus moduli.

Example II: Fuzzy Sphere

Another case of interest is the deformation of the spherical string worldsheet. On the sphere there is no global flat parameterization with $\eta = 1$. It is convenient to represent the two-sphere worldsheet parameters embedded into the three-dimensional Euclidean space:

$$\sigma_1^2 + \sigma_2^2 + \sigma_3^2 = 1 , \qquad (5.42)$$

with the induced metric and volume form η. The (invariant) Poisson bracket is given by the following expression[2]:

$$\{\sigma_i, \sigma_j\} = (1/r)\epsilon_{ijk}\sigma_k . \qquad (5.43)$$

Quantization of the Poisson algebra (5.43) leads to the $su(2)$ Lie algebra commutator,

$$[\hat{\sigma}_i, \hat{\sigma}_j] = i(\theta/r)\epsilon_{ijk}\hat{\sigma}_k , \qquad (5.44)$$

whose unitary irreducible representations are the well-known representations of the $su(2)$ algebra. They are parameterized by the spin of the representation J. The dimensionality of such representation is $N = 2J + 1$. The two-dimensional parameters: the radius of the sphere and the noncommutativity parameter are not independent. They satisfy instead,

$$r^4 = \theta^2 J(J+1) . \qquad (5.45)$$

Again, arbitrary $(2J + 1) \times (2J + 1)$ matrix can be expanded in terms of symmetrized monomials in σ_i – *noncommutative spherical harmonics*, which are the spherical analogues of the Weyl functions.

Turning back to the action one gets exactly the same model as in the previous example with $N = 2J+1$. As a result we get that independently from which geometry one starts one gets basically the same deformed description. The only meaningful parameter is the dimensionality of the matrix and it depends only on the worldsheet area. This is a manifestation of the universality of the matrix description which we plan to explore in the next sections.

5.3.2 Noncommutative Membrane and the BFSS Matrix Model

Let us consider slightly more complex example, namely that of the membrane. For the membrane one can write a Nambu–Goto action too,

$$S_{NG} = T_m \int_{\Sigma_3} d^3\sigma \sqrt{-\det \partial_a X^\mu \partial_b X_\mu} , \qquad (5.46)$$

where T_m is the membrane tension and X are the membrane embedding functions.

---

[2] We drop out the subscript of the invariant Poisson bracket since it creates no confusion while it is the only used from now on

In the case when the topology of the worldvolume Σ_3 is of the type $\Sigma_3 = I \times \mathcal{M}_2$, where $\mathbb{R}^1$ is the time interval $I = [0, t_0]$ and $\mathcal{M}_2$ is a two-dimensional manifold, one has the freedom to choose the worldsheet parameters σ^i, $i = 1, 2, 3$ in such a way that the time-like tangential will be always orthogonal to the space-like tangential,

$$\partial_0 X^\mu \partial_a X_\mu = 0 . \tag{5.47}$$

In this case the Nambu–Goto action takes the following form

$$S_{NG} = T_m \int d\tau \, d^2\sigma \sqrt{\frac{1}{2} \dot{X}^2 \Sigma_{\mu\nu}^2} , \tag{5.48}$$

where

$$\Sigma_{\mu\nu} = \epsilon_{ab} \partial_a X_\mu \partial_b X_\nu . \tag{5.49}$$

In the complete analogy to the case of the string let us rewrite the Nambu–Goto action in the Polyakov form,

$$S_{NGP} = \int d^3\sigma\eta \left[\frac{T_m^2}{2} \dot{X}^2 + \frac{1}{4} \{X_\mu, X_\nu\}^2 \right] , \tag{5.50}$$

where the (invariant) Poisson bracket is defined as

$$\{X, Y\} = \eta^{-1} \epsilon_{ab} \partial_a X \, \partial_b Y . \tag{5.51}$$

Since we partially fixed the reparameterization gauge invariance by choosing the time direction we have the constraint (5.47). This leads to the following constraint,

$$\{\dot{X}^\mu, X_\mu\} = 0 . \tag{5.52}$$

Now, straightforwardly repeating the arguments of the previous subsection one can write down the matrix model action. In the present case the action takes the following form:

$$S_m = \int dt \left(\beta \, \mathrm{tr} \, \frac{1}{2} \dot{X}^2 + \alpha \, \mathrm{tr} \, \frac{1}{4} [X_\mu, X_\nu]^2 \right) , \tag{5.53}$$

where $\beta = 2\pi T^2/\theta$ and $\alpha = 2\pi/\theta$, respectively. The action (5.53) should be supplemented with the following constraint:

$$[\dot{X}_\mu, X_\mu] = 0 . \tag{5.54}$$

The constraint (5.54) can be added to the action (5.53) with the Lagrange multiplier A_0. In this case the action acquires the following form:

$$S_{gi} = \int dt \left(\beta \, \mathrm{tr} \, \frac{1}{2} (\nabla_0 X_\mu)^2 + \alpha \, \mathrm{tr} \, \frac{1}{4} [X_\mu, X_\nu]^2 \right) , \tag{5.55}$$

which is identical (upto definition of parameters α and β) to the bosonic part of the BFSS action (5.11). By the redefinition of the matrix fields and rescaling of the time one can eliminate the constants α and β, so in what follows we can put both to unity.

So far we have considered only the bosonic parts of the membrane. Including the fermions (when they exist) introduces no conceptual changes. Therefore, derivation of the fermionic parts of the IKKT and BFSS matrix model description of the string and membrane is entirely left to the reader.

Exercise 6. Derive the fermionic part of both matrix models starting from the superstring/supermembrane.

5.4 Equations of Motion: Classical Solutions

In this section, we consider two types of theories, namely the string and the membrane in the Nambu–Goto–Polyakov form and the corresponding matrix models. One can write down equations of motion and try to find out some simple classical solutions in order to compare these cases among each other.

The static equations of motion in the membrane case coincide with the string equations of motion. Therefore, it is enough to consider only the last case: Any solution in the IKKT model has also the interpretation as a classical vacuum of the BFSS theory.

5.4.1 Equations of Motion Before Deformation: Nambu–Goto–Polyakov String

Consider first the equations of motion corresponding to the Nambu–Goto–Polyakov string (5.21) in the form one gets just before the deformation procedure.

Variation of X_ν produces the following equations,

$$\{X_\mu, \eta^{-1}\{X_\mu, X_\nu\}\} = 0 , \qquad (5.56a)$$

while the variation of η produces the constraint

$$\eta^2 = \frac{1}{4} \frac{\{X_\mu, X_\nu\}^2}{T^2} . \qquad (5.56b)$$

(As in the Polyakov particle case the last equation can be used to eliminate η from the action (5.21) in order to get the original Nambu–Goto action (5.14).)

The equations of motion (5.56) possess a large symmetry related to the reparameterization invariance (5.23). To find some solutions it is useful (but not necessary!) to fix this gauge invariance. As the use of the model is to describe branes, one may be interested in solutions corresponding to infinitely extended branes, which have the topology of $\mathbb{R}^2$. In this simplest case, one can impose the gauge $\eta = 1/4T^2$. Then, the equations of motion (5.56) are reduced to

$$\{X_\mu, \{X_\mu, X_\nu\}\} = 0, \qquad \{X_\mu, X_\nu\}^2 = 1 . \tag{5.57}$$

In the case of two dimensions ($\mu, \nu = 1, 2$), one can find even the generic solution. It is given by an arbitrary canonical transformation $X_{1,2} = X_{1,2}(\sigma_1, \sigma_2)$. This is easy to see if to observe that the second equation in (5.57) requires that the XX Poisson bracket must be a canonical one. The first equation is then satisfied automatically. One can also see that all the arbitrariness in the solution is due to the remnant of the reparameterization invariance which is given by the *area preserving diffeomorphisms*. This situation is similar to one met in the case of two-dimensional gauge theories where there are no physical degrees of freedom left to the gauge fields beyond the gauge arbitrariness. As we will see later, this similarity is not accidental, in some sense the above matrix model is indeed a two-dimensional gauge theory.

The situation is different in more than two dimensions. In this case we are not able to write down the generic solution, but one can find a significant particular one. The simplest solutions of (5.57) can be obtained by just lifting up the two-dimensional ones to higher dimensions. In particular, one has the following solution

$$X_1 = \sigma_1, \qquad X_2 = \sigma_2, \qquad X_i = 0, \quad i = 3, \ldots, 10 . \tag{5.58}$$

It is not difficult to check that the solution (5.58) satisfy to both (5.57). The physical meaning of this solution is an infinite Euclidean brane extended in the plane (1, 2).

One can see, that by the nature of the model in which fields X_μ are functions of a two-dimensional parameter the solutions to the equations of motion are forced always to describe two-dimensional surfaces, i.e., single brane configurations. One can go slightly beyond this limitation allowing X's to be multivalent functions of σ's. In this case, one is able to describe a certain set of multibrane systems, each sheet of X corresponding to an individual brane. This situation in application to spherical branes was analyzed in more details in [19].

Another question one may ask is whether one can find solutions describing a compact worldsheet. We are not going to give any proof of the fact that such type of solutions do not exist. Rather we consider a simple example of a cylindrical configuration and show that the equations of motion are not satisfied by it. An infinite cylinder as an extremal case of the torus can be given by the following parametric description:

$$X_1 = \sin \sigma_1, \qquad X_2 = \cos \sigma_1, \qquad X_3 = \sigma_2 . \tag{5.59}$$

Equation (5.59) describes a cylinder obtained from moving the circle in the plane (1,2) along the third axis. The parameterization (5.59) satisfy the constraint (5.47), therefore to see whether such surface is a classically stable it is enough to check the first equation of (5.57). The explicit evaluation of the equations of motion gives

$$\{X_\mu, \{X_\mu, X_1\}\} = -X_1 \neq 0 \,, \qquad (5.60)$$
$$\{X_\mu, \{X_\mu, X_2\}\} = -X_2 \neq 0 \,, \qquad (5.61)$$
$$\{X_\mu, \{X_\mu, X_0\}\} = 0 \,. \qquad (5.62)$$

As one can see, only the equation of motion for the third noncompact direction is satisfied. Other equations can be satisfied if one modifies the action of the model by adding mass terms for X_1 and X_2:

$$S \to S + m^2(X_1^2 + X_2^2) \,. \qquad (5.63)$$

Exercise 7. Modify the classical action in a way to allow the spherical brane solutions. Worldsheets quantize this model and compare it to the BMN matrix model.

Another interesting type of solutions is given by singular configurations with trivial Poisson bracket,

$$\{X_\mu, X_\nu\} = 0 \,. \qquad (5.64)$$

Obviously, these configurations satisfy the equations of motion. This solution corresponds to an arbitrary open or closed smooth one-dimensional line embedded in $\mathbb{R}^D$. The problem appears when one tries to make this type of solution to satisfy the constraint (5.47) arising from the gauge fixing $\eta^2 = 1/4T^2$. This configuration, however is still an acceptable solution before the gauge fixing. The degeneracy of the two-dimensional surface into the line results into the degeneracy of the two-dimensional surface reparameterization symmetry into the subgroup of the line reparameterizations. This means in particular that $\eta^2 = 1/4T^2$ is not an acceptable gauge condition in this point, one must impose $\eta = 0$ instead.

Let us now turn to the noncommutative case and see how the situation is changed there.

5.4.2 Equations of Motion After Deformation: IKKT/BFSS Matrix Models

After quantization of the worldsheet/worldvolume we are left with no Polyakov auxiliary field η. The role of this field in the noncommutative theory is played by the choice of the representation. As is most cases we cannot smoothly variate the representation, we have no equations of motion corresponding to this parameter. So, we are left with only equations of motion corresponding to the variation of X's. For the IKKT model these equations read

$$[X_\mu, [X_\mu, X_\nu]] = 0 \,, \qquad (5.65)$$

while for the BFSS model the variation of X leads to the following dynamical equations,

$$\ddot{X}_\mu + [X_\mu, [X_\mu, X_\nu]] = 0 , \qquad (5.66)$$

where we also put the brane tension to unity: $T = 1$. If one is interested in only the static solutions ($\dot{X} = 0$) to the BFSS equations of motion, then the (5.66) is reduced down to the IKKT equation of motion. Therefore, in what follows we consider only the last one.

By the first look at (5.65) it is clear that one can generalize the string solution (5.58) from the commutative case. Namely, one can check that the configuration

$$X_1 = \hat{\sigma}_1, \qquad X_2 = \hat{\sigma}_2, \qquad X_i = 0, \qquad i = 3, \ldots, D , \qquad (5.67)$$

satisfy the equations of motion (5.65). By the analogy with the commutative case we can say that this configuration describes either Euclidean D-string (IKKT) or a static membrane (BFSS). The solution (5.67) corresponds to the Heisenberg algebra

$$[X_1, X_2] = 1 , \qquad (5.68)$$

which allows only the infinite-dimensional representation. The value of X is not bounded, therefore this solution corresponds to a noncompact brane.

What is the role of the η-constraint here? The algebra (5.68) does not completely specify the solution unless the nature of its representation is also given. In particular, the algebra of $\hat{\sigma}$'s can be irreducibly represented on the whole Hilbert space. In the semiclassical limit this can be seen to correspond to the constraint of the previous subsection.

As we discussed in the case of commutative string, any solution to the equations of motion describes a two-dimensional surface and, therefore, has the Poisson bracket of the rank (in indices μ and ν) two or zero. In contrast to this, in the noncommutative case one may have solutions with an arbitrary even rank between zero and D. Indeed, consider a configuration,

$$X_a = p_a, \qquad a = 1, \ldots, p+1, \qquad X_i = 0, \qquad i = p+2, \ldots, D , \qquad (5.69)$$

such that

$$[p_a, p_b] = iB_{ab}, \qquad \det B \neq 0 , \qquad (5.70)$$

where B is the matrix with c-number entries B_{ab}. Such set of operators always exists if the Hilbert space is infinite dimensional separable. The set of operators p_a generates a Heisenberg algebra. Interesting cases are when the Heisenberg algebra (5.70) is represented irreducibly on the Hilbert space of the model, or when this irreducible representation is n-tuple degenerate. We will analyze these cases in the next sections.

How about the compact branes? As we have already discussed in the previous section, the compact worldsheet solution corresponds to finite-dimensional matrices X_μ. As it appears for such matrices the only solution to the equation of motion which exists is one with the trivial commutator,

$$[X_\mu, X_\nu] = 0 . \qquad (5.71)$$

To prove this fact, suppose we find such a solution with $B_{\mu\nu} = [X_\mu^{(0)}, X_\nu^{(0)}] \neq 0$ and satisfying the equations of motion (5.65). The IKKT action (BFSS energy) computed on such a solution is

$$S(X) = -\frac{1}{4} B_{\mu\nu}^2 \operatorname{tr} \mathbb{I} \neq 0 . \tag{5.72}$$

Since this is a solution to the equations of motion the variation of the action should vanish on the solution,

$$\delta S = \operatorname{tr} \frac{\delta S}{\delta X_\mu}(X^{(0)})\delta X_\mu = 0, \qquad \text{for } \forall \delta X_i , \tag{5.73}$$

which is not the case: Take $\delta X_\mu = \epsilon X_\mu^{(0)}$ to find out that $\delta S|_{X^{(0)}} \neq 0$. So, there are no solutions with nontrivial commutator for the finite-dimensional matrix space.

Consider now the extremal case of singular solutions with vanishing commutators,

$$[X_\mu, X_\nu] = 0 . \tag{5.74}$$

Obviously, from (5.74) automatically follows that the equations are satisfied too. This solution exists in both finite as well as infinite-dimensional cases. Since the commutativity of X_μ's allows their simultaneous diagonalization

$$X_\mu = \begin{pmatrix} x_1^\mu & & \\ & x_2^\mu & \\ & & \ddots \end{pmatrix} , \tag{5.75}$$

this means that the branes which are described by the matrix models are localized x_k^μ being the coordinates of the kth brane.

The Symmetry of the Solutions

The various types of solutions have different symmetry properties. Thus, the solution of the type (5.69) with the algebra of p_a's irreducibly represented over the Hilbert space of the model has no internal symmetries. Indeed, by the Schurr's lemma any operator commuting with all p_a is proportional to the identity. In the case when the representation is n-tuple degenerate one has a $U(n)$ symmetry mixing the representations. The degenerate case (5.74), when $B_{\mu\nu} = 0$ give rise to some symmetries too. Indeed, an arbitrary diagonal matrix commute with all X_μ given by (5.75). If no two branes are in the same place: $x_m^\mu \neq x_n^\mu$ for any $m \neq n$, then the configuration breaks the $U(N)$ symmetry group (in the finite-dimensional case) down to the Abelian subgroup $U(1)^N$.

5.5 From the Matrix Theory to Noncommutative Yang–Mills

This and the following section are mainly based on the papers [24–29], the reader is also referred to the lecture notes [30] and references therein.

The main idea is to use the solutions from the previous section both as classical vacua, such that arbitrary matrix configuration is regarded as a perturbation of this vacuum configuration, and as a basic set of operators in terms of which the above perturbations are expanded. Now follow the details.

5.5.1 Zero Commutator Case: Gauge Group of Diffeomorphisms

Consider first the case of the solution with the vanishing commutator (5.74). We are interested in configurations in which the branes form a p-dimensional lattice. Using the rotational symmetry of the model, one can choose this lattice to be extended in the dimensions $1, \ldots, p$:

$$X_a \equiv p_a, \quad a+1, \ldots, p; \qquad X_I = 0, \qquad I = p+1, \ldots D . \qquad (5.76)$$

Then an *arbitrary* configuration can be represented as

$$X_a = p_a + A_a, \qquad X_I = \Phi_I . \qquad (5.77)$$

Let us take the limit $N \to \infty$ and take such a distribution of the branes in which they form an infinite regular p-dimensional lattice:

$$p_a \to \lambda n_a, \qquad n_a \in \mathbb{Z} , \qquad (5.78)$$

such that the Hilbert space can be split in the product of p infinite-dimensional subspaces $\mathcal{H}_a$

$$\mathcal{H} = \otimes_{a=1}^{p} \mathcal{H}_a , \qquad (5.79)$$

such that each eigenvalue λn_a is nondegenerate in $\mathcal{H}_a$. In this case the operators p_a can be regarded as ($-i$ times) partial derivatives on a p-dimensional torus of the size $1/\lambda$,

$$p_a = -i\partial_a . \qquad (5.80)$$

Now, let us turn to the perturbation of the vacuum configuration (5.77) and try to write it in terms of operators p_a. Since the algebra of p_a's is commutative, they alone fail to generate an irreducible representation in terms of which one can expand an arbitrary operator acting on the Hilbert space $\mathcal{H}$. One must instead supplement this set with p other operators x^a, which together with p_a form a Heisenberg algebra irreducibly represented on $\mathcal{H}$,

$$[x^a, x^b] = 0, \qquad [p_a, x^b] = -i\delta_a{}^b . \qquad (5.81)$$

From the algebra (5.81) follows that the operators x^a have a continuous spectrum which is bounded: $-\pi/\lambda \le x^a < \pi/\lambda$. This precisely means that x^a

are operators of coordinates on the p-dimensional torus. Then, an arbitrary matrix X can be represented as a an operator function of the operators p_a and x^a,

$$X = \hat{X}(\hat{p}, \hat{x}).$$

In the "x-picture" this will be a differential operator $X(-i\partial, x)$. There are many ways to represent a particular operator X as a operator function of p_a and x^a which is related to the *ordering*. The *Weyl ordering* we will consider in the next subsection, here let us use a different one in which all operators p_a are on the right to all x^a. In such an ordering prescription one can write down a Fourier expansion of the operator in the following form

$$X = \frac{1}{(2\pi)^p} \int d^p z \, \tilde{X}(z, x) \, e^{i\hat{p} \cdot z} . \tag{5.82}$$

In this parameterization the product of two operators is given by an involution product of the symbols:

$$\widetilde{XY}(z, x) = \tilde{X} * \tilde{Y}(z, x) = \frac{1}{(2\pi)^p} \int d^p y \, \tilde{X}(y, x) \tilde{Y}(z - y, x + y) . \tag{5.83}$$

The trace of an operator can be computed in a standard way, namely

$$\operatorname{tr} X = \int d^p x \, \langle x | \, X \, | x \rangle = \int d^p x \, \tilde{X}(0, x) = \int d^p x \, d^p l \, X(l, x) , \tag{5.84}$$

where in the last part $X(l, x)$ is the normal symbol of which is obtained by the replacement of operator $\hat{p}_a$ by an ordinary variable l_a in the definition (5.82),

$$X(l, x) = \frac{1}{(2\pi)^p} \int d^p z \, \tilde{X}(z, x) \, e^{i l \cdot z} , \tag{5.85}$$

$$\tilde{X}(z, x) = \operatorname{tr} e^{-i\hat{p} \cdot z} X . \tag{5.86}$$

Now we are ready to write down the whole matrix action (5.32) in terms of the normal symbols. It looks as follows,

$$S = \int d^p l \, d^p x \left(-\frac{1}{4} \mathcal{F}_{ab}^2 + \frac{1}{2} (\nabla_a \Phi_I)^2 - \frac{1}{4} [\Phi_I, \Phi_J]_*^2 \right) , \tag{5.87}$$

where

$$\mathcal{F}_{ab}(l, x) = \partial_a A_b(l, x) - \partial_b A(l, x) - [A_a, A_b]_*(l, x) , \tag{5.88}$$

$$\nabla_a \Phi = \partial_a \Phi + [A_a, \Phi]_*(l, x) , \tag{5.89}$$

$$[A, B]_*(l, x) = A * B(l, x) - B * A(l, x) \tag{5.90}$$

and the star product is defined as in (5.83).

The model defined by the action (5.87) has the meaning of Yang–Mills theory with the infinite-dimensional gauge group of diffeomorphism transformations generated by the operators

$$T_f = \mathrm{i}f^a(x)\partial_a \ . \tag{5.91}$$

Because of the noncommutative nature of the products involved in the action (5.87) the local gauge group is not commutative. However, if one tries to write down the group of global gauge symmetry, one finds out that this group is, in fact nothing else that $U(1)$. Changing only slightly the character of the solution one can also get a non-Abelian global group. Indeed, consider the solution as in (5.76) with the exception that the Hilbert space is not just (5.79), but is given by the product of parts $\mathcal{H}_a$ at some (positive integer) power n:

$$\mathcal{H} = (\otimes_{a=1}^{p}\mathcal{H}_a)^{\otimes n} \ . \tag{5.92}$$

Repeating with this solution the same manipulations which lead us to (5.87) with the only exception that in this case an arbitrary matrix is represented by an $(n \times n)$-matrix-valued function instead of just "ordinary" one, we arrive to the action similar to (5.87) with the exception that the fields take their value in the $u(n)$ algebra and the global gauge group is, respectively, $U(n)$. We hope that the things will be clarified a lot when the reader will pass the next subsection.

Ordinary Gauge Model?

A question one may ask oneself is if the fluctuations of the matrix models can be restricted in such a way to get a "normal" Yang–Mills theory with a compact Lie group. In the present case one may restrict the fluctuations around the background (5.76) to depend on $\hat{x}^a$ operators only. This aim can be achieved by imposing the following constraints on the matrices X_μ:

$$[x_a, X_b] = \mathrm{i}\delta_{ab}, \qquad [x_a, X_I] = 0 \ . \tag{5.93}$$

Let us note that X_a and x_a do not form the Heisenberg algebra because the commutator between X_a do not necessarily vanish:

$$[X_a, X_b] \equiv F_{ab} \neq 0 \ . \tag{5.94}$$

Dynamically, the constraint (5.93) can be implemented through the modification of the matrix action by the addition of the constraint (5.93) with the Lagrange multiplier. The modified matrix model action reads:

$$S_c = \mathrm{tr}\left(\frac{1}{4}[X_\mu, X_\nu]^2 + \rho_{\mu\nu}([x_\mu, X_\nu] - \Delta_{\mu\nu}) + T^2\right) , \tag{5.95}$$

where $\rho_{\mu\nu}$ are the Lagrange multipliers, $x_\mu = (x_a, 0)$ and $\Delta_{\mu\nu}$ is equal to δ_{ab} when $(\mu\nu) = (ab)$ and zero otherwise. The limit $N \to \infty$ of the matrix model specified by the action (5.95) produces the Abelian gauge model. Under similar setup one can obtain also non-Abelian gauge models.

5.5.2 Nonzero Commutator: Noncommutative Yang–Mills Model

In this subsection, we consider the matrix action as a perturbation of the background configuration given by (5.69) and (5.70). Here, we plan to give a more detailed approach also partly justifying the result of the previous subsection. The operators p_a generate a $(p + 1)/2$-dimensional Heisenberg algebra. If this algebra is represented irreducibly on the Hilbert space of the model (which is in fact our choice), then an arbitrary operator acting on this space can be represented as an operator function of p_a. Let us consider this situation in more details.

Irreducibility of the representation in particular means that any operator commuting with all p_a is a c-number constant. From this follows that the operators

$$P_a = [p_a, \cdot] \,, \tag{5.96}$$

which are Hermitian on the space of square trace operators equipped with the scalar product $(A, B) = \operatorname{tr} A^* B$, are diagonalizable and have nondegenerate eigenvalues.

Exercise 8. Prove this!

By a direct check one can verify that the operator $e^{i k_a \hat{x}^a}$, where $\hat{x}^a = \theta^{ab} \hat{p}_b$, $\theta \equiv B^{-1}$ is an eigenvector for P_a with the eigenvalue k_a:

$$P_a \cdot e^{i k \cdot \hat{x}} = [p_a, e^{i k \cdot \hat{x}}] = k_a e^{i k \cdot \hat{x}} \,. \tag{5.97}$$

This set of eigenvectors form an orthogonal basis (P_a's are Hermitian). One can normalize the eigenvectors to delta function trace,

$$E_k = c_k e^{i k \cdot \hat{x}}, \qquad \operatorname{tr} E_{k'}^* E_k = \delta(k' - k) \,. \tag{5.98}$$

The normalizing coefficients c_k can be found from evaluating explicitly the trace of $e^{i(k-k')\hat{x}}$ in (5.98) and equating it to the Dirac delta. Let us compute this trace and find the respective quotients. To do this, consider the basis where the set of operators x^μ splits in pairs p_i, q^i satisfying the standard commutation relations: $[p_i, q_j] = -i\theta \delta_{ij}$.

As we know from courses of Quantum Mechanics the trace of the operator

$$e^{-i k' \hat{x}} \cdot e^{i k \hat{x}} = e^{i(k-k')\hat{x}} \, e^{\frac{i}{2} k' \times k} \,, \tag{5.99}$$

can be computed in q-representation as,

$$\operatorname{tr} e^{i(k-k')\hat{x}} \, e^{\frac{i}{2} k' \times k} = \int dq \, \langle q | e^{-i(l'_i - l_i)q^i + (z'^i - z^i)p_i} | q \rangle = 1/|c_k|^2 \delta(k' - k) \,, \tag{5.100}$$

where $|q\rangle$ is the basis of eigenvectors of $\mathbf{q}^i$,

$$\mathbf{q}^i | q \rangle = q^i | q \rangle, \qquad \langle q' | q \rangle = \delta(q' - q) \,, \tag{5.101}$$

and l_i, z^i (l_i, z^i) are components of k_μ (k'_μ) in the in the parameterizations: $x^\mu \to p_i, q^i$. Explicit computation gives,

$$1/|c_k|^2 = \frac{(2\pi)^{\frac{p}{2}}}{\sqrt{\det\theta}} . \tag{5.102}$$

Now, we have the basis of eigenvectors E_k and can write any operator F in terms of this basis,

$$\hat{F} = \int \mathrm{d}k\, \tilde{F}(k)\, \mathrm{e}^{ik\hat{x}} , \tag{5.103}$$

where the "coordinate" $\tilde{F}(k)$ is given by,

$$\tilde{F}(k) = \frac{\sqrt{\det\theta}}{(2\pi)^{\frac{p}{2}}}\,\mathrm{tr}(\mathrm{e}^{-ik\hat{x}} \cdot \hat{F}) . \tag{5.104}$$

Function $\tilde{F}(k)$ can be interpreted as the Fourier transform of an L^2 function $F(x)$,

$$F(x) = \int \mathrm{d}k \tilde{F}(k)\,\mathrm{e}^{ik_\mu x^\mu} = \sqrt{\det\theta}\int \frac{\mathrm{d}k}{(2\pi)^{p/2}}\mathrm{e}^{ikx}\,\mathrm{tr}\,\mathrm{e}^{-ik\hat{x}}\hat{F} . \tag{5.105}$$

And viceversa, to any L^2 function $F(x)$ from one can put into correspondence an L^2 operator $\hat{F}$ by inverse formula,

$$\hat{F} = \int \frac{\mathrm{d}x}{(2\pi)^{p/2}} \int \frac{\mathrm{d}k}{(2\pi)^{p/2}} F(x)\,\mathrm{e}^{ik(\hat{x}-x)} . \tag{5.106}$$

Equations (5.105) and (5.106) providing a one-to-one correspondence between L^2 functions and operators with finite trace,

$$\mathrm{tr}\,\mathbf{F}^\dagger \cdot \mathbf{F} < \infty , \tag{5.107}$$

give in fact formula for the Weyl symbols. By introducing distributions over this space of operators one can extend the above map to operators with unbounded trace.

Exercise 9. Check that (5.105) and (5.106) lead in terms of distributions to the correct Weyl ordering prescription for polynomial functions of p_μ.

Let us note that the map (5.105) and (5.106) can be rewritten in the following form,

$$F(x) = (2\pi)^{p/2}\sqrt{\det\theta}\,\mathrm{tr}\,\hat{\delta}(\hat{x}-x)\hat{F}, \qquad \hat{F} = \int \mathrm{d}^p x\,\hat{\delta}(\hat{x}-x)F(x) , \tag{5.108}$$

where we introduced the operator,

$$\hat{\delta}(\hat{x}-x) = \int \frac{\mathrm{d}^p k}{(2\pi)^p}\,\mathrm{e}^{ik\cdot(\hat{x}-x)} . \tag{5.109}$$

This operator satisfies the following properties,

$$\int d^p x \, \hat{\delta}(\hat{x} - x) = \mathbb{I} \,, \tag{5.110a}$$

$$(2\pi)^{p/2} \sqrt{\det \theta} \, \mathrm{tr} \, \hat{\delta}(\hat{x} - x) = 1 \,, \tag{5.110b}$$

$$(2\pi)^{p/2} \sqrt{\det \theta} \, \mathrm{tr} \, \hat{\delta}(\hat{x} - x) \hat{\delta}(\hat{x} - y) = \delta(x - y) \,, \tag{5.110c}$$

where in the r.h.s. of last equation is the ordinary delta function. Also, operators $\hat{\delta}(\hat{x} - x)$ for all x form a complete set of operators,

$$[\hat{\delta}(\hat{x} - x), \mathbf{F}] \equiv 0 \Rightarrow F \propto \mathrm{i} \,. \tag{5.110d}$$

The commutation relations of $\hat{x}^\mu$ also imply that $\hat{\delta}(\hat{x} - x)$ should satisfy,

$$[\hat{x}^\mu, \hat{\delta}(\hat{x} - x)] = \mathrm{i}\theta^{\mu\nu} \partial_\nu \hat{\delta}(\hat{x} - x) \,. \tag{5.110e}$$

In fact, one can define alternatively the noncommutative plane starting from operator $\hat{\delta}(\hat{x} - x)$ satisfying (5.110), with $\hat{x}^\mu$ defined by,

$$\hat{x}^\mu = \int d^p x \, x^\mu \hat{\delta}(\hat{x} - x) \,. \tag{5.111}$$

In this case (5.110e) provides that $\hat{x}^\mu$ satisfy the Heisenberg algebra (5.69), while the property (5.110d) provides that they form a complete set of operators. Relaxing these properties allows one to introduce a more general noncommutative spaces.

Let us the operator $\hat{\delta}(x)$ in the simplest case of two-dimensional noncommutative plane. The most convenient is to find its matrix elements $D_{mn}(x)$ in the oscillator basis given by,

$$|n\rangle = \frac{(\hat{a}^\dagger)^n}{\sqrt{n!}} |0\rangle \,, \qquad \hat{a} |0\rangle = 0 \,, \tag{5.112}$$

where the oscillator operators $\hat{a}$ and $\hat{a}^\dagger$ are the noncommutative analogues of the complex coordinates,

$$\hat{a} = \sqrt{\frac{1}{2\theta}}(\hat{x}^1 + \mathrm{i}\hat{x}^2) \,, \qquad \hat{a}^\dagger = \sqrt{\frac{1}{2\theta}}(\hat{x}^1 - \mathrm{i}\hat{x}^2) \,; \qquad [\hat{a}, \hat{a}^\dagger] = 1 \,. \tag{5.113}$$

Then the matrix elements read

$$D_{mn}(x) = \langle m| \, \hat{\delta}^{(2)}(\hat{a} - z) \, |n\rangle = \mathrm{tr} \, \hat{\delta}^{(2)}(\hat{a} - z) P_{nm} \,, \tag{5.114}$$

where $P_{nm} = |n\rangle \langle m|$.

As one can see, up to a Hermitian transposition the matrix elements of $\hat{\delta}(\hat{x} - x)$ correspond to the Weyl symbols of operators like $|m\rangle \langle n|$, or so called Wigner functions. The computation of (5.114) gives,[3]

---

[3] For the details of computation see e.g. [31]

$$D^\theta_{mn}(z, \bar z) = (-1)^n \left(\frac{2}{\sqrt\theta}\right)^{m-n+1} \sqrt{\frac{n!}{m!}}\, e^{-z\bar z/\theta} \left(\frac{z^m}{\bar z^n}\right) L^{m-n}_n(2z\bar z/\theta) \,,$$

(5.115)

where $L^{m-n}_n(x)$ are Laguerre polynomials,

$$L^\alpha_n(x) = \frac{x^{-\alpha} e^x}{n!} \left(\frac{\mathrm d}{\mathrm d x}\right)^n (e^{-x} x^{\alpha+n}) \,.$$

(5.116)

It is worthwhile to note that in spite of its singular origin the symbol of the delta operator is a smooth function which is rapidly vanishing at infinity. The smoothness comes from the fact that the operator elements are written in an L^2 basis. In a non-L^2 basis, e.g., in the basis of x_1 eigenfunctions D^θ would have more singular form.

The above computations can be generalized to p-dimensions. Written in the complex coordinates $z_i, \bar z_i$ corresponding to oscillator operators (5.113), which diagonalize the noncommutativity matrix this looks as follows,

$$D_{mn} = D^{\theta(1)}_{m_1 n_1}(z_1, \bar z_1) D^{\theta(2)}_{m_2 n_2}(z_2, \bar z_2) \cdots D^{\theta(p/2)}_{m_{p/2} n_{p/2}}(z_{p/2}, \bar z_{p/2}) \,,$$

(5.117)

where

$$[z_i, \bar z_j]_* = \delta_{ij}, \quad i = 1, \ldots, p/2 \,.$$

(5.118)

Having the above map one can establish the following relations between operators and their Weyl symbols.

1. It is not difficult to derive that,

$$(2\pi)^{p/2} \sqrt{\det\theta}\, \mathrm{tr}\,\mathbf{F} = \int \mathrm d x\, F(x) \,.$$

(5.119)

2. The (noncommutative) product of operators is mapped into the *star* or *Moyal* product of functions,

$$\mathbf{F} \cdot \mathbf{G} \to F * G(x) \,,$$

(5.120)

where $F * G(x)$ is defined as,

$$F * G(x) = e^{-\frac{1}{2}\theta^{\mu\nu} \partial_\mu \partial'_\nu} F(x) G(x') \Big|_{x'=x} \,.$$

(5.121)

In terms of operator $\hat\delta(\hat x - x)$, this product can be written as follows,

$$F * G(x) = \int \mathrm d^p y\, \mathrm d^p z\, K(x; y, z) F(y) G(z) \,,$$

(5.122)

where

$$K(x; y, z) = (2\pi)^{p/2} \sqrt{\det\theta}\, \mathrm{tr}\,\hat\delta(\hat x - x)\hat\delta(\hat x - y)\hat\delta(\hat x - z)$$
$$= e^{\frac{1}{2}\partial^y_\mu \theta^{\mu\nu} \partial^z_\nu} \delta(y - x)\delta(z - x) \,,$$

(5.123)

∂_μ^y and ∂_μ^z are, respectively, $\partial/\partial y^\mu$ and $\partial/\partial z^\mu$, and in the last line one has ordinary delta functions.

On the other hand, the ordinary product of functions was not found to have any reasonable meaning in this context.

3. Another property of the star product is that in the integral it can be replaced by the ordinary product:

$$\int d^p x \, F * G(x) = \int d^p x \, F(x)G(x) \, . \qquad (5.124)$$

4. Interesting feature of this representation is that partial derivatives of Weyl symbols correspond to commutators of respective operators with $i p_\mu$,

$$[\mathbf{ip}_\mu, \mathbf{F}] \rightarrow i(p_\mu * F - F * p_\mu)(x) = \frac{\partial F(x)}{\partial x^\mu} \, , \qquad (5.125)$$

where p_μ is linear function of x^μ: $p_\mu = -\theta_{\mu\nu}^{-1} x^\nu$.

This is an important feature of the star algebra of functions distinguishing it from the ordinary product algebra. In the last one cannot represent the derivative as an *internal automorphism* while in the star algebra it is possible due to its nonlocal character. This property is of great importance in the field theory since, as it will appear later, it is the source of duality relations in noncommutative gauge models which we turn to in the next section.

Exercise 10. Derive (5.119)–(5.125).

Let us turn back to the matrix model action (5.32) and represent an arbitrary matrix configuration as a perturbation of the background (5.69):

$$X_a = p_a + A_a, \quad X_I = \Phi_I, \qquad a = 1, \ldots, p+1, \quad I = p+2, \ldots, D \, . \quad (5.126)$$

Passing from operators A_a and Φ to their Weyl symbols using (5.108), (5.120), and (5.125) one gets following representation for the matrix action (5.32):

$$S = \int d^p x \left(-\frac{1}{4}(\mathcal{F}_{ab} - B_{ab})^2 + \frac{1}{2}(\nabla_a \Phi_I)^2 - \frac{1}{4}[\Phi_I, \Phi_J]_*^2 \right) , \qquad (5.127)$$

where

$$F_{\mu\nu} = \partial_\mu A_\nu - \partial_\nu A_\mu - i[A_\mu, A_\nu]_* \, . \qquad (5.128)$$

In the case of the irreducible representation of the algebra (5.70) this describes the $U(1)$ gauge model.

One can consider an n-tuple degenerate representation in this case as well. As in the previous case the index labeling the representations become an internal symmetry index and the global gauge group of the model becomes $U(n)$. Indeed, the operator basis in which one can expand an arbitrary operator now is given by

$$E_k^\alpha = \sigma^\alpha \otimes e^{ik \cdot \hat{x}} , \qquad (5.129)$$

where σ^α, $\alpha = 1, \ldots, n^2$ are the adjoint generators of the $u(n)$ algebra. These can be normalized to satisfy,

$$[\sigma^\alpha, \sigma^\beta] = i\epsilon^{\alpha\beta\gamma}\sigma^\gamma, \qquad \mathrm{tr}_{su(2)}\,\sigma^\alpha\sigma^\beta = \delta^{\alpha\beta} , \qquad (5.130)$$

where $\epsilon^{\alpha\beta\gamma}$ are the structure constants of the $u(n)$ algebra:

$$\epsilon^{\alpha\beta\gamma} = -i\,\mathrm{tr}_{su(2)}[\sigma^\alpha, \sigma^\beta]\sigma^\gamma , \qquad (5.131)$$

which follows from (5.130). Then an operator $\hat{F}$ is mapped to the following function $F(x)$:

$$
\begin{aligned}
F^\alpha(x) &= \sqrt{\det\theta} \int \frac{d^p k}{(2\pi)^{p/2}} e^{ikx} \,\mathrm{tr}\left\{(\sigma^\alpha \otimes e^{ik\cdot\hat{x}}) \cdot \hat{F}\right\} \\
&= (2\pi)^{p/2}\sqrt{\det\theta}\,\mathrm{tr}\left\{(\sigma^\alpha \otimes \hat{\delta}(\hat{x} - x)) \cdot \hat{F}\right\} . \qquad (5.132)
\end{aligned}
$$

Equation (5.132) gives the most generic map from the space of operators to the space of p-dimensional $u(n)$-algebra-valued functions.

Exercise 11. Prove that p is always even.

Just for the sake of completeness let us give also the formula for the inverse map,

$$\hat{F} = \int d^p x\,(\sigma^\alpha \otimes \hat{\delta}(\hat{x} - x))F^\alpha(x) , \qquad (5.133)$$

Applying the map (5.132) and (5.133) to the IKKT matrix model (5.32) or to the BFSS one (5.55), one gets, respectively, the p or $p + 1$-dimensional noncommutative $u(n)$ Yang–Mills model.

Exercise 12. Derive the p- and $(p + 1)$-dimensional noncommutative supersymmetric gauge model from the matrix actions (5.32) and (5.55), using the map (5.132) and its inverse (5.133).

Some comments regarding both gauge models described by the actions (5.87) and (5.128) are in order. In spite of the fact that both models look very similar to the "ordinary" Yang–Mills models, the perturbation theory of this models are badly defined in the case of noncompact noncommutative spaces. In the first case the nonrenormalizable divergence is due to extra integrations over l in the "internal" space. In the case of noncommutative gauge model the behavior of the perturbative expansion is altered by the IR/UV mixing [32, 33]. The supersymmetry or low dimensionality improves the situation allowing the "bad" terms to cancel (see [34–37]). On the other hand, the compact noncommutative spaces provide both IR and UV cut off and the field theory on such spaces is finite [38]. In the case of zero commutator background the behavior of the perturbative expansions depends on the eigenvalue distribution. Faster the eigenvalues increase, better the expansions converge. However, there is always the problem of the zero modes corresponding to the diagonal matrix excitations (functions of commutative p_a's). There is a hope that integrating over the remaining modes helps to generate a dynamical term for the zero modes too. Indeed, for purely bosonic model one has a repelling potential after the one-loop integration of the nondiagonal modes. The fermions contribute with the attractive potential. In the supersymmetric case the repelling bosonic contribution is cancelled by the attractive fermionic one and diagonal modes remain nondynamical (Y. Makeenko, private communication).

Exercise 13. Consider the Eguchi–Kawai model given by the action (5.9). Write down the equations of motion and find the classical solutions analogous to (5.69). One can have noncommutative solutions even for finite N. Explain, why? Consider arbitrary matrix configuration as a perturbation of the above classical backgrounds and find the resulting models. What is the space on which these models live? How the same space can be obtained from a noncompact matrix model.

We considered exclusively the bosonic models. When the supersymmetric theories are analyzed one has to deal also with the fermionic part. In the case of compact noncommutative spaces which correspond to finite size matrices one has a discrete system with fermions. In the lattice gauge theories with fermions there is a famous problem related to the fermion *doubling* [39]. Concerning the theories on the compact noncommutative spaces it was found that in some cases one can indeed have fermion doubling [25][4] some other cases were reported to be doubling free and giving alternative solutions to the long standing lattice problem [41].

5.6 Matrix Models and Dualities of Noncommutative Gauge Models

In the previous section, we realized that the matrix model from different "points" of the moduli space of classical solutions looks like different gauge models. These models can have different dimensionality or different global gauge symmetry group, but they all are equivalent to the original IKKT or BFSS matrix model. This equivalence can be used to pass from some noncommutative model back to the matrix model and then to a different noncommutative model and vice versa. Thus, one can find a one-to-one map from one model to an equivalent one.

In reality, one can jump the intermediate step by writing a new solution direct in the noncommutative gauge model and passing to Weyl (re)ordered description with respect to the new background. From the point of view of noncommutative geometry this procedure is nothing else that the change of

---

[4] For the case of the unitary Eguchi–Kawai-type model with fermions see [40]

the noncommutative variable taking into account also the ordering. Let us go to the details. Consider two different background solutions given by $p^{(i)}_{\mu_{(i)}}$, where $\mu_{(i)} = 1, \ldots, p_{(i)}$ and the index $i = 1, 2$ labels the backgrounds. Denote the orders of degeneracy of the backgrounds by $n_{(i)}$. The commutator for both backgrounds is given by

$$[p^{(i)}_{\mu_{(i)}}, p^{(i)}_{\nu_{(i)}}] = iB^{(i)}_{\mu_{(i)}\nu_{(i)}} . \tag{5.134}$$

Applying to a $p_{(1)}$-dimensional $u(n_{(1)}$ algebra-valued field $F^{\alpha_{(1)}}(x_{(1)})$ first the inverse Weyl transformation (5.133) which maps it in the operator form and then the direct transformation (5.132) from the operator form to the second background one gets a $p_{(2)}$-dimensional $u(n_{(2)}$ algebra-valued field $F^{\alpha_{(2)}}(x_{(2)})$ defined by

$$F^{\alpha_{(2)}}(x_{(2)}) = \int d^{p_{(1)}} x_{(1)} K^{\alpha_{(2)}\,\alpha_{(1)}}_{(2|1)}(x_{(2)}|x_{(1)}) F^{\alpha_{(1)}}(x_{(1)}) , \tag{5.135}$$

where the kernel $K^{\alpha_{(2)}\,\alpha_{(1)}}_{(2|1)}(x_{(2)}|x_{(1)})$ is given by

$$K^{\alpha_{(2)}\,\alpha_{(1)}}_{(1|2)}(x_{(2)}, x_{(1)}) = (2\pi)^{p_{(2)}/2}\sqrt{\det\theta_{(2)}}$$
$$\times \operatorname{tr}\left\{(\sigma^{\alpha_{(2)}}_{(2)} \otimes \hat{\delta}(\hat{x}_{(2)} - x_{(2)})) \cdot (\sigma^{\alpha_{(1)}}_{(1)} \otimes \hat{\delta}(\hat{x}_{(1)} - x_{(1)}))\right\}, \tag{5.136}$$

where $x_{(i)}$ and $\sigma^{\alpha_{(i)}}_{(i)}$ are the coordinate and algebra generators corresponding to the background $p^{(i)}_{\mu_{(i)}}$.

Equation (5.136) still appeals to the background independent operator form by using the $\hat{\delta}$-operators and trace. This can be eliminated in the following way. Consider the functions $x^{\mu_{(2)}}_{(2)}(x^{\mu_{(1)}}_{(1)}, \sigma^{\alpha_{(1)}}_{(1)}) = x^{\mu_{(2)};\alpha_{(1)}}_{(2)}(x^{\mu_{(1)}}_{(1)})\sigma^{\alpha_{(1)}}_{(1)}$ and $\sigma^{\alpha_{(2)}}_{(2)}(x^{\mu_{(1)}}_{(1)}, \sigma^{\alpha_{(1)}}_{(1)}) = \sigma^{\alpha_{(2)};\alpha_{(1)}}_{(2)}(x^{\mu_{(1)}}_{(1)})\sigma^{\alpha_{(1)}}_{(1)}$ which are the symbols of the second background $\hat{x}^{\mu_{(2)}}_{(2)}$ which are Weyl-ordered with respect to the first background. Namely, they are the solution to the equation,

$$x^{\mu_{(2)}}_{(2)} *_{(1)} x^{\nu_{(2)}}_{(2)} - x^{\nu_{(2)}}_{(2)} *_{(1)} x^{\mu_{(2)}}_{(2)} = \theta^{(2)}_{\mu_{(2)}\nu_{(2)}} , \tag{5.137}$$

and for $\sigma_{(2)}$

$$\sigma^{\alpha_{(2)}}_{(2)} *_{(1)} \sigma^{\beta_{(2)}}_{(2)} - \sigma^{\alpha_{(2)}}_{(2)} *_{(1)} \sigma^{\beta_{(2)}}_{(2)} = i\epsilon^{\alpha_{(2)}\beta_{(2)}\gamma_{(2)}}\sigma^{\gamma_{(2)}}_{(2)} , \tag{5.138}$$

where $*_{(1)}$ includes both the noncommutative with $\theta_{(1)}$ and the $u(n_{(1)})$ matrix products and we did not write explicitly the arguments $(x^{\mu_{(1)}}_{(1)}, \sigma^{\alpha_{(1)}}_{(1)})$ and $u(n_{(1)})$ matrix indices of $x_{(2)}$ and $\sigma_{(2)}$. Then, the kernel (5.136) can be rewritten in the $x_{(1)}$ background as follows,

$$K^{\alpha_{(2)}\,\alpha_{(1)}}_{(1|2)}(x_{(2)}, x_{(1)})$$
$$= \sqrt{\frac{\det 2\pi\theta_{(2)}}{\det 2\pi\theta_{(1)}}} d^{\alpha_{(1)}\beta_{(1)}\gamma_{(1)}}_{(1)} \left(\sigma^{\alpha_{(2)};\beta_{(1)}}_{(2)} *_{(1)} \delta^{\gamma_{(1)}}_{*_{(1)}}(x_{(2)}(x_{(1)}) - x_{(2)})\right), \tag{5.139}$$

where $d_{(1)}^{\alpha\beta\gamma} = \text{tr}_{(1)}\, \sigma_{(1)}^{\alpha}\sigma_{(1)}^{\beta}\sigma_{(1)}^{\gamma}$ and

$$\delta_{*(1)}^{\gamma(1)}(x_{(2)}(x_{(1)}) - x_{(2)}) = \int \frac{d^{p(2)}l}{(2\pi)^{p(2)}}\, \text{tr}_{(1)}\, \sigma_{(1)}^{\gamma(1)}\, e_{*(1)}^{il\cdot(x_{(2)}(x_{(1)})-x_{(2)})}, \qquad (5.140)$$

$e_*^{f(x)}$ is the star exponent computed with the noncommutative structure corresponding to $*$.

General expression for the basis transform (5.135) with the kernel (5.136) or (5.139) looks rather complicate almost impossible to deal with. Therefore, it is useful to consider some particular examples that we take from [13] which show that in fact the objects are still treatable.

5.6.1 Example 1: The $U(1) \longrightarrow U(n)$ Map

Let us present the explicit construction for the map from $U(1)$ to $U(2)$ gauge model in the case of two-dimensional noncommutative space. The map we are going to discuss can be straightforwardly generalized to the case of arbitrary even dimensions as well as to the case of arbitrary $U(n)$ group.

The two-dimensional noncommutative coordinates are

$$[x^1, x^2] = i\theta . \qquad (5.141)$$

As we already discussed, noncommutative analog of complex coordinates is given by oscillator rising and lowering operators,

$$a = \sqrt{\frac{1}{2\theta}}(x^1 + ix^2), \qquad \bar{a} = \sqrt{\frac{1}{2\theta}}(x^1 - ix^2) \qquad (5.142)$$

$$a\,|n\rangle = \sqrt{n}\,|n-1\rangle, \qquad \bar{a}\,|n\rangle = \sqrt{n+1}\,|n+1\rangle, \qquad (5.143)$$

where $|n\rangle$ is the oscillator basis formed by eigenvectors of $N = \bar{a}a$,

$$N\,|n\rangle = n\,|n\rangle . \qquad (5.144)$$

The gauge symmetry in this background is noncommutative $U(1)$.

We will now construct the noncommutative $U(2)$ gauge model. For this, consider the $U(2)$ basis which is given by following vectors,

$$|n', a\rangle = |n'\rangle \otimes e_a, \qquad a = 0, 1 \qquad (5.145)$$

$$e_0 = \begin{pmatrix} 1 \\ 0 \end{pmatrix}, \qquad e_1 = \begin{pmatrix} 0 \\ 1 \end{pmatrix}, \qquad (5.146)$$

where $\{|n'\rangle\}$ is the oscillator basis and $\{e_a\}$ is the "isotopic" space basis.

The one-to-one correspondence between $U(1)$ and $U(2)$ bases can be established in the following way [42, 43],

$$|n'\rangle \otimes e_a \sim |n\rangle = |2n' + a\rangle , \qquad (5.147)$$

where $|n\rangle$ is a basis element of the $U(1)$ Hilbert space and $|n'\rangle \otimes e_a$ is a basis element of the Hilbert space of $U(2)$ theory. (Note that they are two bases of the same Hilbert space.)

Let us note that the identification (5.147) is not unique. For example, one can put an arbitrary unitary matrix in front of $|n\rangle$ in the r.h.s. of (5.147). This in fact describes all possible identifications and respectively maps from $U(1)$ to $U(2)$ model.

Under this map, the $U(2)$-valued functions can be represented as scalar functions in $U(1)$ theory. For example, constant $U(2)$ matrices are mapped to particular functions in $U(1)$ space. To find these functions, it suffices to find the map of the basis of the $u(2)$ algebra given by Pauli matrices σ_α, $\alpha = 0, 1, 2, 3$.

In the $U(1)$ basis Pauli matrices look as follows,

$$\sigma_0 = \sum_{n=0}^{\infty} (|2n\rangle \langle 2n| + |2n+1\rangle \langle 2n+1|) \equiv \mathbb{I} , \qquad (5.148a)$$

$$\sigma_1 = \sum_{n=0}^{\infty} (|2n\rangle \langle 2n+1| + |2n+1\rangle \langle 2n|) , \qquad (5.148b)$$

$$\sigma_2 = -i \sum_{n=0}^{\infty} (|2n\rangle \langle 2n+1| - |2n+1\rangle \langle 2n|) , \qquad (5.148c)$$

$$\sigma_3 = \sum_{n=0}^{\infty} (|2n\rangle \langle 2n| - |2n+1\rangle \langle 2n+1|) , \qquad (5.148d)$$

while the "complex" coordinates a' and $\bar{a}'$ of the $U(2)$ invariant space are given by the following,

$$a' = \sum_{n=0}^{\infty} \sqrt{n}(|2n-2\rangle \langle 2n| + |2n-1\rangle \langle 2n+1|) , \qquad (5.149a)$$

$$\bar{a}' = \sum_{n=0}^{\infty} \sqrt{n+1}(|2n+2\rangle \langle 2n| + |2n+3\rangle \langle 2n+1|) . \qquad (5.149b)$$

One can see that when trying to find the Weyl symbols for operators given by (5.148), (5.149), one faces the problem that the integrals defining the Weyl symbols diverge. This happens because the respective functions (operators) do not belong to the noncommutative analog of L^2 space (are not square-trace).

Let us give an alternative way to compute the functions corresponding to operators (5.148) and (5.149). To do this let us observe that operators

$$\Pi_+ = \sum_{n=0}^{\infty} |2n\rangle \langle 2n| \,, \tag{5.150}$$

and

$$\Pi_- = \sum_{n=0}^{\infty} |2n+1\rangle \langle 2n+1| \,, \tag{5.151}$$

can be expressed as[5]

$$\Pi_+ = \frac{1}{2} \sum_{n=0}^{\infty} \left(1 + \sin \pi \left(n + \frac{1}{2} \right) \right) |n\rangle \langle n| \to \frac{1}{2} \left(1 + \sin_* \pi \left(\bar{z} * z + \frac{1}{2} \right) \right) \,, \tag{5.152}$$

and

$$\Pi_- = i - \Pi_+ = \frac{1}{2} \left(1 - \sin_* \pi \left(\bar{z} * z + \frac{1}{2} \right) \right) = \frac{1}{2} \left(1 - \sin_* \pi |z|^2 \right) \,, \tag{5.153}$$

where $\sin_*$ is the "star" sin function defined by the star Taylor series,

$$\sin_* f = f - \frac{1}{3!} f * f * f + \frac{1}{5!} f * f * f * f * f - \cdots \,, \tag{5.154}$$

with the star product defined in variables $z, \bar{z}$ as follows,

$$f * g(\bar{a}, a) = e^{\partial \bar{\partial}' - \bar{\partial} \partial'} f(\bar{z}, z) g(\bar{z}', z') |_{z'=z} \,, \tag{5.155}$$

where $\partial = \partial/\partial z$, $\bar{\partial} = \partial/\partial \bar{z}$ and analogously for primed z' and $\bar{z}'$. For convenience we denoted Weyl symbols of a and $\bar{a}$ as z and $\bar{z}$.

The easiest way to compute (5.152) and (5.153) is to find the Weyl symbol of the operator,

$$I_k^{\pm} = \frac{1 \pm \sin \left(\bar{a} a + \frac{1}{2} \right)}{(\bar{a} a + \gamma)^k} \,, \tag{5.156}$$

where γ is some constant, mainly $\pm 1/2$.

For sufficiently large k, the operator $I_k^{\pm}$ becomes square trace for which the formula (5.132) defining the Weyl map is applicable. The result can be analytically continued for smaller values of k, using the following recurrence relation,

$$I_{k-m}^{\pm}(\bar{z}, z) = \underbrace{\left(|z|^2 + \gamma - \frac{1}{2} \right) * \cdots * \left(|z|^2 + \gamma - \frac{1}{2} \right)}_{m \text{ times}} * I_k^{\pm}(\bar{z}, z) \,. \tag{5.157}$$

The last equation requires computation of only finite number of derivatives of $I_k^{\pm}(\bar{z}, z)$ arising from the star product with polynomials in $\bar{z}, z$.

Exercise 14. Compute the Weyl symbol for the operator (5.156).

---

[5] Weyl symbols of a and $\bar{a}$ are denoted, respectively, as z and $\bar{z}$. The same rule applies also to primed variables

5.6.2 Example 2: Map Between Different Dimensions

Consider the situation when the dimension is changed. This topic was considered in [26, 28].

Consider the Hilbert space $\mathcal{H}$ corresponding to the representation of the two-dimensional noncommutative algebra (5.141), and $\mathcal{H} \otimes \mathcal{H}$ (which is in fact isomorphic to $\mathcal{H}$) which corresponds to the four-dimensional noncommutative algebra generated by

$$[x^1, x^2] = i\theta_{(1)}, \qquad [x^3, x^4] = i\theta_{(2)} . \qquad (5.158)$$

In the last case noncommutative complex coordinates correspond to two sets of oscillator operators, a_1, a_2 and $\bar{a}_1$, $\bar{a}_2$, where,

$$a_1 = \sqrt{\frac{1}{2\theta_{(1)}}}(x^1 + ix^2), \qquad\qquad \bar{a}_1 = \sqrt{\frac{1}{2\theta_{(1)}}}(x^1 - ix^2) \qquad (5.159a)$$

$$a_1 |n_1\rangle = \sqrt{n_1} |n_1 - 1\rangle , \qquad\qquad \bar{a}_1 |n_1\rangle = \sqrt{n_1 + 1} |n_1 + 1\rangle , \qquad (5.159b)$$

$$a_2 = \sqrt{\frac{1}{2\theta_{(2)}}}(x^3 + ix^4), \qquad\qquad \bar{a}_2 = \sqrt{\frac{1}{2\theta_{(2)}}}(x^3 - ix^4) \qquad (5.159c)$$

$$a |n\rangle_2 = \sqrt{n_2} |n_2 - 1\rangle , \qquad\qquad \bar{a}_2 |n_2\rangle = \sqrt{n_2 + 1} |n_2 + 1\rangle , \qquad (5.159d)$$

and the basis elements of the "four-dimensional" Hilbert space $\mathcal{H} \otimes \mathcal{H}$ are $|n_1, n_2\rangle = |n_1\rangle \otimes |n_2\rangle$.

The isomorphic map $\sigma : \mathcal{H} \otimes \mathcal{H} \to \mathcal{H}$ is given by assigning a unique number n to each element $|n_1, n_2\rangle$ and putting it into correspondence to $|n\rangle \in \mathcal{H}$. So, the problems is reduced to the construction of an isomorphic map from one-dimensional lattice of nonnegative integers into the two-dimensional quarter-infinite lattice. This can be done by consecutive enumeration of the two-dimensional lattice nodes starting from the angle (00). The details of the construction can be found in [26, 28].

As we discussed earlier, this map induces an isomorphic map of gauge and scalar fields from two- to four-dimensional noncommutative spaces.

This can be easily generalized to the case with arbitrary number of factors $\mathcal{H} \otimes \cdots \otimes \mathcal{H}$ corresponding to $p/2$ "two-dimensional" noncommutative spaces. In this way, one obtains the isomorphism σ which relates two-dimensional noncommutative function algebra with a p-dimensional one, for p even.

5.6.3 Example 3: Change of θ

So far, we have considered maps which relate algebras of noncommutative functions in different dimensions or at least taking values in different Lie algebras. Due to the fact that they change considerably the geometry, these maps could not be deformed smoothly into the identity map. In this section, we consider a more restricted class of maps which do not change either dimensionality or the gauge group but only the noncommutativity parameter. Obviously, this can be smoothly deformed into identity map, therefore one may consider infinitesimal transformations.

The new noncommutativity parameter is given by the solution to the equations of motion. In this framework, the map is given by the change of the background solution p_μ by an infinitesimal amount: $p_\mu + \delta p_\mu$. Then, a solution with the constant field strength $F_{\mu\nu}^{(\delta p)}$ will change the noncommutativity parameter as follows,

$$\theta^{\mu\nu} + \delta\theta^{\mu\nu} \equiv (\theta_{\mu\nu}^{-1} + \delta\theta_{\mu\nu}^{-1})^{-1} = (\theta_{\mu\nu}^{-1} + F_{\mu\nu})^{-1} . \tag{5.160}$$

Note that the above equation does not require $\delta\theta$ to be infinitesimal.

Since we are considering solutions to the gauge field equations of motion $A_\mu = \delta p_\mu$ one should fix the gauge for it. A convenient choice would be, e.g., the Lorentz gauge, $\partial_\mu \delta p_\mu = 0$. Then, the solution with

$$A_\mu^{(\delta p)} \equiv \delta p_\mu = (1/2)\epsilon_{\mu\nu}\theta^{\nu\alpha}p_\alpha \tag{5.161}$$

with antisymmetric $\epsilon_{\mu\nu}$ has the constant field strength

$$F_{\mu\nu}^{(\delta p)} \equiv \delta\theta_{\mu\nu}^{-1} = \epsilon_{\mu\nu} + (1/4)\epsilon_{\mu\alpha}\theta^{\alpha\beta}\epsilon_{\beta\nu} = \epsilon_{\mu\nu} + O(\epsilon^2) . \tag{5.162}$$

This corresponds to the following variation of the noncommutativity parameter,

$$\delta\theta^{\mu\nu} = -\theta^{\mu\alpha}\epsilon_{\alpha\beta}\theta^{\beta\nu} - \frac{1}{4}\theta^{\mu\alpha}\epsilon_{\alpha\gamma}\theta^{\gamma\rho}\epsilon_{\rho\beta}\theta^{\beta\nu} = -\theta^{\mu\alpha}\delta\theta_{\alpha\beta}^{-1}\theta^{\beta\nu} + O(\epsilon^2) . \tag{5.163}$$

Let us note that such kind of infinitesimal transformations were considered in a slightly different context in [44].

Let us find how noncommutative functions are changed with respect to this transformation. To do this, let us consider how the Weyl symbol (5.132) transforms under the variation of background (5.161). For an arbitrary operator ϕ after short calculation we have,

$$\delta\phi(x) = \frac{1}{4}\delta\theta^{\alpha\beta}(\partial_\alpha\phi * p_\beta(x) + p_\beta * \partial_\alpha\phi(x)) . \tag{5.164}$$

In obtaining this equation we had to take into consideration the variation of p_μ as well as of the factor $\sqrt{\det\theta}$ in the definition of the Weyl symbol (5.132).

By the construction, this variation satisfies the "star-Leibnitz rule,"

$$\delta(\phi * \chi)(x) = \delta\phi * \chi(x) + \phi * \delta\chi(x) + \phi(\delta*)\chi(x) , \tag{5.165}$$

where $\delta\phi(x)$ and $\delta\chi(x)$ are defined according to (5.164) and variation of the star product is given by

$$\phi(\delta*)\chi(x) = \frac{1}{2}\delta\theta^{\alpha\beta}\partial_\alpha\phi * \partial_\beta\chi(x) . \tag{5.166}$$

The property (5.165) implies that δ provides a homomorphism (which is in fact an isomorphism) of star algebras of functions.

The above transformation (5.164) do not apply, however, to the gauge field $A_\mu(x)$ and gauge field strength $F_{\mu\nu}(x)$. This is the case because the respective fields do not correspond to invariant operators. Indeed, according to the definition $A_\mu = X_\mu - p_\mu$, where X_μ is corresponds to such an operator. Therefore, the gauge field $A_\mu(x)$ transforms in a nonhomogeneous way,[6]

$$\delta A_\mu(x) = \frac{1}{4}\delta\theta^{\alpha\beta}(\partial_\alpha A_\mu * p_\beta + p_\beta * \partial_\alpha A_\mu) + \frac{1}{2}\theta_{\mu\alpha}\delta\theta^{\alpha\beta}p_\beta . \tag{5.167}$$

The transformation law for $F_{\mu\nu}(x)$ can be computed using its definition (5.128) and the "star-Leibnitz rule" (5.165) as well as the fact that it is the Weyl symbol of the operator,

$$F_{\mu\nu} = \mathrm{i}[X_\mu, X_\nu] - \theta_{\mu\nu} . \tag{5.168}$$

Of course, both approaches give the same result,

$$\delta F_{\mu\nu}(x) = \frac{1}{4}\delta\theta^{\alpha\beta}(\partial_\alpha F_{\mu\nu} * p_\beta + p_\beta * \partial_\alpha F_{\mu\nu})(x) - \delta\theta^{-1}_{\mu\nu} . \tag{5.169}$$

The infinitesimal map described above has the following properties:

(i).   It maps gauge equivalent configurations to gauge equivalent ones, therefore it satisfies the Seiberg–Witten equation,

$$U^{-1} * A * U + U^{-1} * dU \rightarrow U'^{-1} *' A' *' U' + U'^{-1} *' d'U' . \tag{5.170}$$

(ii).   It is linear in the fields.

(iii). Any background independent functional is invariant under this transformation. In particular, any gauge invariant functional whose dependence on gauge fields enters through the combination $X_{\mu\nu}(x) = F_{\mu\nu} + \theta^{-1}_{\mu\nu}$ is invariant with respect to (5.164)–(5.169). This is also the symmetry of the action provided that the gauge coupling transforms accordingly.

(iv).   Formally, the transformation (5.164) can be represented in the form,

$$\delta\phi(x) = \delta x^\alpha \partial_\alpha\phi(x) = \phi(x + \delta x) - \phi(x) , \tag{5.171}$$

where $\delta x^\alpha = -\theta^{\alpha\beta}\delta p_\beta$ and no star product is assumed. This looks very similar to the coordinate transformations.

The map we just constructed looks very similar to the famous Seiberg–Witten map, which is given by the following variation of the background p_μ [23],

---

[6] In fact, the same happens in the map between different dimensions

$$\delta_{\text{SW}} p_\mu = -\frac{1}{2}\epsilon_{\mu\nu}\theta^{\nu\alpha}A_\alpha \ . \tag{5.172}$$

In (5.161), we have chosen δp_μ independent of gauge field background. (In fact, the gauge field background was switched-on later, after the transformation.) An alternative way would be to have nontrivial field $A_\mu(x)$ from the very beginning and to chose δp_μ to be of the Seiberg–Witten form. Then, the transformation laws corresponding to such a transformation of the background coincide exactly with the standard SW map. This appears possible because the function $p_\mu = -\theta_{\mu\nu}^{-1}x^\nu$ has the same gauge transformation properties as $-A_\mu(x)$,

$$p_\mu \to U^{-1} * p_\mu * U(x) - U^{-1} * \partial_\mu U(x) \ . \tag{5.173}$$

5.7 Discussion and Outlook

This lecture notes was designed as a very basic and very subjective introduction to the field. Many important things were not reflected and even not mentioned here. Among these, very few was said about the brane dynamics and interpretation which was the main motivation for the development of the matrix models, while the literature on this topic is enormously vast. For this we refer the reader to other reviews and lecture notes mentioned in the introduction (as well as to the references one can find inside these papers).

Recently, the role of the matrix models in the context of AdS/CFT correspondence became more clear. Some new matrix models arise in the description of the anomalous dimensions of composite super-Yang–Mills operators (see e.g., [45, 46].

Another recent progress even not mentioned here but which is related to matrix models is their use for the computation of the superpotential of $\mathcal{N} = 1$ supersymmetric gauge theories [47–49].

Acknowledgments

This lecture notes reflects my experience that I gained due to the communication with many persons. My thanks are directed most of all to my collaborators and colleagues from Bogoliubov lab in Dubna, Physics Department of University of Crete, Laboratori Nazionali di Frascati. I am grateful to the friends who helped me with various problems a while this text was being written: Giorgio Pagnini, Carlo Cavallo, Valeria, and Claudio Minardi. This work was supported by the INTAS-00-00262 grant, the Alexander von Humboldt research fellowship.

References

1. A. Morozov: arXiv:hep-th/0502010
2. J.J.M. Verbaarschot: arXiv:hep-th/0502029

3. T.A. Brody, J. Flores, J.B. French, P.A. Mello, A. Pandey, S.S.M. Wong: Rev. Mod. Phys. **53**, 385 (1981)

4. T. Guhr, A. Muller-Groeling, H.A. Weidenmuller: Phys. Rep. **299**, 189 (1998) [arXiv:cond-mat/9707301]

5. J.C. Osborn, D. Toublan, J.J.M. Verbaarschot: Nucl. Phys. B **540**, 317 (1999) [arXiv:hep-th/9806110]

6. J.J.M. Verbaarschot: Phys. Rev. Lett. **72**, 2531 (1994) [arXiv:hep-th/9401059]

7. E.V. Shuryak, J.J.M. Verbaarschot: Nucl. Phys. A **560**, 306 (1993) [arXiv:hep-th/9212088]

8. N. Ishibashi, H. Kawai, Y. Kitazawa, A. Tsuchiya: Nucl. Phys. B **498**, 467 (1997) [arXiv:hep-th/9612115]

9. T. Banks, W. Fischler, S.H. Shenker, L. Susskind: Phys. Rev. D **55**, 5112 (1997) [arXiv:hep-th/9610043]

10. J. Polchinski: Phys. Rev. Lett. **75**, 4724 (1995) [arXiv:hep-th/9510017]

11. O. Aharony, S.S. Gubser, J.M. Maldacena, H. Ooguri, Y. Oz: Phys. Rep. **323**, 183 (2000) [arXiv:hep-th/9905111]

12. J. Polchinski, *String Theory*, 2 volumes, (Cambridge University Press, Cambridge, 1998)

13. E. Kiritsis: arXiv:hep-th/9709062

14. T. Eguchi, H. Kawai: Phys. Rev. Lett. **48**, 1063 (1982)

15. D. Berenstein, J.M. Maldacena, H. Nastase: JHEP **0204**, 013 (2002) [arXiv:hep-th/0202021]

16. D. Berenstein, E. Gava, J.M. Maldacena, K.S. Narain, H. Nastase: arXiv:hep-th/0203249

17. D. Berenstein, H. Nastase: arXiv:hep-th/0205048

18. P. Valtancoli: Int. J. Mod. Phys. A **18**, 967 (2003) [arXiv:hep-th/0206075]

19. C. Sochichiu: Phys. Lett. B **574**, 105 (2003) [arXiv:hep-th/0206239]

20. Y.K. Cheung, M. Krogh: Nucl. Phys. B **528**, 185 (1998) [arXiv:hep-th/9803031]

21. C.S. Chu, P.M. Ho: Nucl. Phys. B **550**, 151 (1999) [arXiv:hep-th/9812219]

22. C.S. Chu, P.M. Ho: Nucl. Phys. B **568**, 447 (2000) [arXiv:hep-th/9906192]

23. N. Seiberg, E. Witten: JHEP **9909**, 032 (1999) [arXiv:hep-th/9908142]

24. C. Sochichiu: JHEP **0005**, 026 (2000) [arXiv:hep-th/0004062]

25. C. Sochichiu: Phys. Lett. B **485**, 202 (2000) [arXiv:hep-th/0005156]

26. C. Sochichiu: JHEP **0008**, 048 (2000) [arXiv:hep-th/0007127]

27. C. Sochichiu: arXiv:hep-th/0010149

28. C. Sochichiu: arXiv:hep-th/0012262

29. E. Kiritsis, C. Sochichiu: arXiv:hep-th/0202065

30. C. Sochichiu: arXiv:hep-th/0202014

31. J.A. Harvey: arXiv:hep-th/0102076

32. S. Minwalla, M. Van Raamsdonk, N. Seiberg: JHEP **0002**, 020 (2000) [arXiv:hep-th/9912072]

33. M. Van Raamsdonk, N. Seiberg: JHEP **0003**, 035 (2000) [arXiv:hep-th/0002186]

34. S. Sarkar: JHEP **0206**, 003 (2002) [arXiv:hep-th/0202171]

35. W. Bietenholz, F. Hofheinz, J. Nishimura: JHEP **0209**, 009 (2002) [arXiv:hep-th/0203151]

36. A.A. Slavnov: Phys. Lett. B **565**, 246 (2003) [arXiv:hep-th/0304141]

37. M. Buric, V. Radovanovic: arXiv:hep-th/0305236

38. M.M. Sheikh-Jabbari: JHEP **9906**, 015 (1999) [arXiv:hep-th/9903107]

39. H.B. Nielsen, M. Ninomiya: Phys. Lett. B **105**, 219 (1981)

40. N. Kitsunezaki, J. Nishimura: Nucl. Phys. B **526**, 351 (1998) [arXiv:hep-th/9707162]
41. A.P. Balachandran, G. Immirzi: Phys. Rev. D **68**, 065023 (2003) [arXiv:hep-th/0301242]
42. V.P. Nair, A.P. Polychronakos: Phys. Rev. Lett. **87**, 030403 (2001) [arXiv:hep-th/0102181]
43. D. Bak, K.M. Lee, J.H. Park, "Comments on noncommutative gauge theories", Phys. Lett. B **501**, 305–312 (2001) [arXiv:hep-th/0011244]
44. T. Ishikawa, S.I. Kuroki, A. Sako: J. Math. Phys. **43**, 872 (2002) [arXiv:hep-th/0107033]
45. A. Agarwal, S.G. Rajeev: Mod. Phys. Lett. A **19**, 2549 (2004) [arXiv:hep-th/0405116]
46. S. Bellucci, C. Sochichiu: Nucl. Phys. B **726**, 233 (2005) [arXiv:hep-th/0410010]
47. R. Dijkgraaf, C. Vafa: arXiv:hep-th/0208048
48. R. Dijkgraaf, C. Vafa: Nucl. Phys. B **644**, 21 (2002) [arXiv:hep-th/0207106]
49. R. Dijkgraaf, C. Vafa: Nucl. Phys. B **644**, 3 (2002) [arXiv:hep-th/0206255]

Index

Lecture Notes in Physics

For information about earlier volumes
please contact your bookseller or Springer
LNP Online archive: springerlink.com